Design, Development and Processing of Aluminium Alloys and Their Composite Materials

Design, Development and Processing of Aluminium Alloys and Their Composite Materials

Guest Editors

Peng Tang
Mingyi Zheng
Kang Wang
Xingzhi Pang

MDPI

Basel • Beijing • Wuhan • Barcelona • Belgrade • Novi Sad • Cluj • Manchester

Guest Editors
Peng Tang
School of Resources,
Environment and Materials
Guangxi University
Nanning
China

Mingyi Zheng
School of Materials Science
and Engineering
Harbin Institute of
Technology
Harbin
China

Kang Wang
School of Materials Science
and Engineering
Dongguan University of
Technology
Dongguan
China

Xingzhi Pang
School of Resources,
Environment and Materials
Guangxi University
Nanning
China

Editorial Office
MDPI AG
Grosspeteranlage 5
4052 Basel, Switzerland

This is a reprint of the Special Issue, published open access by the journal *Crystals* (ISSN 2073-4352), freely accessible at: https://www.mdpi.com/journal/crystals/special_issues/3ZM674FEOO.

For citation purposes, cite each article independently as indicated on the article page online and as indicated below:

Lastname, A.A.; Lastname, B.B. Article Title. *Journal Name* **Year**, *Volume Number*, Page Range.

ISBN 978-3-7258-7897-0 (Hbk)
ISBN 978-3-7258-7898-7 (PDF)
https://doi.org/10.3390/books978-3-7258-7898-7

Contents

About the Editors

Peng Tang

Peng Tang is an Associate Professor and Master's Supervisor in Materials Science and Engineering at Guangxi University. He earned his Ph.D. in Materials Processing Engineering from South China University of Technology, following a Bachelor's and Master's degree at Guangxi University. He conducted postdoctoral research in Civil Engineering at Guangxi University and was a visiting scholar at Nanyang Technological University (Singapore). His research focuses on lightweight recyclable structural materials, interface self-organization, programmable flexible smart materials, and laser-based intelligent manufacturing. He has led 18 research projects, including NSFC grants, published 76 papers (43 SCI), and holds 7 authorized patents. He was honored as "Teacher in My Heart" at Guangxi University (2023) and served as session chair for ICCES 2026. He serves on the editorial boards of several prestigious journals, including Rare Metals, China Foundry, Crystals, China Foundry, etc. He is also a frequent reviewer for over 30 international journals and a national expert reviewer for the National Natural Science Foundation of China.

Mingyi Zheng

Mingyi Zheng is currently a Professor in the School of Materials Science and Engineering at the Harbin Institute of Technology, China. In 2008, he was awarded New Century Excellent Talent in University in China. He has published approximately 230 peer-reviewed papers, garnering over 12,000 citations with an H-index of 68. He ranks among the World's Top 2% Most-Cited Scientists for both career-long and single-year impact (2021–2025), and was listed among Elsevier's Most Cited Chinese Researchers in 2020 and 2021. Dr. Zheng received the International Magnesium Science & Technology Award - Innovative Research & Application Award in 2020, and the Person of the Year Award in 2024. He serves on the editorial boards of several international materials science journals, including the renowned *Materials Science and Engineering: A* and the *Journal of Magnesium and Alloys*. Additionally, he serves as the Vice President of the Chinese Committee of Magnesium Alloys within the Chinese Society for Materials. His research primarily focuses on magnesium alloys, metal matrix composites, and bulk nano-metals processed by severe plastic deformation.

Kang Wang

Kang Wang is affiliated with the School of Materials Science and Engineering, Dongguan University of Technology. Research interests center on the preparation and performance study of light alloys and their composites. He has served as the principal investigator of two provincial and municipal Natural Science Foundation projects and participated in the National Key Research and Development Program as well as two projects of the National Natural Science Foundation of China. Achievements include ten SCI-indexed papers published as first or corresponding author.

Xingzhi Pang

Xingzhi Pang is a Master's supervisor at the School of Resources, Environment and Materials, Guangxi University. His primary research areas include novel non-ferrous metal materials and composite materials. He has served as the Principal Investigator for one key research and development project funded by Guangxi Zhuang Autonomous Region, one project funded by the Guangxi Natural Science Foundation, and one industry-commissioned project. He has also participated in several national-level research projects. Dr. Pang has published over ten SCI-indexed papers as the first or corresponding author and holds more than twenty authorized patents as the first inventor, two of which have been successfully commercialized. As a core team member, he has received one first prize, one second prize, and one third prize for provincial/ministerial-level teaching achievements. As the lead supervisor, he has guided students to win two bronze awards at the China International College Students' Innovation Competition.

crystals MDPI

Editorial

The Design, Development and Processing of Aluminium Alloys and Their Composite Materials

Peng Tang [1,*], Mingyi Zheng [2], Kang Wang [3,*] and Xingzhi Pang [1,*]

1 Guangxi Key Laboratory of Processing for Non-Ferrous Metals and Featured Materials, School of Resources, Environment and Materials, Guangxi University, Nanning 530004, China
2 School of Materials Science and Engineering, Harbin Institute of Technology, Harbin 150001, China; zhenghe@hit.edu.cn
3 School of Materials Science and Engineering, Dongguan University of Technology, Dongguan 523808, China
* Correspondence: tp@gxu.edu.cn (P.T.); wangk@dgut.edu.cn (K.W.); pxzh2006@126.com (X.P.)

1. Introduction

Aluminium, as the most abundant metallic element in the Earth's crust, holds a pivotal position in modern industry due to its exceptional combination of a low density, high specific strength, excellent workability, superior corrosion resistance, and remarkable recyclability. From aerospace and automotive engineering to construction and transportation, aluminium alloys have become indispensable materials that enable lightweight design and sustainable manufacturing. However, with the ever-increasing demand for higher performance, enhanced energy efficiency, and reduced environmental impact, the continuous advancement of aluminium alloy design, development, and processing technologies remains a critical pursuit for materials scientists and engineers worldwide.

This Special Issue, titled "The Design, Development and Processing of Aluminium Alloys and Their Composite Materials", brings together a collection of 14 original research articles that explore a wide spectrum of topics, ranging from fundamental microstructure–property relationships to advanced processing techniques and innovative material designs. Collectively, these contributions provide valuable insights into the complex interplay between composition, processing, microstructure, and the resulting mechanical, thermal, and corrosion properties of aluminium-based materials.

The published articles in this Special Issue can be broadly categorized into several key thematic areas, each addressing critical challenges and opportunities in aluminium alloy research.

2. Welding and Joining Technologies for Aluminium Alloys

The joining of aluminium alloys remains a critical industrial challenge, particularly in the context of lightweight structural fabrication. Several contributions in this Special Issue address this topic through both experimental and numerical approaches. In the work by Peng et al. [1], the laser welding process of 6056 aluminium alloy T-joints was systematically investigated using Simufact Welding numerical simulation. The authors validated their finite element model against experimental results and explored the effects of bilateral synchronous versus asynchronous laser welding on molten pool stability. Through orthogonal testing, optimal process parameters were identified, and the influence of welding sequence on post-weld equivalent stress and deformation was analyzed. This study provides a robust framework for process optimization in laser welding of aluminium alloy structures. Complementing this, Guan et al. [2] conducted an experimental investigation into the effects of reaming on the fatigue life of cold-expanded 7050-T7451 aluminium alloy.

The split-sleeve cold expansion technology is widely employed to enhance the fatigue performance of fastening holes in aerospace applications. Their results demonstrated that reaming by 0.2–0.4 mm effectively eliminates minor defects and improves fatigue life, while excessive reaming beyond 0.5 mm leads to stress relief and removal of the refined grain layer, thereby reducing fatigue performance. These findings offer essential guidance for optimizing post-expansion processing in aerospace fastener applications. A comprehensive review by Borah et al. [3] consolidates recent advancements in friction stir spot welding (FSSW) of aluminium alloys, with particular emphasis on tool design, thermal characterization, and mechanical properties. The review critically synthesizes comparative insights into temperature distribution, weld strength, hardness variation, and metallurgical transformations reported across numerous studies. By identifying knowledge gaps and potential design improvements, this work provides a valuable roadmap for developing more efficient and robust FSSW processes for industrial applications.

3. Alloy Composition Optimization and Microstructural Control

The relationship between alloy composition, microstructure, and mechanical properties is a central theme in aluminium alloy development. Several studies in this Special Issue explore the effects of trace element additions and processing conditions on microstructural evolution and performance. Qin et al. [4] investigated the influence of rare earth yttrium (Y) on the microstructure and heat-resistant properties of ZL109 alloys. The results revealed that Y addition refines α-Al dendrites and modifies eutectic Si morphology, while also forming a heat-resistant Al_2Si_2Y phase. The optimal Y content was determined to be 0.2 wt.%, which significantly improved both room-temperature and high-temperature tensile properties. This study highlights the potential of rare earth elements in enhancing the high-temperature performance of cast Al–Si–Cu–Mg–Ni alloys. Similarly, Wu et al. [5] examined the effect of yttrium on iron-rich phases and mechanical properties of as-cast AA8176 (Al–1Fe) alloy with low silicon concentration. The addition of 0.5 wt.% Y transformed coarse needle-like $Al_{13}Fe_4$ phases into short rod-like morphologies, while forming fine $Al_{10}Fe_2Y$ phases, resulting in a substantial reduction in the average length of iron-rich phases and an improvement in mechanical properties. This work provides important insights into the modification of detrimental iron-rich phases in Al–Fe alloys through rare earth microalloying. Addressing the challenge of iron impurity control in recycled aluminium, Cao et al. [6] systematically investigated the effects of Fe content on the microstructure, mechanical properties, and corrosion resistance of as-cast 5083 aluminium alloy. Their results showed that increasing Fe content induces a transition from compact Chinese-script α-Al(Fe,Mn)Si phases to coarse needle-like β-AlFeSi phases, with optimal mechanical properties achieved at 0.45 wt.% Fe. This study establishes an experimental basis for impurity tolerance control in recycled 5083 alloys, contributing to the sustainable utilization of secondary aluminium resources. In another contribution, Huang et al. [7] explored the micro-alloying effect of samarium (Sm) on the thermal conductivity and mechanical properties of Al–5Ni cast alloys. The addition of 0.1 wt.% Sm refined both α-Al grains and eutectic structures while simultaneously improving thermal conductivity through the reduction of Fe and Si impurity solubility in the α-Al matrix. This work demonstrates a promising approach for developing Al–Ni alloys with combined high thermal conductivity and satisfactory mechanical properties.

4. Composite Materials and Interface Engineering

The development of aluminium matrix composites and the understanding of interface phenomena are crucial for achieving enhanced material performance. Several contributions in this Special Issue focus on these aspects. Pang et al. [8] employed first-principles calcu-

https://doi.org/10.3390/cryst16040237

lations to investigate the interface characteristics between $TiB_2(0001)$ and $L1_2$-$Al_3Zr(001)$. Considering twelve different interface models with various terminations and stacking positions, the authors identified the most stable interface configurations based on adhesion work, interfacial energy, and electronic structure analysis. This theoretical study provides fundamental insights into the "poisoning" phenomenon observed in Al–Ti–B grain refiners when excess Zr is present, offering a deeper understanding of heterogeneous nucleation mechanisms in aluminium alloys. Jiang et al. [9] reported on the toughening mechanism of $CaAl_{12}O_{19}$ in red mud–Al_2O_3 composite ceramics. By incorporating 1.5 wt.% Cr_2O_3 and sintering at 1500 °C, the authors achieved a composite with flexural strength of 297.03 MPa and fracture toughness of 6.57 $MPa \cdot m^{1/2}$. The formation of $CaAl_{12}O_{19}$ phases that create "endo-crystalline" structures with Al_2O_3 grains was identified as the key toughening mechanism, changing the fracture mode from intergranular to transgranular. This study not only demonstrates an effective approach for utilizing red mud resources but also establishes the potential of these composites for electronic packaging applications. El Hachemi et al. [10] presented a dual approach combining spectroscopic ellipsometry and wave optics simulations to characterize the optical properties of nano-designed TiN/AlN composite dielectrics. Their methodology enabled the retrieval of effective refractive index and extinction coefficient for this complex nanostructured composite, with strong agreement between experimental and numerical results. This work provides a rational framework for designing material models for optical property studies across a broad spectrum.

5. Advanced Processing and Characterization

Innovations in processing techniques and characterization methods are essential for advancing aluminium alloy technology. Several contributions in this Special Issue explore such advancements. Mo et al. [11] investigated the interactions between Sc_2O_3 ceramics and calcium–magnesium–alumina–silicate (CMAS) at elevated temperatures, a topic relevant to thermal barrier coating applications. Their study revealed that a dense reaction layer forms on the ceramic surface, with composition and thickness dependent on temperature. At 1300 °C, the garnet phase-dominated reaction layer thickens according to parabolic kinetics, providing effective resistance to Mg^{2+} and Al^{3+} infiltration. These findings contribute to the understanding of CMAS corrosion resistance in rare earth oxide-based thermal barrier coatings. The study by Liu et al. [12] investigated the typical microstructure of twin-roll cast (TRC) 2139 aluminium alloy and its impact on mechanical properties, comparing with conventional mold casting (MC). The TRC process resulted in a denser dendritic structure and, after heat treatment, yielded a 10%+ improvement in both strength and ductility compared to MC sheets. This work establishes TRC as a superior industrial route for producing high-performance aluminium sheets, with significant implications for automotive and aerospace applications. Shi et al. [13] examined the effect of Al–5Ti–2B grain refiner on the microstructure and mechanical properties of recycled Al–7Si–0.3Mg–1Fe alloy. The optimal refinement was achieved at 1% addition, yielding tensile strength of 149.4 MPa and elongation of 4.3%. The study provides insights into the synergistic effects of grain refinement on α-Al, eutectic Si, and Fe-rich intermetallics, contributing to the high-value utilization of recycled aluminium alloys. Hu et al. [14] conducted research on the efficient desilication process of low-grade bauxite in Guangxi using barium hydroxide pre-desilication technology. Under optimal conditions (3 g $Ba(OH)_2$, 25 °C, 50 min), a desilication rate of 97.53% was achieved, significantly improving the Al/Si ratio and reducing subsequent processing costs. This work addresses the critical challenge of silicon removal from low-grade bauxite, supporting the sustainable development of the alumina industry.

https://doi.org/10.3390/cryst16040237

6. Concluding Remarks

The 14 articles collected in this Special Issue represent a diverse and comprehensive exploration of the current frontiers in aluminium alloy research. Collectively, they highlight several key trends and future directions:

Sustainable Processing: The growing emphasis on recycled aluminium, red mud utilization, and low-grade bauxite processing reflects the industry's commitment to circular economy principles and environmental sustainability.

Advanced Joining Technologies: The continued development of laser welding, friction stir spot welding, and cold expansion techniques demonstrates the importance of robust joining solutions for lightweight structures.

Microalloying and Composition Optimization: The strategic use of rare earth elements and transition metals to refine microstructures, modify harmful phases, and enhance both mechanical and functional properties remains a powerful strategy for alloy development.

Interface Engineering: The fundamental understanding of interfaces—whether in composite materials, grain refiner interactions, or coating systems—is critical for achieving superior material performance.

Multiscale Characterization and Simulation: The integration of experimental characterization with numerical simulation and first-principles calculations enables deeper insights into structure–property relationships and accelerates materials development.

We express our sincere gratitude to all authors for their valuable contributions to this Special Issue. Their rigorous research and insightful findings have significantly advanced the field of aluminium alloy design, development, and processing. We also extend our appreciation to the reviewers for their thorough and constructive evaluations, which ensured the high quality of the published works. Finally, we thank the editorial staff of Crystals for their professional support throughout the preparation of this Special Issue.

We hope that this collection of articles will serve as a valuable resource for researchers, engineers, and students working in the field of aluminium alloys and composite materials, inspiring further innovations and advancements in this dynamic and critically important area.

Author Contributions: All authors contributed equally to this work. All authors have read and agreed to the published version of the manuscript.

Funding: This work was supported by National Natural Science Foundation of China (52261024), Guangxi Natural Science Foundation (2025GXNSFAA069657, 2023GXNSFAA026374), Guangxi Key Technologies R&D Program (AB24010286), and Baise City Science and Technology Program (Baike ZQ252811).

Conflicts of Interest: The authors declare no conflicts of interest.

References

1. Peng, J.; Xie, S.; Chen, T.; Wang, X.; Yu, X.; Yang, L.; Ni, Z.; Ling, Z.; Yuan, Z.; Shi, J.; et al. Numerical Simulation and Process Optimization of Laser Welding in 6056 Aluminum Alloy T-Joints. *Crystals* **2025**, *15*, 35. [CrossRef]
2. Guan, M.; Xue, Q.; Zhuang, Z.; Hu, Q.; Qi, H. Experimental Study of Reaming Sizes on Fatigue Life of Cold-Expanded 7050-T7451 Aluminum Alloy. *Crystals* **2024**, *14*, 803. [CrossRef]
3. Borah, M.J.; Sengupta, K.; Hossain, B.; Mondal, A.K.; Ling, B.; Jain, H.; Nath, Y.; Ling, B. A Critical Review on Friction Stir Spot Welding of Aluminium Alloys: Tool, Mechanical, and Micro-Structural Characteristics. *Crystals* **2025**, *15*, 755. [CrossRef]
4. Qin, X.; Li, A.; Li, X. Effect of Y on the Microstructures and Heat-Resistant Property of ZL109 Alloys. *Crystals* **2025**, *15*, 75. [CrossRef]
5. Wu, W.; Lai, W.; Cao, Z.; Li, C.; Zhao, M. Effect of Yttrium on Iron-Rich Phases and Mechanical Properties of As-Cast Al-Fe Alloy with Low Si Concentration. *Crystals* **2025**, *16*, 28. [CrossRef]
6. Cao, J.; Zhao, W.; Li, J.; Tang, H.; Zheng, X.; He, K.; Zhao, Q.; Yang, H.; Lu, X.; Lei, S.; et al. Effect of Fe Content on the Microstructure and Properties of 5083 Aluminum Alloy. *Crystals* **2025**, *16*, 192. [CrossRef]

7. Huang, Y.; Wang, K.; Zhu, Z.; Cheng, W.; Li, W. Micro-Alloying with Samarium to Enhance the Thermal Conductivity and Mechanical Properties of Al-5Ni Cast Alloys. *Crystals* **2025**, *15*, 609. [CrossRef]
8. Pang, X.; Yang, L.; Nong, H.; Pang, M.; Wang, G.; Li, J.; Chen, Z.; Zeng, W.; Xiao, Z.; Yang, Z.; et al. Understanding the Interface Characteristics Between TiB_2(0001) and $L1_2$-Al_3Zr(001): A First-Principles Investigation. *Crystals* **2024**, *14*, 979. [CrossRef]
9. Jiang, S.; Li, A.; Wang, Z. Toughening Mechanism of $CaAl_{12}O_{19}$ in Red Mud–Al_2O_3 Composite Ceramics. *Crystals* **2024**, *14*, 924. [CrossRef]
10. El Hachemi, M.; Khanna, N.; Barborini, E. Spectroscopic Ellipsometry and Wave Optics: A Dual Approach to Characterizing TiN/AlN Composite Dielectrics. *Crystals* **2025**, *15*, 143. [CrossRef]
11. Mo, Z.; Mo, Z.; Yu, Z.; Cheng, Y.; Miao, Y.; Liang, T. Interactions Between Sc_2O_3 Ceramics and Calcium–Magnesium–Alumina–Silicate (CMAS) at Elevated Temperature. *Crystals* **2025**, *15*, 198. [CrossRef]
12. Liu, Z.; Wang, Y.; Cheng, Q.; Ma, L.; Yang, Z.; Xu, Y.; Luo, Y.; Qin, X.; Liang, T. The Typical Microstructure of Twin-Roll Cast 2139 Alloy and Its Impact on Mechanical Properties. *Crystals* **2025**, *15*, 13. [CrossRef]
13. Shi, W.; Chen, L.; He, B.; Lu, B.; Yang, J. Effect of Al-5Ti-2B on the Microstructure and Mechanical Properties of Recycled Al-7Si-0.3Mg-1Fe Alloy. *Crystals* **2025**, *15*, 584. [CrossRef]
14. Hu, G.; Li, A.; Xia, A.; Zhang, D.; Pan, L.; Zhao, X.; Pang, X. Research on the Efficient Desilication Process of Low-Grade Bauxite in Guangxi. *Crystals* **2025**, *15*, 675. [CrossRef]

https://doi.org/10.3390/cryst16040237

crystals

MDPI

Article

Numerical Simulation and Process Optimization of Laser Welding in 6056 Aluminum Alloy T-Joints

Jin Peng [1,2,3,*], Shihua Xie [1], Tiejun Chen [1], Xingxing Wang [1,*], Xiaokai Yu [4], Luqiang Yang [4], Zenglei Ni [1], Zicheng Ling [1], Zhipeng Yuan [1], Jianjun Shi [1] and Zhibin Yang [5]

[1] Henan International Joint Laboratory of High-Efficiency Special Green Welding, Material College, North China University of Water Resources and Electric Power, Zhengzhou 450045, China; chentiejun_ncwu@163.com (T.C.); lingzicheng@ncwu.edu.cn (Z.L.)
[2] Ningbo Intelligent Machine Tool Research Institute Co., Ltd. of China National Machinery Institute Group, Ningbo 315700, China
[3] National United Engineering Laboratory for Advanced Bearing Tribology, Henan University of Science and Technology, Luoyang 471023, China
[4] Luoyang Bearing Research Institute Co., Ltd., Luoyang 471039, China; yxk0359@163.com (X.Y.); yangluqiang135@163.com (L.Y.)
[5] School of Materials Science and Engineering, Dalian Jiaotong University, Dalian 116028, China
* Correspondence: pengjin1985120@163.com (J.P.); wangxingxing@ncwu.edu.cn (X.W.); Tel./Fax: +86-0371-67120189 (J.P.)

Abstract: This paper conducts a numerical simulation of the laser welding process for 6056 aluminum alloy stringers and skin T-joints using Simufact Welding. Initially, the accuracy of the finite element simulation is validated, followed by an exploration of the impact of bilateral asynchronous and bilateral synchronous laser welding on molten pool stability. Process parameters, including laser power, welding speed, fixture clamping force, and preheat temperature, are optimized through orthogonal testing. Furthermore, the influence of welding sequences on post-weld equivalent stress and deformation in three stringers' T-joints is analyzed. The numerical simulation results indicate that the stability of the molten pool is superior in bilateral synchronous welding compared to asynchronous welding. Optimized process parameters were obtained through orthogonal testing, and subsequent experiments demonstrated that the welding sequence of welding both sides first, followed by the middle, produced lower post-weld equivalent stress and reduced overall joint deformation.

Keywords: laser welding; numerical simulation; 6056 aluminum alloy; T-joint; optimized process parameters

1. Introduction

Aluminum alloys are widely recognized for their unique combination of low density and high strength [1–3], making them a key material in industries such as maritime, aerospace, and high-speed rail. In particular, aluminum alloy stringer and skin T-joints are commonly employed in structural applications [4,5]. Traditionally, these joints have been assembled using riveting, but this method increases the overall structural weight, posing design and efficiency challenges [6]. With the continuous advancement of welding technologies, laser welding has emerged as a promising alternative. Notably, the dual laser beam synchronous welding technology, developed by Airbus in Germany, is considered one of the most advanced methods for aluminum alloy joints [7–9].

https://doi.org/10.3390/cryst15010035

Recent research efforts have increasingly focused on numerical simulations to optimize welding processes and understand underlying thermal and mechanical behaviors [10]. One major reason is the high cost and time consumption associated with experimental work, especially when dealing with complex parameters and large numbers of trials [11,12]. Numerical simulations allow researchers to conduct extensive parametric studies in a relatively short time, enabling the exploration of a wide range of welding conditions without the need for costly material consumption and labor-intensive experiments. Moreover, simulations can provide detailed insights into temperature distribution, stress analysis, and deformation, which would be difficult or impossible to capture in physical experiments [13,14]. This makes simulations not only cost-effective but also more flexible, as they can model various welding conditions under different material properties and geometric configurations [15].

Vrtiel et al. [16] conducted numerical simulations of the laser welding process on 4 mm thick S650MC high-strength steel plates using ANSYS software (Release 18.1). They developed a comprehensive simulation model for temperature field analysis, leading to the determination of optimal welding parameters for S650MC steel. Similarly, Khoshroyan et al. [17] applied ANSYS to develop a three-dimensional finite element model to study the temperature distribution, residual stresses, and deformations in MIG welding of reinforced aluminum alloy Al6061-T6 plates. Their work demonstrated the impact of welding sequences and parameters on the residual stress and deformation in T-joints.

Duggirala et al. [18] explored the thermo-fluid dynamics during the laser deep fusion welding of AA2024 aluminum alloy, using a transient three-dimensional simulation model. Through finite element modeling, they quantitatively analyzed the effects of temperature-driven surface tension and its influence on melt pool dynamics and geometry. Further contributions by Alghamdi et al. [19] introduced a two-dimensional finite element model to simulate the phase transformation and its effect on residual stress distribution in multiple welding passes of thick T-joints. Additionally, Lu et al. [20] modeled oscillating laser welding in 6061 aluminum alloy, investigating the influence of beam oscillation on temperature fields, melt flow, and defect suppression mechanisms.

With advancements in computational tools, numerical simulation has become an indispensable technique in welding research. It plays a critical role in optimizing process parameters, ensuring weld quality, minimizing testing costs, and guiding the development of new materials and structures. Particularly for the dual-beam synchronous laser welding of stringer-skin T-joints, the application of numerical simulations is of significant practical value [21–24]. Xia et al. [25] conducted a numerical study on the molten pool flow behavior and pore evolution during the dual-laser beam oscillating bilateral synchronous welding (DLBOW) of T-joints. The results indicated that DLBOW technology effectively reduces porosity in T-joint welds. A comprehensive numerical CFD model of the melting process was developed, which aligned well with experimental results. The study also explored the tracking of pore morphology under different laser incidence angles and the application of the Fresnel multiple reflection model to capture pore evolution during the welding process. Yang et al. [26] conducted a numerical simulation study to analyze heat transfer and fluid flow during double-sided laser beam welding of T-joints for aluminum aircraft fuselage panels. The study focused on the thermal and fluid dynamic behaviors within the molten pool during the welding process. The results indicated that the heat transfer characteristics and fluid flow significantly influence the weld quality, including penetration and bead formation. The authors developed a comprehensive numerical model to simulate the process, which was validated by experimental results. This work provided valuable insights into optimizing welding parameters for improved joint integrity and enhanced performance of aluminum fuselage structures in aerospace applications.

Stringer–skin T-joints are commonly used in critical structural applications such as aerospace and high-speed rail, where stringent quality standards are required. These joints often experience complex loading conditions, making the optimization of their welding process crucial for ensuring long-term structural integrity and reliable performance under operational stresses. Numerical simulations of stringers and skin T-joints provide an effective method for optimizing key welding parameters such as laser power, welding speed, and fixture clamping force, ultimately enhancing joint quality [19]. In this study, we use Simufact Welding software(version 5.0) to simulate the laser welding process of 6056 aluminum alloy stringers and skin T-joints, aiming to optimize the process parameters and enhance joint quality.

2. Experimental Materials and Models

2.1. *Material of T-Joints and Dimensions*

The materials utilized for the T-joint base plate and stringer in this study are all 6056 aluminum alloy, which belongs to the aluminum–magnesium–silicon series (6××× series) [26], and the chemical composition of the 6056 aluminum alloy is presented (shown) in Table 1.

Table 1. Chemical composition of 6056 aluminum alloy (mass fraction).

Alloying Element	Mg	Si	Cu	Mn	Zn	Al
Content (wt%)	0.9	1.0	0.8	0.6	0.4	balance

The model and dimensions of the three T-joints studied in this paper are shown in Figures 1 and 2.

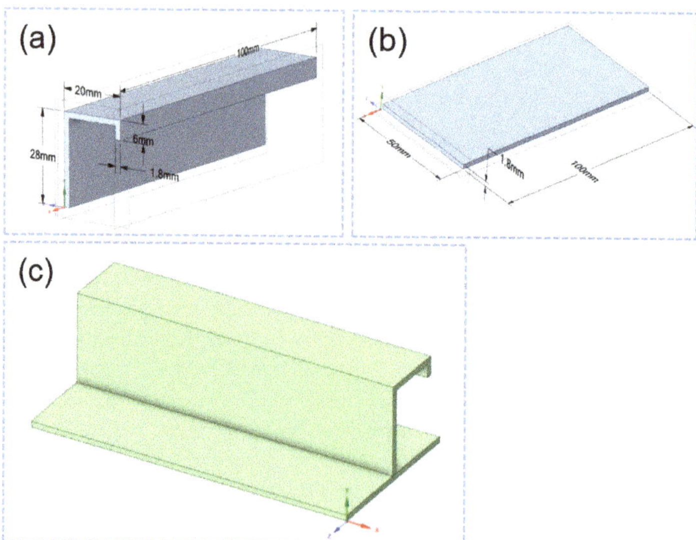

Figure 1. Dimensions of the stringer T-joints. (**a**) Stringer; (**b**) base plate; (**c**) joint.

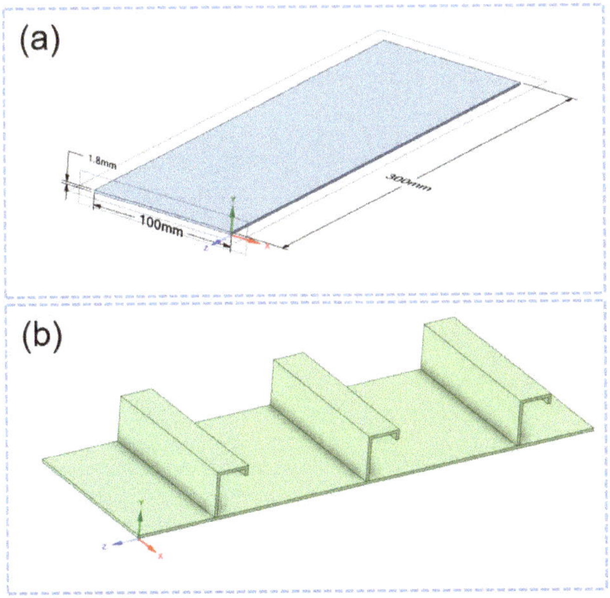

Figure 2. Model of three stringer T-joints. (**a**) Base plate; (**b**) joint.

2.2. Finite Element Meshing of T-Joints

Simufact Welding, a software program specifically designed for welding applications, was selected for the welding simulation in this study. It offers predefined process templates, material libraries, and advanced algorithms for simulating complex heat sources and multi-pass welds. These features make it particularly effective in modeling the dynamic effects of welding, such as temperature distribution, residual stresses, and distortion. Additionally, its integration with meshing tools like HyperMesh simplifies the preprocessing stage, ensuring that the simulation setup is both efficient and accurate.

Hexahedral elements were used for the finite element meshing in the HyperMesh software(version 2020). These elements provide higher accuracy in capturing complex thermal gradients and mechanical deformations around the weld pool, making them particularly effective in predicting the residual stresses and distortion typically associated with welding processes.

A transition mesh strategy was employed to refine the mesh around the weld joint and its adjacent areas. Finer elements were used near the weld, while coarser elements were applied in regions farther from the weld, which are less affected by the heat source [27,28]. The mesh size in the seam area of the stringer was 0.9 mm × 0.9 mm, and in the surrounding areas, it was 1.8 mm × 1.8 mm, resulting in a total of 2722 mesh elements. The base plate was meshed in a similar manner, generating a total of 3889 mesh elements. At the connection between the base plate and the three stringers, the mesh size was set to 2 mm × 2 mm, while 4 mm × 4 mm was used in other areas, resulting in a total of 2575 mesh elements, as shown in Figure 3.

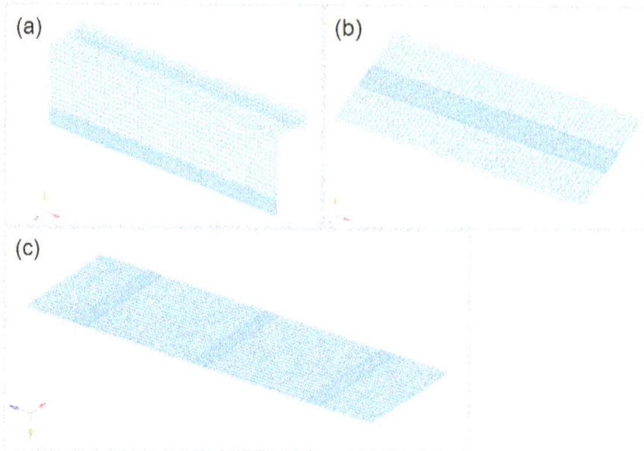

Figure 3. Mesh generation. (**a**) Stringer; (**b**) base plate; (**c**) base plate for the three stringer T-joints.

2.3. Finite Element Simulation Welding Heat Source Modeling

Welding finite element simulations typically employ various heat source models, including the Gaussian heat source model, double elliptic heat source model, uniform body heat source model, and cone heat source model [29–31].

In this study, the heat source model utilized combines a Gaussian surface heat source with an adaptive body heat source model. Within the Simufact Welding simulation software, this combined heat source model is simplified. The plasma heat source situated above the small hole exhibits a heat flow density that follows a Gaussian distribution, while the columnar heat source below the small hole maintains a constant heat flow density. This configuration can be represented by a uniformly distributed Gaussian cylindrical heat source model. A schematic representation of the heat source model is illustrated in Figure 4.

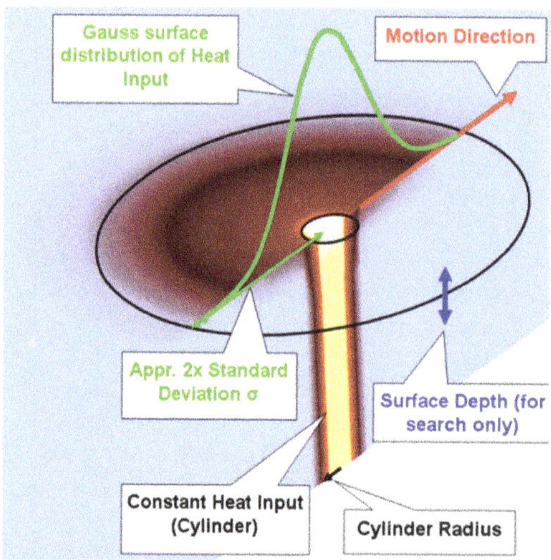

Figure 4. Schematic of the laser welding heat source model.

The heat source model used for laser welding T-joints combines a Gaussian surface heat source and an adaptive body heat source. The parameters for this model are as follows. The radius of the column heat source is 0.22 mm, and its depth is 1.95 mm. The radius of the surface heat source is 1.2 mm, with a depth of 0.1 mm. The cylindrical heat source accounts for 86% of the total heat input, with an energy conversion efficiency of 19% [32].

These parameters are calibrated for the specific laser welding conditions of the T-joint configuration.

2.4. Thermophysical Parameters of Materials

The constant thermophysical property parameters of the 6056 aluminum alloy utilized in this study are provided in Table 2 [33], which are crucial for accurately characterizing the material's behavior in the context of laser welding simulations. Furthermore, the temperature-dependent variations of specific thermophysical properties for the 6056 aluminum alloy are described by Equations (1)–(3) [34]. These equations define the relationships for specific heat, thermal conductivity, and kinematic viscosity as functions of temperature.

Table 2. Selected thermophysical property parameters of 6056 aluminum alloy.

Physical Quantity	Expressed Symbol	Unit	Numerical Value
Solid phase density	ρ_s	kg·m^{-3}	2720
Solid phase line temperature	T_S	K	860
Liquid phase line temperature	T_L	K	917
Latent heat of vaporization	L_v	J·kg^{-1}	1.08×10^7
Coefficient of thermal expansion	β_k	K^{-1}	1.92×10^{-5}

The mathematical expression for the variation of specific heat with temperature for 6056 aluminum alloy is:

$$C_p\left(\mathrm{J \cdot kg^{-1}}\right) = \begin{cases} -0.001 \times T^2 + 1.1609 \times T + 267.71 & T \leq 573 \\ 0.0009 \times T^2 - 0.3901 \times T + 514.45 & 573 < T \leq 913 \\ -0.0001 \times T^2 + 0.5832 \times T + 435.14 & 913 < T \end{cases} \quad (1)$$

The mathematical expression for the variation of thermal conductivity with temperature for 6056 aluminum alloy is:

$$k\left(\mathrm{W \cdot m^{-1} \cdot K^{-1}}\right) = \begin{cases} 0.0001 \times T^2 - 0.0697 \times T + 95.334 & T \leq 860 \\ -0.0048 \times T^2 + 9.2812 \times T - 4275.6 & 860 < T \leq 917 \\ -0.00001 \times T^2 + 0.0582 \times T + 148.74 & 917 < T \leq 2740 \\ -0.00001 \times T^2 + 0.085 \times T + 71.111 & 2740 < T \end{cases} \quad (2)$$

The mathematical expression for the variation of kinetic viscosity with temperature for 6056 aluminum alloy:

$$\mu\left(\mathrm{kg \cdot m^{-1} \cdot s^{-1}}\right) = \begin{cases} 1 \times 10^{-7} \times T^2 - 0.0002 \times T + 0.1202 & 897 \leq T \leq 937 \\ 2 \times 10^{-11} \times T^2 - 5 \times 10^{-7} \times T + 0.0038 & 937 < T \leq 2650 \\ -6 \times 10^{-8} \times T^2 + 0.0003 \times T - 0.4151 & 2650 < T \leq 2720 \end{cases} \quad (3)$$

3. Analysis of Results

3.1. Validation of the Heat Source Model for Numerical Simulation

Numerical simulation involves the use of computational methods to solve mathematical models, thereby replicating experimental processes. Typically, discrepancies exist between simulated and experimental results, requiring calibration of the heat source model. The simulation in this study used the same parameters as those in Yang Zhibin's experiment [35]. Therefore, a comparison was made between the weld seam surface morphology and melt pool morphology from both the experimental and simulated results, as shown in Figures 5 and 6. In Figure 5, the left portion and in Figure 6, the upper portion represent the experimental results from Yang Zhibin, while the comparison in the figures corresponds to the simulated results from this study.

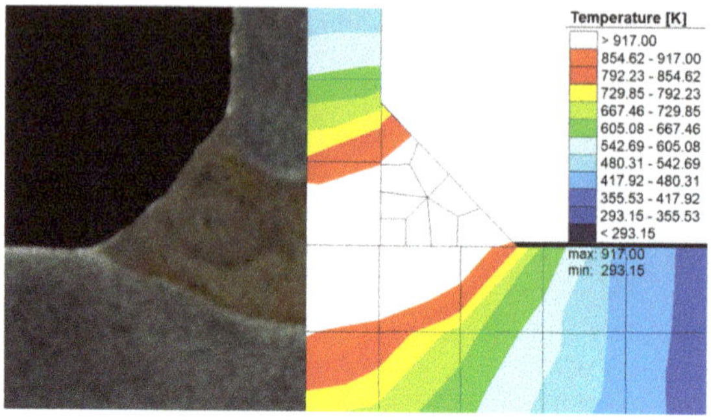

Figure 5. Comparison between experimental and simulated weld seam cross-section.

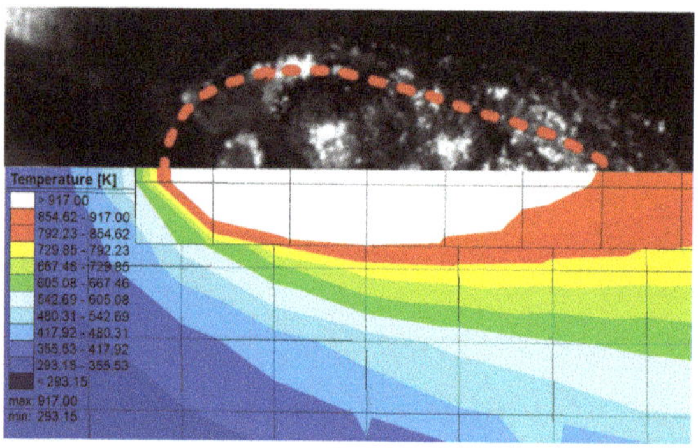

Figure 6. Comparison between experimental and simulated melt pool.

Table 3 provides a quantitative comparison of the surface morphology parameters for the weld cross-section and molten pool in the dual-side synchronous laser welding of aluminum alloy T-joints. The experimental data obtained from Yang Zhibin's work and the simulated results from this study are in close agreement. The smallest relative error, 1.57%, was observed for the molten pool length, while the largest relative error, 6.75%, was found in the molten pool width. The average relative error across all parameters is 4.61%. These values are within acceptable engineering tolerances, indicating a high degree of agreement between the simulated and experimental results [36]. This consistency confirms that the heat source model developed in this study is effective and reliable for simulating the laser welding process of 6056 aluminum alloy T-joints.

Table 3. Comparison of weld cross-section and melt pool surface topography parameters.

	Welding Depth/mm	Melt Pool Length/mm	Pool Width/mm
analog value	1.034	4.331	1.921
experimental value	0.980	4.400	2.060
relative error	5.51%	1.57%	6.75%

3.2. *Bilateral Synchronous and Bilateral Asynchronous Laser Welding*

To determine the most suitable laser welding method, a series of simulation experiments were conducted using Simufact Welding software. The stability of the molten pool and variations in melt depth for both bilateral synchronous and asynchronous laser welding were compared and analyzed to identify the optimal approach [2].

For these simulations, the laser power (P) was set to 1800 W, the welding speed (v) to 3.8 m/min, and the ambient temperature (T) to 293.15 K. In the case of the bilateral asynchronous laser welding, two welding robots were used. The first robot performed welding on one side of the T-joint and paused for 50 s to allow the weldment to cool to room temperature before the second robot welded the opposite side of the T-joint.

The schematic representations of the melt pool shape and weld depth, along with the fusion depth curves for both welding methods at different time intervals, are shown in Figures 7–10, where t represents the time interval after welding completion.

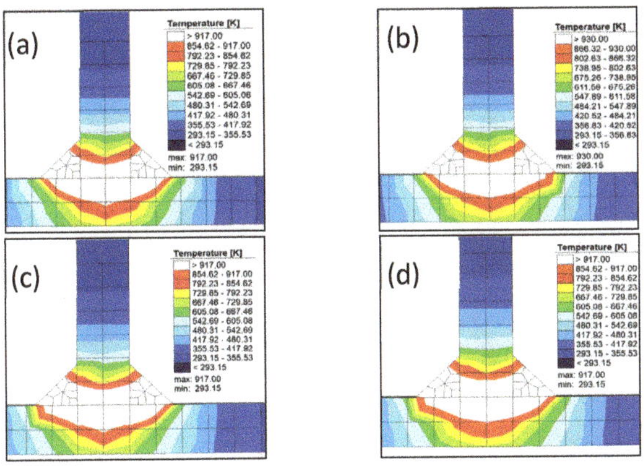

Figure 7. Melt pool at different time instances during double-sided synchronous laser welding. (**a**) t = 0.1078 s; (**b**) t = 0.5286 s; (**c**) t = 0.8451 s; (**d**) t = 1.266 s.

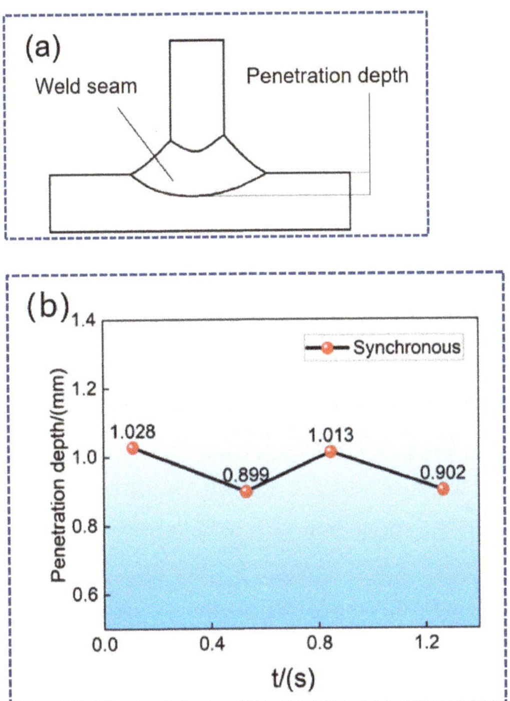

Figure 8. Double-sided synchronous laser welding. (**a**) Schematic of weld penetration, (**b**) penetration depth variation curve.

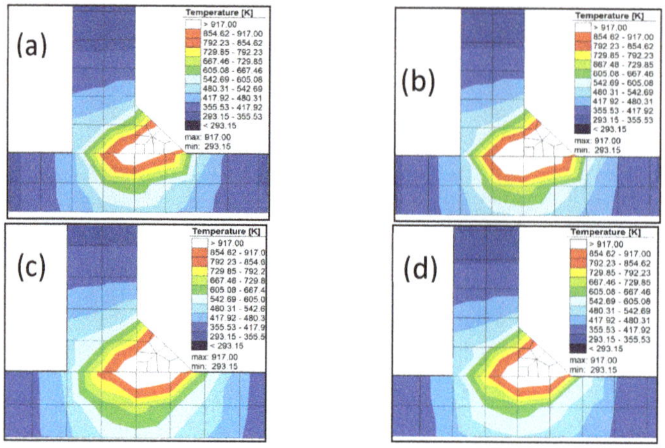

Figure 9. Melt pool at different time instances during double-sided asynchronous laser welding. (**a**) t = 0.1078 s; (**b**) t = 0.4243 s; (**c**) t = 1.266 s; (**d**) t = 1.475 s.

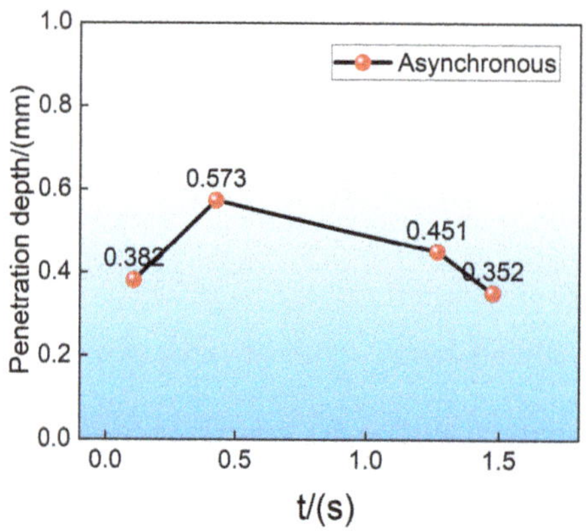

Figure 10. Penetration depth variation curve of double-sided asynchronous laser welding.

A comparison of experimental results indicates that the molten pool stability during the bilateral asynchronous laser welding is inadequate, leading to poor control over melt depth and increasing the likelihood of issues such as excessive melt depth or keyhole closure. In contrast, the bilateral synchronous laser welding exhibits a more stable molten pool shape.

Based on the data, the sample standard deviation of melt depth for the bilateral asynchronous welding was calculated to be 0.0695, while for the bilateral synchronous welding, it was 0.0981. These results suggest that the bilateral synchronous laser welding provides a more stable process, with better heat distribution and improved precision in controlling melt depth.

In summary, considering factors such as molten pool stability, melt depth control accuracy, and overall joint quality, bilateral synchronous laser welding offers distinct advantages for welding aluminum alloy T-joints. Consequently, the subsequent experiments in this study will adopt the bilateral synchronous laser welding method.

3.3. Orthogonal Experimental Design and Result Analysis of Single Stringer T-Joints

This study utilizes an orthogonal experimental design to evaluate the impact of various parameter combinations on the distribution of welding stress and deformation, aiming to identify the optimal process parameter configuration. Through a review of the literature, four key variables were identified for the orthogonal experimental design: laser power, welding speed, fixture clamping force, and preheating temperature. Preliminary simulations were conducted to establish the approximate ranges for these parameters, resulting in the experimental conditions outlined in Table 4.

Table 4. Table of welding process factor levels.

Level	A. Laser Power/W	B. Welding Speed/(m·min)$^{-1}$	C. Fixture Clamping Force/N	D. Preheating Temperature/K
1	1300	3.6	80	333.15
2	1400	3.8	100	353.15
3	1500	4.0	120	373.15

In this study, the four variables are denoted as A, B, C, and D, representing laser power, welding speed, fixture clamping force, and preheating temperature, respectively. The levels of these variables are numbered as 1, 2, and 3. For example, the combination A2B1C3D1 corresponds to a laser power of 1400 W, welding speed of 3.6 m/min, fixture clamping force of 120 N, and preheating temperature of 333.15 K.

Using these four parameters, a standard orthogonal array, L9(34), was applied to simulate the double-sided synchronous laser welding of single stringer T-joints. Following the completion of the simulations, post-processing was conducted to extract the equivalent stress and deformation values for each parameter combination, as shown in Table 5. Both the equivalent stress and deformation were measured at the same sampling point on the weld seam, 150 s after the welding process was completed.

Table 5. Orthogonal test program and test results.

Experiment Number	A Laser Power/W	B Welding Speed/(m min)$^{-1}$	C Fixture Clamping Force/N	D Preheating Temperature/K	Equivalent Stress/MPa	Deformation /mm
Case 1	1300	3.6	80	333.15	143.680	0.123
Case 2	1300	3.8	100	353.15	156.170	0.145
Case 3	1300	4.0	120	373.15	146.470	0.120
Case 4	1400	3.6	100	373.15	144.102	0.176
Case 5	1400	3.8	120	333.15	159.737	0.105
Case 6	1400	4.0	80	353.15	144.952	0.146
Case 7	1500	3.6	120	353.15	121.700	0.152
Case 8	1500	3.8	80	373.15	129.319	0.175
Case 9	1500	4.0	100	333.15	139.310	0.135

Range analysis is a commonly used method for analyzing data from orthogonal experiments. By calculating the mean and range of experimental results at different levels for each factor, the influence of each factor on the outcomes can be evaluated. By comparing the R-values of different factors, the relative impact of each factor on the experimental results can be determined. Typically, a larger R-value indicates a greater influence of that factor. The basic calculation steps are as follows:

1. Calculate *Ki*:

$$Ki = \sum_{i=1}^{i}(yi) \tag{4}$$

2. Calculate *ki*:

$$ki = \frac{Ki}{\mathrm{N}} \tag{5}$$

3. Calculate *R*:

$$R = \max(ki) - \min(ki) \tag{6}$$

where Ki represents the sum of all the experimental results for a given factor at a specific level; yi represents the i-th experimental result for the given factor at that specific level; ki represents the average of experimental results for the given factor at a specific level;

N represents the number of experiments conducted at this level; and R represents the range of the mean values for a given factor at different levels, indicating the degree of influence that factor has on the experimental results.

The range analysis provides insights into the primary and secondary influencing factors for equivalent stress and deformation, as well as the optimal combination of parameters. The results are shown in Table 6.

Table 6. Analysis of orthogonal test results.

Norm		A	B	C	D
Equivalent stress/MPa	K_1	446.32	409.482	417.951	442.727
	K_2	448.791	445.226	439.582	422.822
	K_3	390.329	430.732	427.907	419.891
	k_1	148.773	136.494	139.317	147.576
	k_2	149.597	148.407	146.527	140.941
	k_3	130.110	143.577	142.636	139.964
	R	19.487	11.913	7.21	7.612
	prioritize factors			A B D C	
	preferred option			$A_3B_1C_1D_3$	
Deformation/mm	K_1	0.388	0.451	0.444	0.363
	K_2	0.427	0.425	0.456	0.443
	K_3	0.462	0.401	0.377	0.471
	k_1	0.129	0.150	0.148	0.121
	k_2	0.142	0.142	0.152	0.148
	k_3	0.154	0.134	0.126	0.157
	R	0.025	0.016	0.026	0.036
	prioritize factors			D C A B	
	preferred option			$A_1B_3C_3D_1$	

The trend graphs illustrating the relationship between each factor and equivalent stress are presented in Figure 11, while Figure 12 displays the trend graphs correlating each factor with deformation.

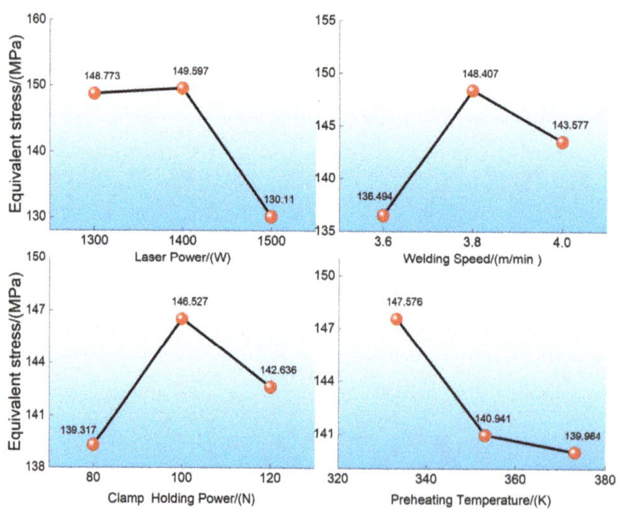

Figure 11. Trends of various factors with equivalent stress.

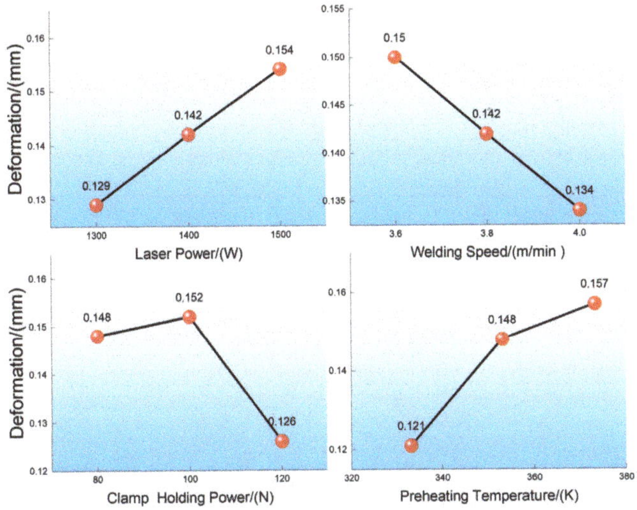

Figure 12. Trends of various factors with deformation.

For multi-indicator orthogonal tests, it is essential to achieve a comprehensive balance among the roles of various factors across different indicators to identify the optimal solution. The comprehensive balance method typically adheres to four guiding principles: consideration of the main influencing factors, the majority principle, minimizing losses while enhancing efficiency, and prioritizing the primary indicators.

In accordance with these principles, the impacts of laser power, welding speed, fixture clamping force, and preheating temperature on the equivalent force and total deformation of the weld are analyzed comprehensively. The optimal combination of process parameters is determined to be a laser power of 1500 W, a welding speed of 3.6 m/min, a fixture clamping force of 120 N, and a preheating temperature of 333.15 K.

The cross-sectional morphology of the molten pool for orthogonal experiments, ranging from Case 1 to Case 9, is illustrated in Figure 13. It is evident that the morphology of the molten pool varies significantly across different levels of parameter combinations. Burn-through defects are observed in Case 4, Case 7, and Case 8, primarily attributed to the combination of higher laser power, slower welding speeds, and elevated preheating temperatures. In the remaining cases, the molten pool morphology is satisfactory, with no burn-through defects present, although slight variations in the depth of the molten pool are noted.

The optimized parameter combinations at various levels obtained from the orthogonal experiments are presented in Table 7. To validate the reliability of these optimized parameter combinations, simulation tests are conducted using the parameters listed in Table 7, while maintaining all other conditions identical to those of the orthogonal experiments. The resulting cross-sectional morphology of the molten pool is depicted in Figure 14, revealing that under the optimized parameters, the cross-sectional morphology of the molten pool is satisfactory, with no burn-through defects observed.

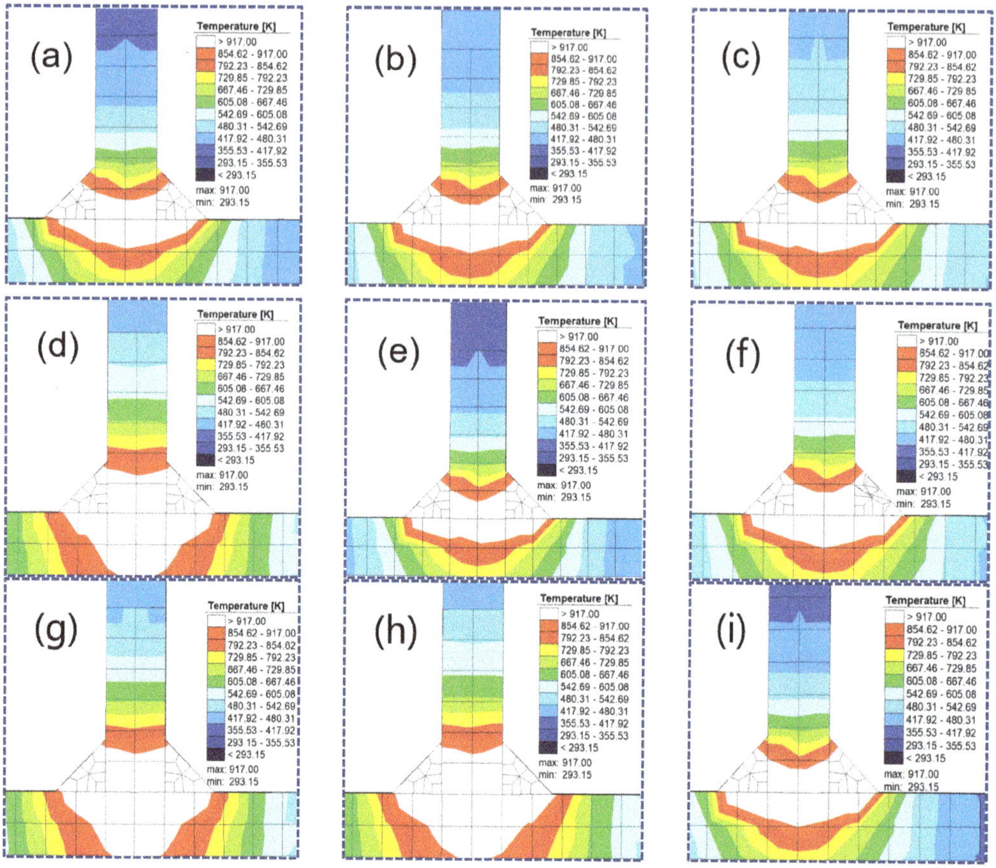

Figure 13. Melt pool morphology of orthogonal test schemes. Figures (**a**–**i**) correspond to case 1 to case 9.

Table 7. Table of optimization levels for each factor.

Factor	A Laser Power/W	B Welding Speed/(m min)$^{-1}$	C Clamping Force/N	D Preheating Temperature/K
value	1500	3.6	120	333.15 K

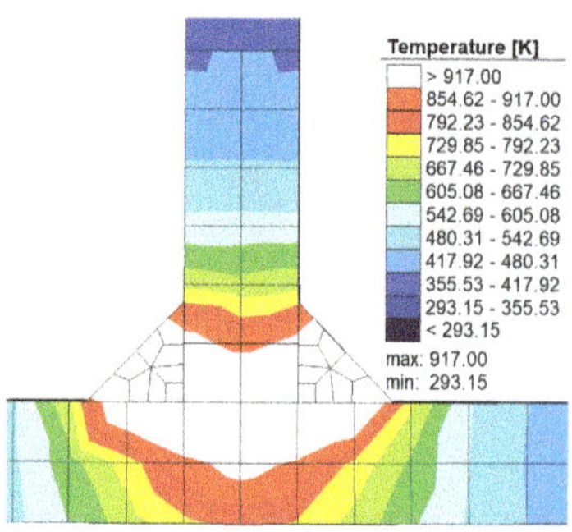

Figure 14. Cross-sectional morphology of the molten pool with optimized parameter combinations.

The variation curves of post-weld equivalent force and total deformation at the weld, corresponding to the optimized parameter combination, are illustrated in Figure 15a,b. At 150 s post-welding, the equivalent force at the weld is measured at 138.110 MPa, while the deformation is recorded at 0.138 mm. A comparison with the orthogonal test schemes indicates that, under the optimized process parameters, the post-weld equivalent force and the amount of deformation are relatively minimal. These optimized parameter combinations can be implemented in the actual welding process to achieve a more favorable distribution of post-weld equivalent force and deformation, ultimately enhancing the overall quality of the welded joints.

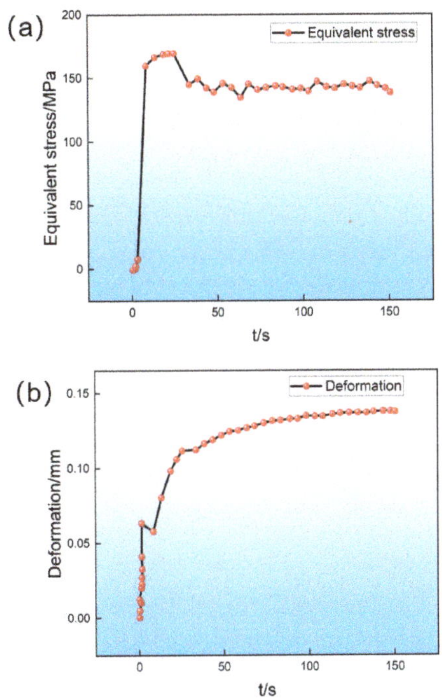

Figure 15. Variation curves of equivalent force and deformation at welded joints. (**a**) Equivalent force (**b**) deformation.

3.4. Study of Welding Sequence of Three Stringer T-Joints

This study builds upon the optimized process parameters for single stringer T-joints, which were derived from orthogonal tests. Using these optimized parameters, we further investigate the impact of different welding sequences on post-weld equivalent stresses and deformations in three stringer T-joints. The simulation model used for this analysis is illustrated in Figure 16.

The primary focus of this section is to examine the effects of three distinct welding sequences on the post-weld equivalent stresses and deformations of the T-joints, with the specific sequences outlined in Table 8. The time interval between the welding of each stringer is set at 3 s.

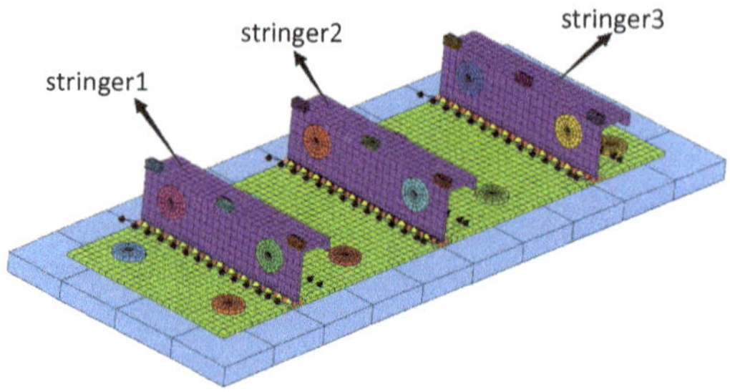

Figure 16. Simulation model of three stringer T-joints.

Table 8. Stringer welding sequence table.

Welding Sequence	1	2	3
Case 1	stringer 2	stringer 1	stringer 3
Case2	stringer 1	stringer 3	stringer 2
Case3	stringer 1	stringer 2	stringer 3

The cloud plots illustrating the distribution of post-weld equivalent stress and total deformation for the three welding sequences are presented in Figures 17 and 18, respectively.

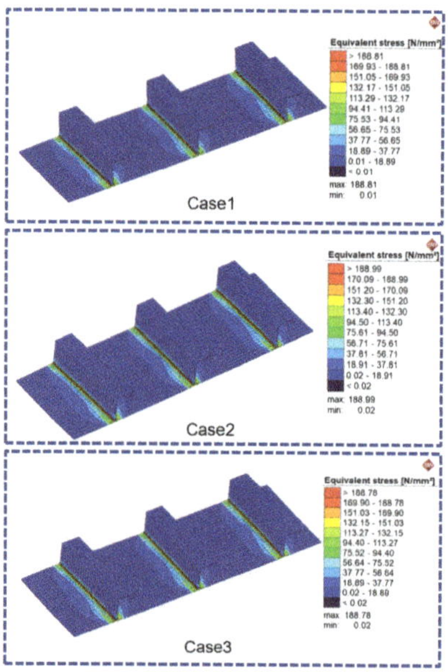

Figure 17. Cloud plot of equivalent stress distribution.

As shown in Figure 17, the maximum post-weld equivalent stresses for the three different welding sequences are 188.81 MPa, 188.99 MPa, and 188.78 MPa, respectively. There is no significant variation in the values, indicating that the three welding sequences have minimal impact on the magnitude and distribution of post-weld equivalent stresses.

In contrast, Figure 18 presents the maximum total post-weld deformations for the three welding sequences, which are 0.95 mm, 1.70 mm, and 0.97 mm, respectively. The welding sequence in Case 2, where the two outer stringers are welded first followed by the center stringer, results in a noticeably higher total deformation compared to the other two sequences. Conversely, the welding sequence in Case 1, where the center stringer is welded first followed by the two outer stringers, shows the smallest total deformation, making it the most optimal welding sequence.

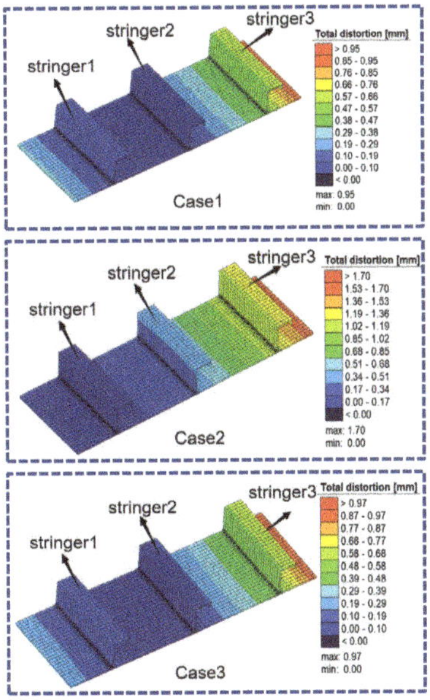

Figure 18. Cloud plot of total deformation distribution.

A plausible explanation for this is that some deformation is inevitably generated during welding. With an optimal welding sequence, such as Case 1, the deformation produced during subsequent welding operations compensates for the deformation from earlier stages. In contrast, a less effective welding sequence, such as Case 2, leads to the accumulation of deformation from each welding stage, resulting in a significantly higher total deformation after welding.

Therefore, it can be concluded that welding the center stringer first, followed by the two outer stringers (as in Case 1), yields more favorable results. This sequence minimizes both post-weld equivalent stress and total deformation, making it the optimal welding order for three-stringer T-joints.

4. Conclusions

This study investigates the laser welding process of 1.8 mm-thick 6056 aluminum alloy stringer and skin T-joints using Simufact Welding simulations. The findings highlight the significant role of welding parameters and sequencing in determining post-weld joint quality, with the following key conclusions:

1. Simulation Validation: The finite element simulation was validated with a maximum error of 6.75% in the molten pool morphology, confirming its accuracy for further analyses.
2. Melt Pool Stability: Bilateral synchronous laser welding demonstrated superior molten pool stability compared to bilateral asynchronous welding, making it the preferred welding method for this study.
3. Effect of Welding Parameters: Among the parameters tested, laser power significantly influenced post-weld equivalent stress, showing a non-linear relationship. Preheating temperature had a major impact on deformation, with a peak deformation reduction at 333.15 K. The optimal welding parameters identified are a laser power of 1500

W, welding speed of 3.6 m/min, fixture clamping force of 120 N, and preheating temperature of 333.15 K.
4. Influence of Welding Sequence: While welding sequence had a minimal effect on post-weld equivalent stress, it notably influenced deformation. The most effective welding sequence involved starting with the middle stringer, followed by the two side stringers, to minimize deformation.

These findings provide a foundation for optimizing welding processes in the manufacture of aluminum alloy T-joints, ensuring improved joint quality and performance.

Author Contributions: Conceptualization, J.P.; methodology, J.P. and S.X.; software, J.P., S.X. and Z.Y. (Zhibin Yang); validation, J.P., S.X. and Z.Y. (Zhibin Yang); formal analysis, J.P., S.X. and X.W.; investigation, S.X., T.C.; resources, J.P., X.W., X.Y. and L.Y.; data curation, J.P., S.X. and T.C.; writing—original draft preparation, J.P. and S.X.; writing—review and editing, J.P., X.W., Z.N., Z.L., Z.Y. (Zhipeng Yuan), J.S. and Z.Y. (Zhibin Yang); visualization, S.X.; supervision, J.P. and X.W.; project ad-ministration, J.P.; funding acquisition, X.W. All authors have read and agreed to the published version of the manuscript.

Funding: This research was funded by Henan Province Science and Technology Research and Development Joint Fund (Industry Category): Key Technology Research and Industrialization of High speed Oil free Lubrication Shaft Pump and Motor for Hydrogen Use (No. 225101610002), the Key Scientific Research Projects of Higher Education Institutions in Henan Province (No. 24A460017).

Data Availability Statement: Data is contained within the article.

Acknowledgments: The authors acknowledge the support provided by the Henan Engineering Research Center of Special Materials and Application of Water Conservancy and hydropower Engineering.

Conflicts of Interest: Jin Peng was employed by Ningbo Intelligent Machine Tool Research Institute Co., Ltd. of China National Machinery Institute Group. Xiaokai Yu and Luqiang Yang was employed by Luoyang Bearing Research Institute Co., Ltd. The remaining authors declare that the research was conducted in the absence of any commercial or financial relationships that could be construed as a potential conflict of interest.

References

1. Starke, E.A.; Staley, J.T. Application of Modern Aluminum Alloys to Aircraft. *Prog. Aerosp. Sci.* **1996**, *32*, 131–172. [CrossRef]
2. Zhou, Z.; Zhan, X.; Gao, Z.; Yan, T.; Liu, J. Effect of Welding Direction on Deformation of Ti6Al4V Alloy Coplanar Double Lap-Joint Produced by Dual Laser Beam Bilateral Synchronous Welding. *Opt. Laser Technol.* **2020**, *131*, 106447. [CrossRef]
3. Liu, T.; Zhan, X.; Zhao, Y.; Bai, M.; Gong, X. Study on 2219 Aluminum Alloy T-Joint during Dual Laser-Beam Bilateral Synchronous Welding: Effect of the Welding Speed and Incident Beam Angle on Grain Morphology. *Opt. Laser Technol.* **2019**, *119*, 105594. [CrossRef]
4. Zhang, X.; Li, L.; Chen, Y.; Yang, Z.; Zhu, X. Experimental Investigation on Electric Current-Aided Laser Stake Welding of Aluminum Alloy T-Joints. *Metals* **2017**, *7*, 467. [CrossRef]
5. Bhagatji, J.D.; Morris, C.; Sridhar, Y.; Bhattacharjee, B.; Kaipa, K.N.; Kravchenko, O.G. Effect of Post-Cured through Thickness Reinforcement on Disbonding Behavior in Skin–Stringer Configuration. *Materials* **2024**, *17*, 3389. [CrossRef]
6. Zhao, H.; Xi, J.; Zheng, K.; Shi, Z.; Lin, J.; Nikbin, K.; Duan, S.; Wang, B. A Review on Solid Riveting Techniques in Aircraft Assembling. *Manuf. Rev.* **2020**, *7*, 40. [CrossRef]
7. Zhan, X.; Xia, L.; Wu, Y.; Yu, H.; Feng, X.; Xia, P. Fracture Mechanism on Al-Li Alloy T-Joint Welded by Dual Laser-Beam Bilateral Synchronous Welding. *Proc. Inst. Mech. Eng. Part B J. Eng. Manuf.* **2019**, *233*, 2074–2088. [CrossRef]
8. Chen, S.; Zhao, Y.; Tian, S.; Gu, Y.; Zhan, X. Study on Keyhole Coupling and Melt Flow Dynamic Behaviors Simulation of 2219 Aluminum Alloy T-Joint during the Dual Laser Beam Bilateral Synchronous Welding. *J. Manuf. Process.* **2020**, *60*, 200–212. [CrossRef]
9. Han, B.; Tao, W.; Chen, Y.; Li, H. Double-Sided Laser Beam Welded T-Joints for Aluminum-Lithium Alloy Aircraft Fuselage Panels: Effects of Filler Elements on Microstructure and Mechanical Properties. *Opt. Laser Technol.* **2017**, *93*, 99–108. [CrossRef]
10. Zhang, X.; Guo, Y.; Fan, X.; Zhao, J.; Yan, B.; Zhang, L. Multi-Objective Optimization of Laser Welding Process Parameters of Steel/Al Based on BO-RF and MOJS. *J. Mech. Sci. Technol.* **2024**, *38*, 861–871. [CrossRef]
11. Wang, Z.; Gao, M. Numerical Simulations of Oscillating Laser Welding: A Review. *J. Manuf. Process.* **2024**, *119*, 744–757. [CrossRef]

12. Jiménez-Xamán, M.; Hernández-Hernández, M.; Tariq, R.; Landa-Damas, S.; Rodríguez-Vázquez, M.; Aranda-Arizmendi, A.; Cruz-Alcantar, P. Numerical Simulations and Mathematical Models in Laser Welding: A Review Based on Physics and Heat Source Models. *Front. Mech. Eng.* **2024**, *10*, 1325623. [CrossRef]
13. Zhen, W.; Li, H.; Wang, Q. Simulation of Residual Stress in Aluminum Alloy Welding Seam Based on Computer Numerical Simulation. *Optik* **2022**, *258*, 168785. [CrossRef]
14. Efa, D.A. Enhancing the Efficiency of Laser Beam Welding: Multi-Objective Parametric Optimization of Dissimilar Materials Using Finite Element Analysis. *Int. J. Adv. Manuf. Technol.* **2024**, *133*, 4525–4541. [CrossRef]
15. Zain-ul-abdein, M.; Nélias, D.; Jullien, J.-F.; Deloison, D. Experimental Investigation and Finite Element Simulation of Laser Beam Welding Induced Residual Stresses and Distortions in Thin Sheets of AA 6056-T4. Mater. *Sci. Eng. A* **2010**, *527*, 3025–3039. [CrossRef]
16. Vrtiel, Š.; Behúlová, M. Analysis of Laser Beam Welding of the S650MC High Strength Steel Using Numerical Simulation. *IOP Conf. Ser. Mater. Sci. Eng.* **2019**, *465*, 012010. [CrossRef]
17. Khoshroyan, A.; Darvazi, A.R. Effects of Welding Parameters and Welding Sequence on Residual Stress and Distortion in Al6061-T6 Aluminum Alloy for T-Shaped Welded Joint. *Trans. Nonferrous Met. Soc. China* **2020**, *30*, 76–89. [CrossRef]
18. Duggirala, A.; Kalvettukaran, P.; Acherjee, B.; Mitra, S. Numerical Simulation of the Temperature Field, Weld Profile, and Weld Pool Dynamics in Laser Welding of Aluminium Alloy. *Optik* **2021**, *247*, 167990. [CrossRef]
19. Alghamdi, A.; Alharthi, H.A. Finite Element Simulation of the Effect of Phase Transformation on Residual Stress in a Thick Section T-Joint. *Crystals* **2022**, *12*, 1422. [CrossRef]
20. Lu, Y.; Deng, Y.; Shi, L.; Jiang, L.; Gao, M. Numerical Simulation of Thermal Flow Dynamics in Oscillating Laser Welding of Aluminum Alloy. *Opt. Laser Technol.* **2023**, *159*, 109003. [CrossRef]
21. Sudnik, W.; Radaj, D.; Breitschwerdt, S.; Erofeew, W. Numerical Simulation of Weld Pool Geometry in Laser Beam Welding. *J. Phys. D Appl. Phys.* **2000**, *33*, 662–671. [CrossRef]
22. Lu, S.P.; Dong, W.C.; Li, D.Z.; Li, Y.Y. Numerical Simulation for Welding Pool and Welding Arc with Variable Active Element and Welding Parameters. *Sci. Technol. Weld. Join.* **2009**, *14*, 509–516. [CrossRef]
23. Kazemi, K.; Goldak, J.A. Numerical Simulation of Laser Full Penetration Welding. *Comput. Mater. Sci.* **2009**, *44*, 841–849. [CrossRef]
24. Tsirkas, S.A.; Papanikos, P.; Kermanidis, T. Numerical Simulation of the Laser Welding Process in Butt-Joint Specimens. *J. Mater. Process. Technol.* **2003**, *134*, 59–69. [CrossRef]
25. Xia, P.; Wang, C.; Mi, G.; Zhang, M.; Xiong, L.; Zhang, X.; Zhai, C.; Feng, X.; Hu, Y. Numerical Simulation of Molten Pool Flow Behavior and Keyhole Evolution Behavior in Dual-Laser Beam Oscillating Bilateral Synchronous Welding of T-Joints. *Int. J. Heat Mass Transf.* **2023**, *209*, 124114. [CrossRef]
26. Yang, Z.; Tao, W.; Li, L.; Chen, Y.; Shi, C. Numerical Simulation of Heat Transfer and Fluid Flow during Double-Sided Laser Beam Welding of T-Joints for Aluminum Aircraft Fuselage Panels. *Opt. Laser Technol.* **2017**, *91*, 120–129. [CrossRef]
27. Zaeh, M.F.; Langhorst, M. A Modelling Approach for the Manufacturing Process Chain of Composite Lightweight Structures. *Adv. Mater. Res.* **2008**, *43*, 157–166. [CrossRef]
28. Wang, L.; Miao, Y.; Hu, G.; Zhao, Y.; Deng, Q. Numerical Simulations and Experimental Verification of T-Structure Welding Deformation Using the Step-by-Step Loading Inherent Strain Method. *Metals* **2023**, *13*, 1312. [CrossRef]
29. Pyo, C.; Kim, J.; Kim, Y.; Kim, M. A Study on a Representative Heat Source Model for Simulating Laser Welding for Liquid Hydrogen Storage Containers. *Mar. Struct.* **2022**, *86*, 103260. [CrossRef]
30. Goldak, J.; Chakravarti, A.; Bibby, M. A New Finite Element Model for Welding Heat Sources. *Metall. Trans. B* **1984**, *15*, 299–305. [CrossRef]
31. Kik, T. Heat Source Models in Numerical Simulations of Laser Welding. *Materials* **2020**, *13*, 2653. [CrossRef] [PubMed]
32. Unni, A.K.; Vasudevan, M. Determination of Heat Source Model for Simulating Full Penetration Laser Welding of 316 LN Stainless Steel by Computational Fluid Dynamics. *Mater. Today Proc.* **2021**, *45*, 4465–4471. [CrossRef]
33. Altay, M.; Aydin, H. Parameter Optimization, Microstructural and Mechanical Properties of Fiber Laser Lap Welds of DP1200 Steel Sheets. *Measurement* **2024**, *235*, 114822. [CrossRef]
34. Peng, J.; Wang, X.; Xu, H.; Yang, J.; Zhang, F. Effect of Droplet Filling Position on Dynamic Behavior of Molten Pool in Laser Welding. *Infrared Laser Eng.* **2020**, *49*, 20200025. [CrossRef]

35. Yang, Z. Study on the Characteristics and Molten Pool Behavior of Double-Sided Laser Welding for Aluminum Alloy Aircraft Fuselage Panels. Ph.D. Thesis, Harbin Institute of Technology, Harbin, China, 2013.
36. Du, Y.; Wang, D.; Liu, X.; Gong, B.; Deng, C. Crack Propagation Angle at the Root of Fillet Welds in T-Joints under Load. *Weld. J.* **2021**, *42*, 21–27+98.

crystals

MDPI

Article

Experimental Study of Reaming Sizes on Fatigue Life of Cold-Expanded 7050-T7451 Aluminum Alloy

Muyu Guan, Qichao Xue *, Zixin Zhuang, Quansheng Hu and Hui Qi

College of Aerospace and Civil Engineering, Harbin Engineering University, Harbin 150000, China
* Correspondence: xueqichao@hrbeu.edu.cn

Abstract: The split-sleeve cold expansion technology is widely used in the aerospace industry, particularly for fastening holes, to enhance the fatigue life of components. However, to ensure proper assembly and improve surface integrity, reaming of the cold-expanded holes is necessary. This study investigates the effects of cold expansion and reaming processes on the fatigue performance of 7050-T7451 aluminum alloy. Fatigue tests, residual stress measurements, and microstructural analyses of the hole edges were conducted on specimens with four different hole diameters after cold expansion and reaming. It was found that the depth of reaming significantly affects fatigue life. During the cold expansion process, the compressive residual stress formed around the hole effectively improves fatigue performance. The experiments demonstrated that reaming by 0.2 mm to 0.4 mm helps eliminate minor defects, thereby improving fatigue life. However, reaming beyond 0.5 mm may lead to stress relief and the removal of dense grains at the hole edges, reducing fatigue life. Therefore, determining the optimal reaming size is crucial for enhancing the reliability of aerospace fasteners.

Keywords: split-sleeve cold expansion; fatigue; reaming; aluminium alloy; residual stress

1. Introduction

With the advancement of the aerospace industry, the study of the reliability of flight vehicles has become highly significant. Most connection methods for flight vehicles are mechanical, such as riveted and bolted connections. The most critical parts of the aircraft involve mechanical connections like rivets and bolts [1–3]. Parts that need frequent replacement, like wings and landing gear, are inevitably subjected to cyclic loads such as mechanical vibrations and centrifugal forces [4,5]. Fatigue cracks can develop in fastening holes with stress concentrations, and fatigue damage originating from these holes is the primary form of fatigue damage in aircraft [6,7]. The 7XXX aluminum alloy of the Al-Zn-Mg-Cu is widely used in the aerospace industry due to its excellent mechanical properties and is a key material in the study of metal fatigue properties [8]. Cold expansion is now widely used in the aircraft industry, particularly for fastening holes [9–11]. In this technique, an oversized ball or mandrel is forced through the fastening holes to locally yield the material, inducing a tangential compressive residual stress field around the holes. This counteracts external loads and reduces stress concentration at the hole edges, improving fatigue resistance [12]. It is generally recognized that cold expansion is the most effective way to improve fatigue resistance. Compressive residual stresses introduced during cold expansion are widely recognized for extending the fatigue life of hole structures [13–15]. Cold expansion methods have been developed over the last decade, with the main techniques being direct cold expansion and split-sleeve cold expansion. The open-seam cold expansion method significantly reduces damage to the hole wall, maintains the correct radial expansion, and allows access from either side [16,17]. Cold expansion of split-sleeve has been widely used to extend the lifespan of fasteners in aircraft [18–21]. Cold expansion of the split-sleeve leaves a ridge in the hole wall after expansion due to cracks in the split-sleeve [19]. The plastic deformation is not axisymmetrically distributed around the hole, and cracking may occur, usually near the crack in the sleeve [22]. In order to avoid material defects caused by the convex ridge after cold expansion and to ensure proper assembly of the fastener, the fastening holes are usually treated with a reaming process [23].

The principle of cold expansion to improve fatigue life is relatively complex, and many scholars have studied the factors and mechanisms involved. Yan et al. [24] conducted experimental studies showing that the cold expansion of split-sleeve resulted in a 1.5 to 3 times increase in the lifespan of TC4 titanium alloys. Li et al. [25] investigated residual stresses around the holes of 7050 aluminum alloy plates at different expansion rates, finding that maximum residual stresses occurred at a 4% expansion rate. Wang et al. [26] found that cold expansion can form gradient fine grains on the hole surface, inhibiting fatigue cracking. Yang et al. [27] found that cold expansion causes a hardened layer to form around the hole, with hardness decreasing up to 450 μm from the hole. Tang et al. [28] proposed a two-scale modeling approach combining the macro-scale cyclic ontology model with crystal plasticity theory, finding that residual stress and the plastic layer together determine the onset. The cold-expanded holes need to undergo a reaming process before they can be put into service. Reaming removes minor defects on the hole edges caused by cold expansion and adapts the fasteners accordingly. Liu et al. [29] analyzed through simulation that the fatigue life of cold-expanded holes increases with greater reaming depth. Alexandre et al. [30] analyzed the change in the stress state of the hole during reaming. However, there is a lack of research on the experimental study of the fatigue life of reamed holes after cold expansion, the changes in surface integrity, and the reasons for decreased fatigue resistance due to oversize reaming.

This paper investigates the 7050-T7451 alloy to study the effects of the reaming process on cold-expanded plates with holes. Four different diameters of cold-expanded holes were reamed to varying depths. Fatigue tests were conducted to assess the impact of different reaming sizes on the fatigue life of these holes. Scanning electron microscope (SEM) analysis examined the microscopic fracture morphology of specimens with varying fatigue lives under different conditions. Electron Backscatter Diffraction (EBSD) analyzed local deformation in reamed and unreamed holes after cold expansion, aiming to reveal the relationship between reaming behavior and fatigue life at the microscopic level. Residual stress before and after reaming was measured using X-ray diffraction. The study identified variations in fatigue life at different reaming depths, the effect of reaming on crystal refinement after cold expansion, and changes in residual stress distribution. These findings provide important guidance for the installation of fasteners in aircraft.

2. Materials and Methods

2.1. Experimental Materials

The experimental material selected for this study is 7050-T7451 aluminum alloy, part of the Al-Zn-Mg-Cu series, widely used in aerospace applications [8]. Its chemical composition and main mechanical properties are shown in Tables 1 and 2. The tensile properties, determined by an INSTRON 8801 (Instron, Norwood, MA, USA) electronic universal testing machine, are illustrated in Figure 1. The alloy is supplied as a 10 mm thick sheet and is designed according to the Chinese standard HB/Z 170-2005 [31] for the cold expansion process of aerospace metal parts. Fatigue specimens with holes are shown in Figure 2, featuring four hole diameters: 4.3 mm, 5.3 mm, 7.18 mm, and 11.1 mm. A split-sleeve from a domestic company was used in this experiment. The sleeve material is 301 stainless steel, with localized material 12Cr7Ni7 as per GB/T 3280-2015 [32]. The thickness of the split-sleeve used for cold expansion is 0.152 mm for Figure 2a,b, 0.203 mm for Figure 2c, and 0.256 mm for Figure 2d. The cold expansion technique described in this paper involves the cold expansion of a split-sleeve; the cold expansion process and cold-expanded specimens as shown in Figure 3. The split-sleeve is secured in the hole, with a mandrel positioned at the center. The mandrel is swiftly pulled through the end of the hole to generate residual stresses. After removing the mandrel and split-sleeve, the hole is reamed at a machining center to achieve the required diameter for the experiment.

Table 1. Chemical composition of 7050-T7451 aluminum alloy.

Elements	Si	Zn	Mg	Cu	Fe	Mn	Cr	Zr	Al
Wt.%	0.03	6.02	2.31	2.04	0.11	0.003	0.009	0.1	Rest

Table 2. Mechanical properties of 7050-T7451 aluminum alloy.

Density ρ (kg/m^3)	Young's Modulus E (Gpa)	Poisson's Radio ν	Tensile Strength σ_b (MPa)	Yield Strength σ_s (MPa)	Elongation A (%)	Fracture Toughness K_{IC} (MPa·m$^{1/2}$)
2830	70.3	0.33	564.7	479.4	13.3	30

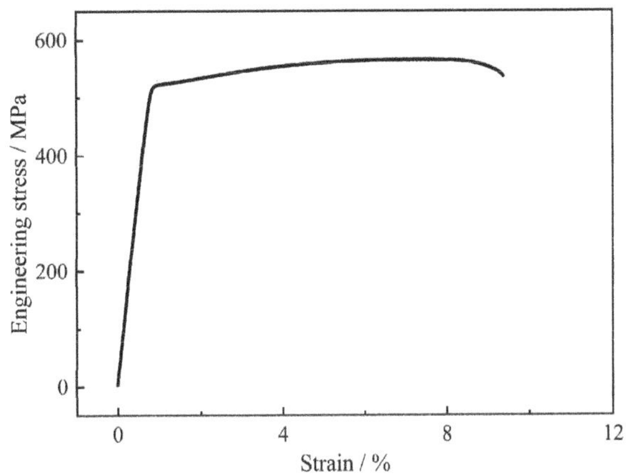

Figure 1. Stress–strain curve of 7050-T7451 aluminum alloy.

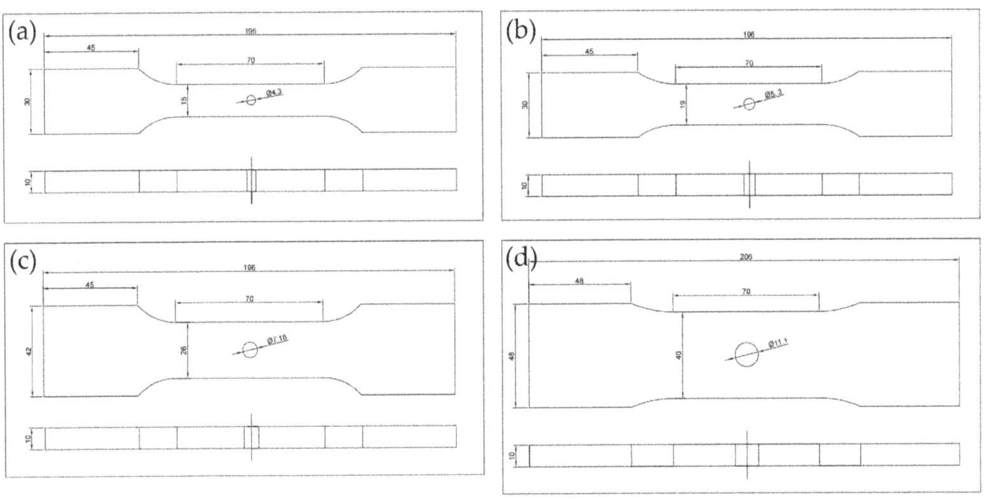

Figure 2. Fatigue specimen size with different hole diameters: (**a**) 4.3 mm, (**b**) 5.3 mm, (**c**) 7.18 mm, (**d**) 11.1 mm.

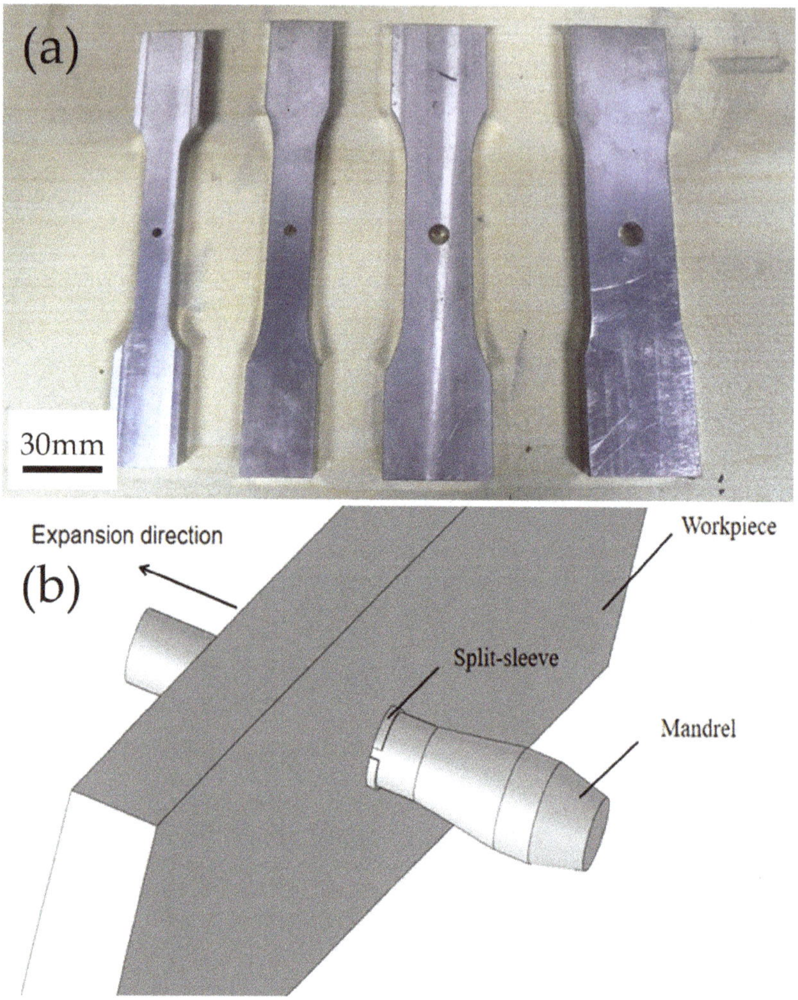

Figure 3. (**a**,**b**) Cold-expanded specimens and the illustration of the split-sleeve cold expansion process.

2.2. Fatigue Test

To compare the effects of different reaming depths on the fatigue life of cold-expanded 7050-T7451 aluminum alloy plates, various specimens were cold-expanded with approximately 4% expansion. The expansion E was calculated as follows:

$$E_r = \frac{(D_0 + 2d) - D}{D} \tag{1}$$

In the equation, D_0 is the diameter of the cold-expanded mandrel, D is the diameter of the initial hole before expansion, and d is the thickness of the split-sleeve. The extrusion rate is calculated in Table 3.

Table 3. The size of the cold expansion tool and cold expansion rate.

Initial Diameter (mm)	Mandrel Diameter (mm)	Split-Sleeve Thickness (mm)	Expansion Ratio E_r (%)
4.3	4.17	0.152	4.04
5.3	5.21	0.152	4.03
7.18	7.06	0.203	3.98
11.1	11.04	0.253	4.01

The specimens of each size were reamed with five different final hole diameters and each reaming depth was equipped with four experiments. The fatigue tests were carried out on INSTRON 8801. The uniaxial fatigue tests were performed with constant load amplitude, sinusoidal loading Stress ratio R = 0.1, stress amplitude 108 MPa, and loading frequency 50 Hz. The experimental matrices for fatigue tests are given in Table 4.

Table 4. Parameters of fatigue specimens with different initial and final hole diameters.

Specimen	Quantity (PCS)	Initial Diameter (mm)	Expansion Ratio E_r (%)	Final Diameter (mm)	Load Spectra Stress Amplitude (MPa)	Stress Ratio R
A1-01	4	4.3	4.04	4.6	108	0.1
A1-02	4	4.3	4.04	4.7	108	0.1
A1-03	4	4.3	4.04	5.0	108	0.1
A1-04	4	4.3	4.04	5.2	108	0.1
A1-05	4	4.3	4.04	5.4	108	0.1
A2-01	4	5.3	4.03	5.4	108	0.1
A2-02	4	5.3	4.03	5.9	108	0.1
A2-03	4	5.3	4.03	6.1	108	0.1
A2-04	4	5.3	4.03	6.3	108	0.1
A2-05	4	5.3	4.03	6.5	108	0.1
A3-01	4	7.18	3.98	7.6	108	0.1
A3-02	4	7.18	3.98	7.8	108	0.1
A3-03	4	7.18	3.98	8.0	108	0.1
A3-04	4	7.18	3.98	8.2	108	0.1
A3-05	4	7.18	3.98	8.4	108	0.1
A4-01	4	11.1	4.01	11.7	108	0.1
A4-02	4	11.1	4.01	12.0	108	0.1
A4-03	4	11.1	4.01	12.3	108	0.1
A4-04	4	11.1	4.01	12.6	108	0.1
A4-05	4	11.1	4.01	12.9	108	0.1
A1-00	4	4.3	0%	4.3	108	0.1
A2-00	4	5.3	0%	5.3	108	0.1
A3-00	4	7.18	0%	7.18	108	0.1
A4-00	4	11.1	0%	11.1	108	0.1

2.3. Characterization

An Electron Backscatter Diffraction (EBSD) system was used to observe the microstructure near the pore walls of specimens after cold expansion without reaming, as well as those cold-expanded and then reamed to 5.0 mm. The EBSD samples were prepared by vibratory polishing with silica suspension, and the data were analyzed using HKL-Channel 5 software. A scanning electron microscope (SEM, SU5000, HITACHI, Hitachi, Japan) was used to analyze the fracture after the fatigue test. X-ray diffraction was employed to measure eight points along the radial direction starting from the hole edge, with intervals of 0.5 mm, using a Cr-Kα X-ray source (wavelength λ = 2.291 Å).

3. Results and Discussion

3.1. Fatigue Life Behavior

In the study by Alexandre et al. [30], the reaming process of cold-expanded holes was simulated. However, the modeling strategy did not consider the thermal effects during reaming, focusing solely on the impact of material removal on the stress state of the part. As a result, there is still a deviation from real experimental conditions. This paper investigates the relationship between reaming size and fatigue life through extensive fatigue experiments. The cold expansion specimen after fatigue testing is shown in Figure 4.

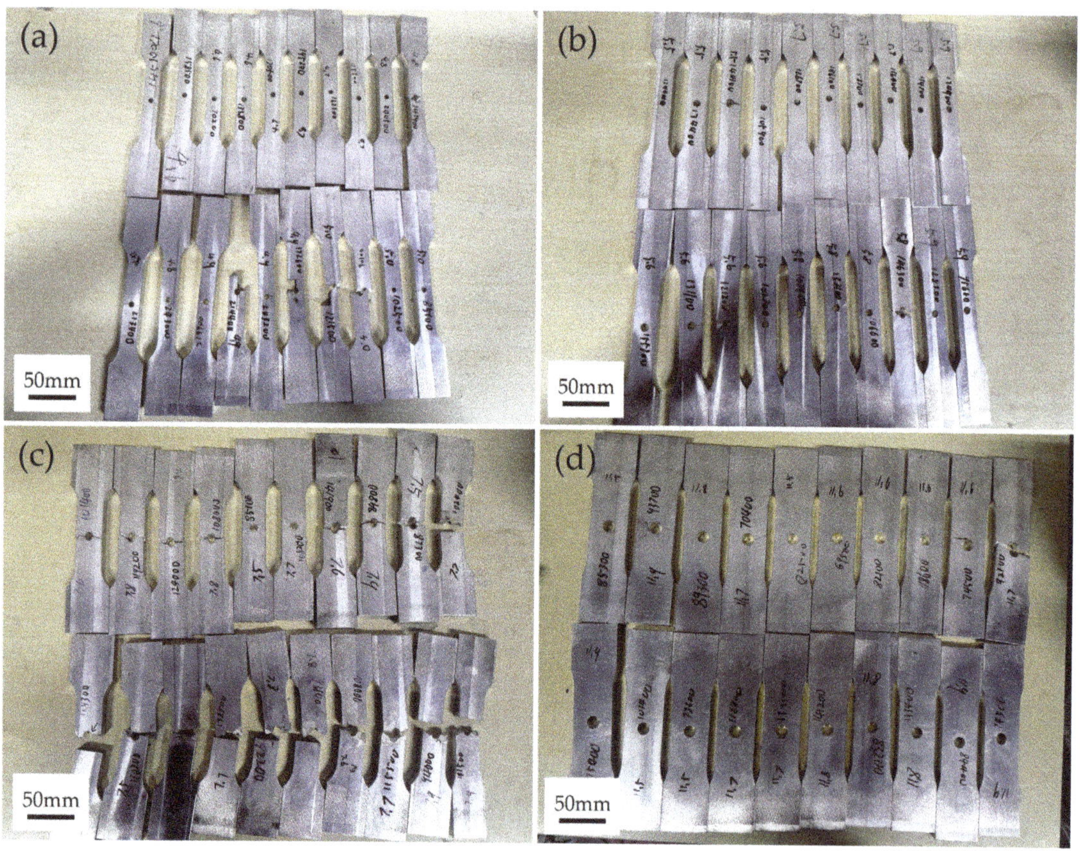

Figure 4. Cold expansion specimens with different initial hole diameters after fatigue testing: (**a**) 4.3 mm; (**b**) 5.3 mm; (**c**) 7.18 mm; (**d**) 11.1 mm.

Figure 5 shows the fatigue life of four different sizes at various final hole diameters. Initially, the fatigue life increases with the reamed size, consistent with the research results of Liu et al. [28]. The trend was similar for specimens with initial hole diameters ranging from 4.3 mm to 11.1 mm. The specimen with an initial hole diameter of 4.3 mm has the highest fatigue life when reamed to 4.8 mm to 4.9 mm; the specimen with an initial hole diameter of 5.3 mm has the highest fatigue life when reamed to 5.7 mm; the specimen with an initial hole diameter of 7.18 mm has the highest fatigue life when reamed to 7.6 mm; and the specimen with an initial hole diameter of 11.1 mm has the highest fatigue life when reamed to 11.8 mm. However, as the reamed hole diameter continues to grow, the fatigue life then decreases. This may be because the reaming process removes the fine-grain hardened layer formed by cold expansion, reducing the material's strength at the hole edge and making fatigue cracks more likely to propagate. Additionally, if the reaming size is too large, it can damage the hole edge plastic layer, reducing constraint in the deeper elastic deformation layer. This weakens the residual stress field's strength, diminishing its positive effect on fatigue life.

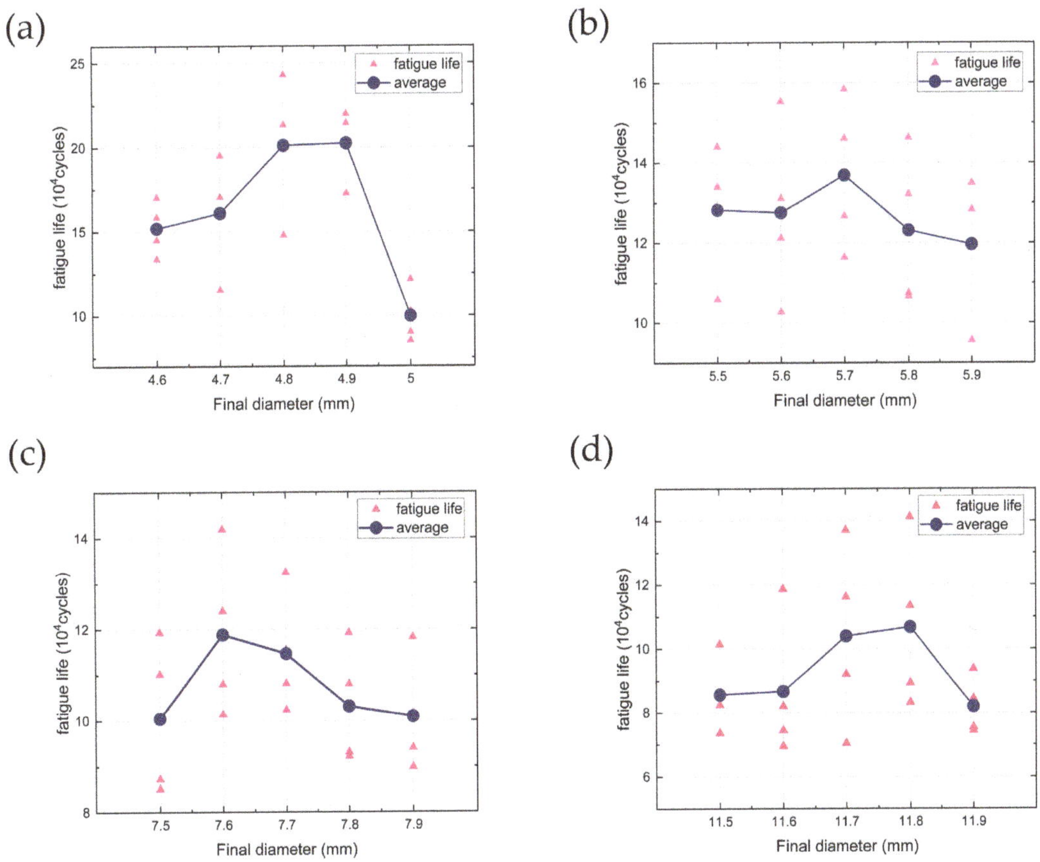

Figure 5. Fatigue life of specimens with different initial hole diameters after being reamed to different final hole diameters: (**a**) 4.3 mm, (**b**) 5.3 mm, (**c**) 7.18 mm, (**d**) 11.1 mm.

In Figure 6, the fatigue life of the reamed specimen with the optimal reamed dimension is compared to that of the untreated specimen. It can be observed that the fatigue life of the properly reamed cold-expanded specimen is approximately 4.4 to 8.7 times higher than that of the untreated specimen.

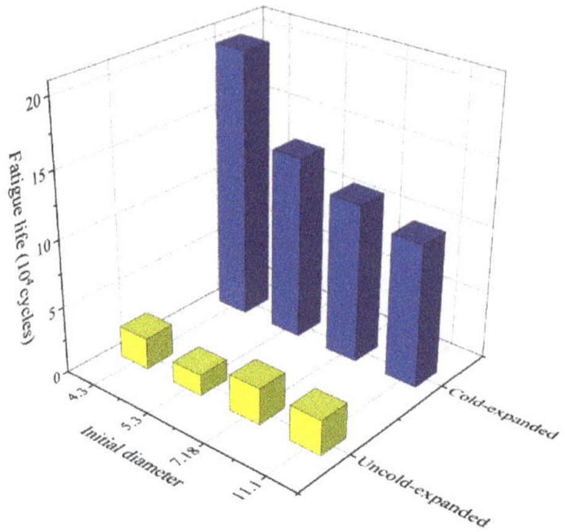

Figure 6. Fatigue life of optimally reamed size specimens and uncold-expanded specimens.

3.2. Observation of Port Morphology

A specimen with an initial hole diameter of 4.3 mm was selected, and microelectron microscopy fracture analysis was performed on the uncold-expanded specimen, the specimen reamed to 4.8 mm after cold expansion, and the specimen reamed to 5.0 mm after cold expansion.

Figure 7 illustrates the typical fracture morphology of the uncold-expanded specimen, which has a fatigue life of 26,500 cycles. The crack initiation point, indicated by an arrow in Figure 7a, shows that the specimen failed in a multi-source fatigue mode. The fatigue crack originated from machining marks and spread from the middle surface of the hole wall. In Figure 7b, the crack initiation zone is clearly visible, along with the crack extension and transient fracture zones. The fatigue expansion zone, shown in Figure 7c, is relatively flat with numerous fatigue bands spaced about 3.14 μm apart. Compared to the crack extension zone, the transient fracture zone is rugged and irregular. As the fatigue crack grows, the effective bearing area decreases, leading to specimen failure after reaching tensile strength. This accounts for about three-fourths of the fracture surface, with numerous plastic deformation features visible in this region, as shown in Figure 7d.

Figure 8 shows the fracture morphology of a specimen cold-expanded and reamed to 4.8 mm, with a fatigue life of 213,400 cycles. The crack extension zone is larger than in uncold-expanded specimens due to residual stresses. The crack originated at the hole entrance, where residual stress reinforcement is minimal, consistent with most studies [4,27]. Tire indentation marks, shown in Figure 8c, indicate effective involvement of compressive residual stresses in the fatigue process. The fatigue strip spacing is about 1.64 μm. The area of the transient fracture zone is compressed, with numerous plastic deformation dimples, as shown in Figure 8d.

Figure 9 depicts the fracture morphology of a specimen reamed to 5.0 mm, with a fatigue life of 90,100 cycles. Compared to the 4.8 mm specimen, this specimen shows more crack initiation sources, not originating at the hole entrance. The expansion zone is shorter, and the transient fracture zone is larger, covering more than one-third of the area, with an average fatigue band spacing of about 2.04 μm. The fatigue expansion rate is higher for the 5.0 mm specimen than for the 4.8 mm specimen, but still lower than that of uncold-expanded specimens. This suggests that larger reaming sizes increase fatigue crack growth rates, highlighting the need to carefully consider reaming depths.

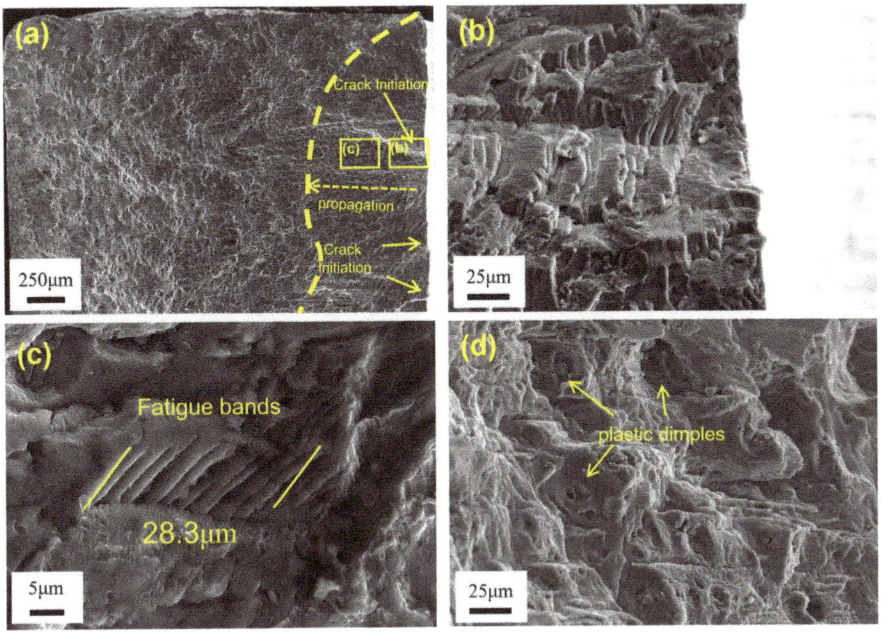

Figure 7. Fracture morphologies of uncold-expanded specimens: (**a**) overall morphology of the port, (**b**) crack source, (**c**) amplified area of the crack source, (**d**) transient fracture area.

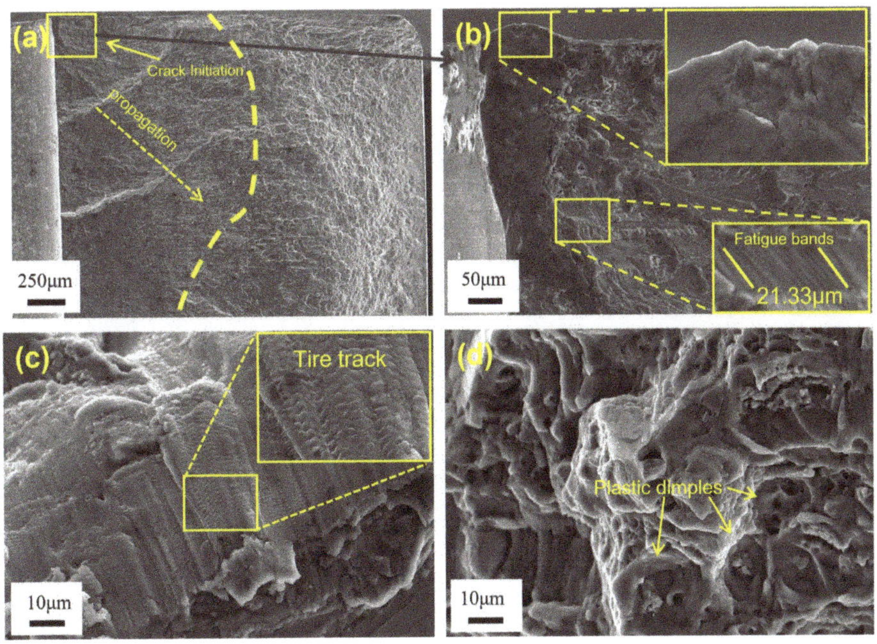

Figure 8. Fracture morphologies of specimens reamed to 4.8 mm after cold expansion: (**a**) overall morphology of the port, (**b**) crack source, crack extension zone, (**c**) tire indentation in the crack extension zone, (**d**) instantaneous fracture zone.

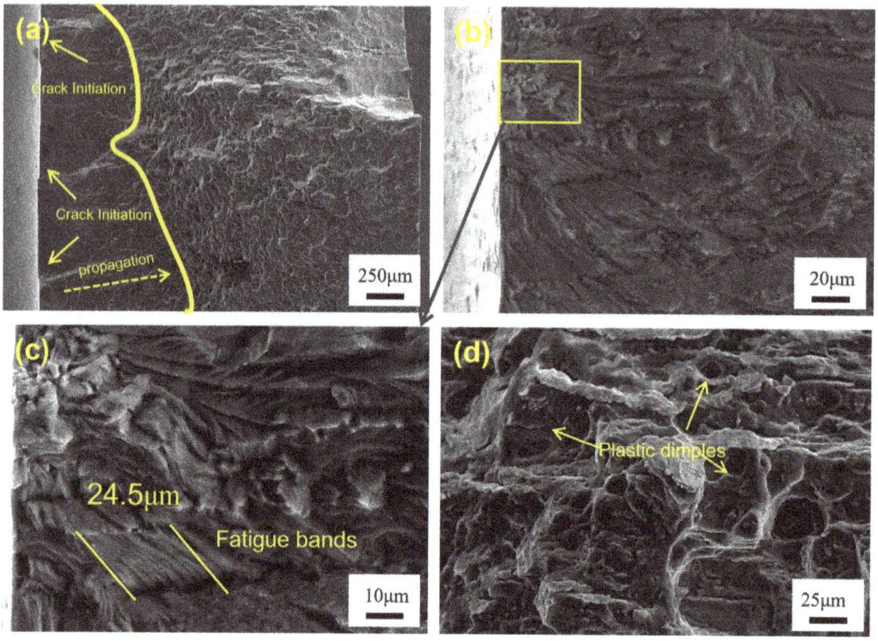

Figure 9. Fracture morphologies of specimens reamed to 5.0 mm after cold extrusion: (**a**) overall morphology of the port, (**b**) crack source, crack extension zone, (**c**) amplified area of the crack source, (**d**) instantaneous fracture zone.

3.3. EBSD Analysis

In this analysis, we examined two specimens. One was cold-expanded with an initial hole diameter of 4.3 mm without reaming, and the other was cold-expanded with an initial hole diameter of 4.3 mm and then reamed to 5.0 mm. Both specimens were analyzed using EBSD. Figure 10a,d show the inverse pole figures (IPF) before and after reaming. Figure 10a indicates that strong crystal refinement formed in the hole wall after cold expansion, with a layer of about 20 μm. It is widely believed that these refined crystals can prevent crack propagation, as cracks must traverse longer paths through the grains, reducing the rate of crack extension [33]. However, this phenomenon disappears after reaming, as shown in Figure 10d, where the crystal refinement layer has been removed. Since the depth of the reaming is much greater than that of the crystal refinement layer, it also removes the ridges of the cold-expanded holes, giving them a more regular shape.

The plastic deformation of the hole is evaluated by the kernel average misorientation (KAM) shown in Figure 10b,e. Significant plastic deformation is observed around the hole after cold expansion, with a deformation layer of about 360 μm. As the hole wall expands outward, the inner material layer deforms more, reaching the yield strength first, while the deeper layer does not reach yield strength but is restrained by the plastic layer, forming residual stresses [25]. In the specimen reamed to 5.0 mm, only a thin layer of deformation remains, which is due to the reaming process, confirming that excessive reaming harms the plastic deformation layer, reducing the constraint on residual stresses and thereby diminishing their beneficial effect on fatigue life.

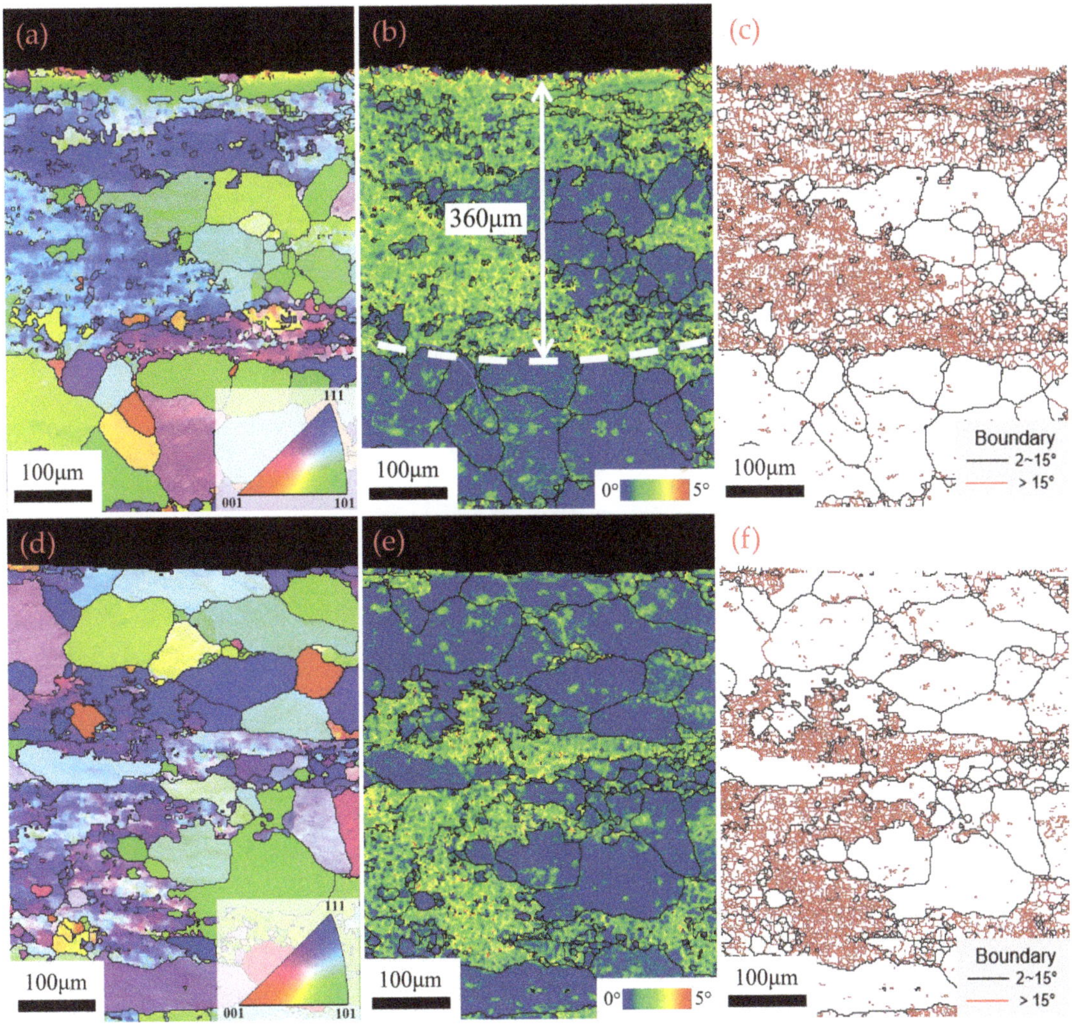

Figure 10. Inverse pole figure, local misorientation average angle map, and grain boundary map of (**a–c**) specimen cold-expanded without reaming and (**d–f**) specimen cold-expanded and reamed to 5.0 mm.

The crystal orientation and grain boundary diagrams are shown in Figure 10c,f. Orientation deviations between 2 and 15° are indicated by black lines, and those above 15° by red lines. The crystal refinement at the hole edges is evident before reaming, but not after. Excessive reaming destroys the crystal refinement layer formed by cold expansion, reducing the fatigue strength of the cold-expanded holes, similar to the findings of Dang et al. [34].

Proper evaluation of the depth of the crystal refinement crystal layer and plastic deformation layer formed around the hole during the cold expansion process is crucial for determining the optimal reaming amount. Wang et al. [26] found that cold expansion of 7075-T6 aluminum alloy with a 4% expansion resulted in a plastic layer about 1200 μm thick. In contrast, Wang et al. [4] observed that cold expansion of TC4 titanium alloy with a 2% expansion formed a much thinner fine-grained layer. Other studies [35,36] observed varying depths of the plastic deformation layer depending on the material and expansion amount. Accurately predicting this thickness is key to guiding reaming dimensions.

3.4. Residual Stress Analysis

Figure 11 shows the distribution of residual stresses along the radial direction for the cold-expanded specimen with an initial hole diameter of 7.18 mm for the unreamed and reamed to 8.4 mm specimens. It can be observed that the intensity of compressive residual stresses in the plate reamed to 8.4 mm is lower than that of the unreamed specimen. The maximum compressive residual stress of the unreamed specimen is 252 MPa, while the maximum residual stress of the specimen reamed to 8.4 mm is only 150 MPa.

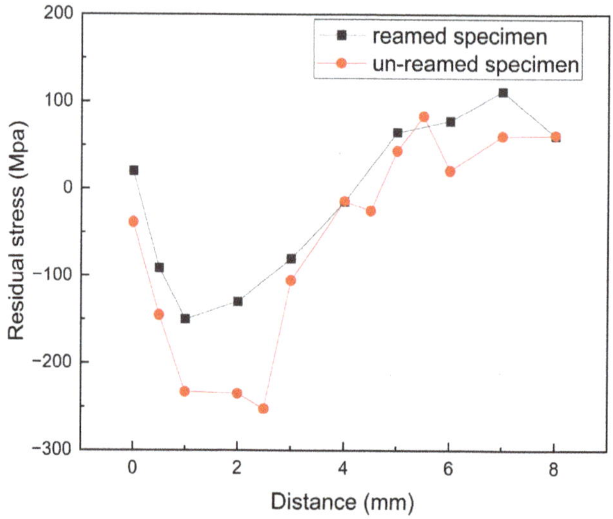

Figure 11. Residual stress distribution in reamed and unreamed specimens after cold expansion.

In the study by Liu et al. [29], it was also found by finite element simulation that the residual compressive stress decreases with the increase in reaming amount. The reason for the decrease in residual stress may be that the reaming size is too large, which reduces or even completely disappears the plastic zone, resulting in the loss of constraints on the residual stress in the plastic zone. The measurement of the residual stresses also confirms the conjecture that the high reaming depth in the fatigue test leads to a reduction in fatigue life. After the residual stresses generated by cold expansion are released, the fatigue gain effect of the residual stresses on the fastening holes is reduced, resulting in the reduction of the fatigue life of the over-reamed specimens with holes.

4. Conclusions

This study investigated the effect of the reaming process on cold-expanded holes by reaming cold-expanded plates with four different hole diameters. By analyzing fatigue life, fracture morphology, microstructure, and residual stress under various reaming conditions, the following conclusions were drawn:

1. Fatigue testing after approximately 4% cold expansion of plates with four different sizes of perforations showed that fatigue life first increases and then decreases with greater reaming. This conclusion applies to apertures ranging from 4.3 mm to 11.1 mm. The maximum fatigue life was about 4.4 to 8.7 times higher than that of the uncold-expanded plate. Excessive reaming results in a significantly smaller crack extension zone in the fatigue specimen, weakening its ability to withstand fatigue.

2. Cold expansion forms a dense grain structure around the hole, creating a crystal refinement layer of approximately 20 μm. Because the height of the convex ridge is much greater than its depth, the reaming process removes this crystal refinement layer.
3. A plastic deformation layer forms at the hole wall, constraining residual stresses in the deeper layers. The reaming process disrupts this plastic layer, releasing the residual stresses. The extent of this effect depends on the reaming depth. Proper assessment of the plastic deformation layer's depth, formed by cold expansion, is crucial in determining the optimal reaming amount.
4. Reaming the cold-expanded plate reduces radial residual stresses around the hole by up to 102 MPa compared to the unreamed plate, due to the loss of stress restraint by the innermost plastic layer. This reduction in residual stresses also decreases the fatigue resistance of the cold-expanded split-sleeve.

Author Contributions: Conceptualization, M.G.; methodology, H.Q.; software, M.G.; validation, H.Q.; formal analysis, Q.H.; investigation, M.G. and Z.Z.; resources, H.Q.; data curation, Z.Z.; writing—original draft preparation, M.G.; writing—review and editing, Q.X. and Q.H.; visualization, Z.Z.; supervision, Q.X.; project administration, Q.X.; funding acquisition, Q.X. All authors have read and agreed to the published version of the manuscript.

Funding: This research was funded by the Natural Science Foundation of Heilongjiang Province (LH2020E070) and the Yantai Science and Technology Innovation Development Program (2022ZDCX002).

Data Availability Statement: In this study, all data are true and reliable. All data are obtained from experimental measurements. The original contributions presented in the study are included in the article.

Conflicts of Interest: The authors declare no conflicts of interest.

References

1. Wang, H. Riveting Sequence Study of Horizontal Stabilizer Assembly Using Finite-Element Analysis and Riveting Equivalent Unit. *J. Aerosp. Eng.* **2014**, *27*, 04014040. [CrossRef]
2. Liu, J.T.; Zhao, A.A.; Ke, Z.Z.; Zhu, Z.D.; Bi, Y.B. Influence of Rivet Diameter and Pitch on the Fatigue Performance of Riveted Lap Joints Based on Stress Distribution Analysis. *Materials* **2020**, *13*, 3625. [CrossRef] [PubMed]
3. Zhao, H.W.; Xi, J.J.; Zheng, K.L.; Shi, Z.S.; Lin, J.G.; Nikbin, K.; Duan, S.H.; Wang, B.W. A review on solid riveting techniques in aircraft assembling. *Manuf. Rev.* **2020**, *7*, 40. [CrossRef]
4. Wang, Q.T.; Li, B.B.; Zheng, Y.M.; Chen, X. Effects of Stress Amplitude and Cold Expansion Strengthening on the Fatigue Property of TC4 Titanium Alloy. *J. Mater. Eng. Perform.* **2023**, *10*, 1–10. [CrossRef]
5. Pitt, S.; Jones, R.J.E.F.A. Multiple-site and widespread fatigue damage in aging aircraft. *Eng. Fail. Anal.* **1997**, *4*, 237–257. [CrossRef]
6. Kumar, S.A.; Babu, N.C.M. Influence of Induced Residual Stresses on Fatigue Performance of Cold Expanded Fastener Holes. In Proceedings of the 5th International Conference on Materials Processing and Characterization (ICMPC), Gokaraju Rangaraju Institution of Engineering & Technology, Hyderabad, India, 12–13 March 2016; pp. 2397–2402.
7. Masserey, B.; Fromme, P. In-situ monitoring of fatigue crack growth using high frequency guided waves. *NDT E Int.* **2015**, *71*, 1–7. [CrossRef]
8. Wang, B.; Liu, Z.Q.; Song, Q.H.; Wan, Y.; Ren, X.P. A Modified Johnson-Cook Constitutive Model and Its Application to High Speed Machining of 7050-T7451 Aluminum Alloy. *J. Manuf. Sci. Eng. Trans. ASME* **2019**, *141*, 011012. [CrossRef]
9. Chakherlou, T.N.; Vogwell, J. The effect of cold expansion on improving the fatigue life of fastener holes. *Eng. Fail. Anal.* **2003**, *10*, 13–24. [CrossRef]
10. Andrew, D.L.; Clark, P.N.; Hoeppner, D.W. Investigation of cold expansion of short edge margin holes with pre-existing cracks in 2024-T351 aluminium alloy. *Fatigue Fract. Eng. Mater. Struct.* **2014**, *37*, 406–416. [CrossRef]
11. Liu, J.; Shao, X.J.; Liu, Y.S.; Yue, Z.F. Effect of cold expansion on fatigue performance of open holes. *Mater. Sci. Eng. A Struct. Mater. Prop. Microstruct. Process.* **2008**, *477*, 271–276. [CrossRef]
12. Yan, W.Z.; Wang, Y.W.; Liang, S.J.; Huo, S.H. Study on the Residual Stress Distribution of Bi-Directional Cold Expansion Process Performed on Open Holes. *J. Press. Vessel Technol. Trans. ASME* **2023**, *145*, 041501. [CrossRef]
13. Lin, Z.L.; Bai, Q.S.; Wang, H.F.; Wu, B.Q.; Li, W.S. Research on the Residual Stress Field of a Compression Bushing-Lug Plate in Cold Expansion Strengthening. *Appl. Sci.* **2023**, *13*, 11694. [CrossRef]
14. Liu, Y.S.; Shao, X.J.; Liu, J.; Yue, Z.F. Finite element method and experimental investigation on the residual stress fields and fatigue performance of cold expansion hole. *Mater. Des.* **2010**, *31*, 1208–1215. [CrossRef]

15. Liu, Y.S.; Liu, J.; Shao, X.J. Study on the Residual Stress Fields, Surface Quality, and Fatigue Performance of Cold Expansion Hole. *Mater. Manuf. Process.* **2011**, *26*, 294–303. [CrossRef]
16. Dey, M.K. *Finite Element Parametric Investigation on the Residual Stresses Induced by Split Sleeve Cold Expansion Process*; Washington State University: Washington, DC, USA, 2021.
17. Karabin, M.E.; Barlat, F.; Schultz, R.W. Numerical and experimental study of the cold expansion process in 7085 plate using a modified split sleeve. *J. Mater. Process. Technol.* **2007**, *189*, 45–57. [CrossRef]
18. Lv, Y.; Dong, M.E.; Zhang, T.; Wang, C.K.; Hou, B.; Li, C.F. Finite Element Analysis of Split Sleeve Cold Expansion Process on Multiple Hole Aluminum Alloy. *Materials* **2023**, *16*, 1109. [CrossRef]
19. Nigrelli, V.; Pasta, S. Finite-element simulation of residual stress induced by split-sleeve cold-expansion process of holes. *J. Mater. Process. Technol.* **2008**, *205*, 290–296. [CrossRef]
20. Guan, D.; Sun, Q. Numerical Simulation of the Cold Expansion and Residual Stress Relaxation for Aluminum Alloy 7050. In Proceedings of the International Conference on Applied Mechanics, Materials and Manufacturing (ICAMMM 2011), Shenzhen, China, 18–20 November 2011; pp. 1656–1661.
21. Özdemir, A.T. Experimental assessment of the redistribution of 3D residual stresses during early fatigue at split-sleeve cold expanded reamed A/C fastener holes. *Sci. Iran.* **2018**, *25*, 1153–1168. [CrossRef]
22. Barlat, F.; Karabin, M.E.; Schultz, R.W. A modified split sleeve for the cold expansion process and application to 7085 plate. In Proceedings of the 10th ESAFORM Conference on Material Forming, Zaragoza, Spain, 18–20 April 2007; pp. 343–348.
23. Huang, Y.J.; Guan, Z.D.; Li, Z.S.; Liu, F.Q.; He, Y.M.; Dan, H. An Numerical Investigation on the Effect of the Combination of Cold Expansion and Interference Fitting on Fatigue Life Improvement of a 7075-T6 Aluminum Alloy Single Plate. In Proceedings of the 8th International Conference on Mechanical and Aerospace Engineering (ICMAE), Prague, Czech Republic, 22–25 July 2017; pp. 632–638.
24. Yan, W.Z.; Wang, X.S.; Gao, H.S.; Yue, Z.F. Effect of split sleeve cold expansion on cracking behaviors of titanium alloy TC4 holes. *Eng. Fract. Mech.* **2012**, *88*, 79–89. [CrossRef]
25. Li, Q.; Xue, Q.C.; Hu, Q.S.; Song, T.; Wang, Y.H.; Li, S.Y. Cold Expansion Strengthening of 7050 Aluminum Alloy Hole: Structure, Residual Stress, and Fatigue Life. *Int. J. Aerosp. Eng.* **2022**, *2022*, 4057898. [CrossRef]
26. Wang, C.G.; Zou, F.; Zhou, E.T.; Fan, Z.L.; Ge, E.D.; An, Q.L.; Ming, W.W.; Chen, M. Effect of split sleeve cold expansion on microstructure and fatigue performance of 7075-T6 aluminum alloy holes. *Int. J. Fatigue* **2023**, *167*, 107339. [CrossRef]
27. Yang, K.; Chen, X.H.; Lv, X.M.; Ding, F.; Dang, B.; Li, F.K.; Wei, D.B.; Zhang, P.Z. Effect of split-sleeve cold hole expansion on the fatigue performance of an additive manufactured TC18 alloy. *J. Mater. Res.* **2023**, *12*, 841–852. [CrossRef]
28. Li, K.S.; Wang, R.Z.; Zhang, X.C.; Yao, S.L.; Cheng, L.Y.; Lei, X.L.; Tu, S.T. Process-performance-prediction integration for fatigue life improvement technologies: An implementation in cold expansion of hole structures. *Int. J. Fatigue* **2023**, *170*, 107507. [CrossRef]
29. Liu, K.Y.; Yang, X.S.; Zhou, L.; Li, M.; Zhu, W.J. Numerical investigation of the effect of hole reaming on fatigue life by cold expansion. *Trans. Can. Soc. Mech. Eng.* **2022**, *46*, 400–411. [CrossRef]
30. Lacombe, A.; Landon, Y.; Paredes, M.; Chirol, C.; Benaben, A. Numerical simulation of the reaming process in aluminium 2024-t351-analysis of the evolution of the material stress state. *Dyna* **2019**, *94*, 215–220. [CrossRef]
31. *HB/Z 170-2005*; Cold Expansion Process for Hole of Aircraft Metal Part. Commission of Science, Technology and Industry for National Defence: Beijing, China, 2005.
32. *GB/T 3280-2015*; Cold Rolled Stainless Steel Plate, Sheet and Strip. General Administration of Quality Supervision, Inspection and Quarantine of the People's Republic of China: Beijing, China, 2015.
33. Jamali, A.; Ma, A.X.; Llorca, J. Influence of grain size and grain boundary misorientation on the fatigue crack initiation mechanisms of textured AZ31 Mg alloy. *Scr. Mater.* **2022**, *207*, 114304. [CrossRef]
34. Dang, Z.C.; Peng, Y.H.; Yang, R.J.; Yan, L.; Gan, X.D.; Ge, D.Y. Experimental study on the effect of reaming on the fatigue life of split-sleeve cold-expanded Ti-6Al-4V alloy components. *Int. J. Fatigue* **2024**, *188*, 108521. [CrossRef]
35. Faghih, S.; Shaha, S.K.; Behravesh, S.B.; Jahed, H. Split sleeve cold expansion of AZ31B sheet: Microstructure, texture and residual stress. *Mater. Des.* **2020**, *186*, 108213. [CrossRef]
36. Su, R.; Li, J.Y.; Liu, W.G.; Xu, C.Z.; Gao, L.J.; Liang, X.Z.; Wu, D.Y.; Huang, X.; Dong, H.C.; Ma, H.K. Investigation on fatigue failure of split-sleeve cold expansion holes of 7075-T651 aluminum alloy. *Mater. Today Commun.* **2023**, *35*, 106290. [CrossRef]

crystals

MDPI

Review

A Critical Review on Friction Stir Spot Welding of Aluminium Alloys: Tool, Mechanical, and Micro-Structural Characteristics

Manash J. Borah [1], Kanta Sarma [1], Yadaiah Nirsanametla [2,3], Barun Haldar [4,*], Arpan K. Mondal [5,*], Borhen Louhichi [6] and Hillol Joardar [7]

1 Programme of Mechanical Engineering, Assam Down Town University, Guwahati 781026, India; manashppp@gmail.com (M.J.B.)
2 Department of Mechanical Engineering, North Eastern Regional Institute of Science and Technology, Itanagar 791109, India
3 Chitkara Centre for Research and Development, Chitkara University, Rajpura 174103, India
4 Industrial Engineering Department, College of Engineering, Imam Mohammad Ibn Saud Islamic University (IMSIU), Riyadh 11432, Saudi Arabia
5 Department of Mechanical Engineering, National Institute of Technical Teachers' Training and Research, Kolkata 700106, India
6 Deanship of Scientific Research, Imam Mohammad Ibn Saud Islamic University (IMSIU), Riyadh 11432, Saudi Arabia
7 Department of Mechanical Engineering, C. V. Raman Global University, Bhubaneswar 752054, India
* Correspondence: bhaldar@imamu.edu.sa (B.H.); arpan@nitttrkol.ac.in (A.K.M.)

Abstract

Aluminum spot welding is extensively applied in automotive, aerospace, and rail sectors due to its favorable strength-to-weight ratio. While resistance spot welding (RSW) has been the traditional method, its high residual stresses, electrode wear, and limited performance with high-strength aluminum alloys have driven interest toward alternative techniques. Friction stir spot welding (FSSW) offers significant advantages over RSW, linear friction welding (LFW), and hybrid processes, including solid-state joining that minimizes porosity, lower energy consumption, and the elimination of consumable electrodes. Compared to LFW, FSSW requires simpler fixturing and is more adaptable for localized repairs, while offering superior joint surface quality over hybrid laser-assisted methods. Despite these advantages, gaps remain in understanding the influence of process parameters on heat generation, microstructural evolution, and mechanical performance. This review consolidates advancements in tool design, thermal characterization, and weld property for aluminum alloys. It presents comparative insights into temperature distribution, weld strength, hardness variation, and metallurgical transformations reported across studies. By critically synthesizing the earlier works, this work identifies knowledge gaps and potential design improvements, aiming to support the development of more efficient and robust FSSW processes for industrial application.

Keywords: friction stir spot welding; FSSW tool; temperature variation; microstructure; ultimate tensile load

1. Introduction

After the development by The Welding Institute, UK Friction stir welding (FSW) has been established as the most innovative and highly effective solid-state welding process to join the materials without melting them. In the FSW process, a non-consumable is used which rotates to generate a sufficient amount of heat due to the friction between the tool and workpiece metals. As the tool travels between the two metal pieces, stirring action takes place on the heated softer material. The tool pin is mainly responsible for the stirring effect [1], allowing the materials of both the metal parts to be mixed and become fused into one another, and after cooling, it behaves like a mechanical locking. A strong defect-free joint is produced by the FSW process, which has been becoming more popular as a preferred method in the fields of aerospace, automotive, and shipbuilding, mainly for the softer metals like aluminum and other lightweight metals. The FSW process has certain advantages over the traditional fusion welds. It is one of the green welding processes due to its energy-efficient process and non-involvement of fumes like the fusion welds. As a solid-state welding process, FSW has the potential to eliminate the common issues of the traditional welding processes like the porosity, heat-inducted cracks, and residual stresses, thus producing strong superior high-quality welding.

Friction stir spot welding (FSSW) has been introduced as a variant of traditional friction stir welding [1–3] which is mostly used for lap joining, as shown in Figure 1, without the translation of the welding tool [4]. It is also a modification of friction stir welding to replace resistance welding or riveting or any other welding joints [5] as FSSW has excellent potential as an alternative to rivet joints and resistance spot welding. The schematic of the friction stir spot welding is shown in Figure 1.

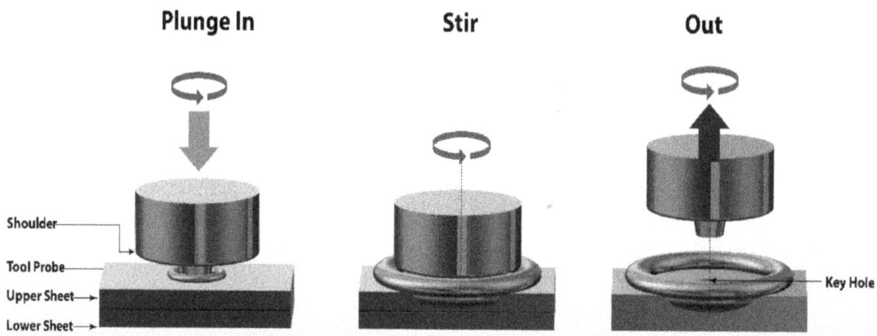

Figure 1. Schematic representation of friction stir spot welding (FSSW).

The traditional resistance spot welding (RSW) is also widely used in the automotive industry; however, RSW can produce larger distortion and is susceptible to cracking. In the case of aluminum and other lightweight alloys, traditional RSW generates extensive amount of heat which affects the materials in and around the spot. On the other hand, the FSSW process does not require any filler material which is the same as the RSW process and it does not produce any hard full emission. Sound joints can be produced by controlling the process parameters such as tool rotational speed, plunge depth, tool shape, and dwell time. Though the FSSW process is less effective for the thicker materials, the FSSW process has the potential to reduce the defects seen in the RSW process.

In recent years, FSSW has been commonly used in automobile, aviation, and aerospace fields, and a lot of industrial and commercial applications [6–10]. To maximize the application of FSSW and to minimize the defects, a different derivative of FSSW [11–25] such as refill FSSW is also being applied [26,27]. Though recent study shows that the joining of high-temperature similar and dissimilar alloys is possible through friction stir welding [10–26], the majority of FSW and FSSW is used for the lighter materials. Thinner plates joined by the fusion welds tends to become distorted, and therefore, makes it difficult to use. Tailor-made blanks are used in automatics to use lighter metal parts. Thinner plates of lightweight materials like aluminum can be joined very easily by the FSSW process, thus making FSSW a potential candidate for the automotive industries. Lighter material parts can help to reduce the overall weight of the motor vehicle and can increase fuel economy. In fact, unlike the fusion welds, FSSW is a clean and green weld without having any fumes or gases. Typically, there are three stages in a friction stir sport welding process: the plunge-in stage, stirring or welding stage, and lastly, the out stage. Figure 2 shows the different stages of the FSSP processes.

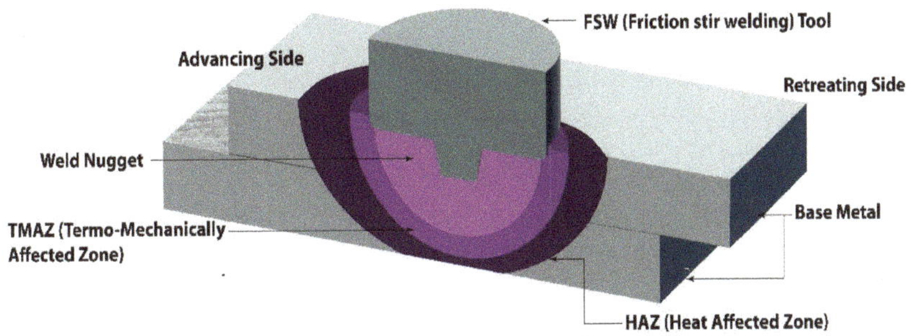

Figure 2. A schematic of the different stages of the friction stir spot welding process.

The automobile industries have also started to use this process to reduce emissions and to improve fuel economy since the characteristics of aluminum alloys (low density, high strength, high ductility) make them the right candidate for it [28]. Moreover, this method is also used for the overlap joining of aluminum alloys which are unable to weld by traditional techniques [3]. With fusion welding of aluminum alloys, several weld defects such as porosity and cracking can be observed which is usually caused by the generated heat that supports the joints of the metals, later resulting in micro segregation of the alloying elements [29].

The amount of heat generated around the pin can be controlled by optimization of the process parameters of FSSW such as rotating velocity, dwell timing, and plunge depth [8,30,31]. With the increasing tool rotational speed and dwell timing, a better tensile shear strength can be observed, which is directly correlated to weld strength and weld width [32]. Due to the presence of a pin in most of the FSSW processes, a keyhole is generally present, which is considered a disadvantage, which is why researchers are trying to alleviate this defect by using pinless or problem tools [33]. Researchers like Vacchi et al. [34] have tried to modify the existing FSSW mechanism and proposed a novel method of spot welding. The traditional FSSW is modified to a novel method named refill friction stir spot welding (RFSSW). The new method has prompted a microstructure refinement and enhanced corrosion behavior [34]. Moreover, it is possible to achieve an optimum weld strength for welded lightweight alloys by selecting appropriate weld parameters for the material combination and adopted tool geometry [35–37]. Thus, the FSSW process is applicable for a favorable microstructure with a minimal number of defects in the case of aluminum alloys [38]. The advances on aluminum alloys help in predicting the promising future of FSSW applied to aluminum alloys. Along with the traditional FSSW works, the micro-friction stir spot welding of aluminum alloys has also been reported [39,40].

Spot welding plays a major role in joining thinner material parts in various industries; therefore, critical study and understanding of developments in friction stir spot welding (FSSW) is the only focus of this work. This review encapsulates the recent research outcomes and advancements of FSSW, underlining its potential for future applications. Through examination, various tools used for FSSW are being discussed and considered as one of the important parameters to decide the welding quality. In this study, results of various authors were discussed and compared at different parameters like dwell time and spindle speed. The comparative analysis of the results gives a clear and comprehensive understanding of the range of resulting temperature gain, microstructural changes, hardness value at the welding zone, and ultimate tensile strength at different welding parameters. Finally, the objective is to extant a complete impression of the recent progress made in FSSW of aluminum alloys, highlighting the technological spreads and thereby promising a future direction for FSSW in this domain.

The performance of FSSW is highly dependent on key process parameters such as tool rotational speed, plunge depth, dwell time, and tool geometry. Proper control of these parameters can significantly influence the heat generation, material flow, and final microstructure of the weld zone. In particular, the design and material of the tool are critical in determining weld quality. Various sections of this review delve into these aspects, starting with the types of tools used in FSSW, followed by an overview of commonly used tool materials including tool steels and polycrystalline cubic boron nitride (PCBN) and their influence on temperature distribution, material flow, and joint strength.

This review compiles and analyzes the latest developments in FSSW of aluminum alloys, focusing on the correlation between tool design, mechanical response, and microstructural evolution. By synthesizing results from different studies, the review aims to provide a comprehensive understanding of the FSSW process and highlight promising directions for further research and industrial application. The novelty of this review lies in its comprehensive correlation of tool design, process parameters, and microstructural evolution in FSSW of aluminum alloys, alongside a critical synthesis of recent findings on mechanical behavior, emerging tool materials, and design guidelines, which have not been collectively analyzed in the prior literature.

2. Tools Used in FSSW

In the FSSW process, the rotating tool is usually plunged with force into the overlapping sheets or plates, and then it is pulled out after a dwell time [41]. The FSSW process utilizes heat as well as frictional deformation to create a good weld joint between the workpiece and tool, causing plastic deformation [33]. Due to this, the metals become joined below the melting point of the base material. With FSSW being a variant of the FSW process, the tools used for FSW are useable for FSSW as well. Figure 3 shows the basic terminology of the FSSW tool.

2.1. Common Tool Materials

The tools used for FSSW usually undergo high temperatures and severe stress [42]. In these case, the characteristics of the material decay gradually, and the tool life decreases for the whole process [43]. Therefore, for selecting the tool material, the workpiece material and tool life is considered to be the critical factor. Moreover, the user's preference and experience can also play an essential role in selecting the tool to weld [44]. Since aluminum alloys do not produce that much tool wear, in such cases steel tools can be used. But materials which are having high melting points such as steel, titanium, or metal matrix composites (MMC) usually wear out in such a way that it can be a matter of worry in certain circumstances [45]. For obtaining a high-quality weld joint with friction stir spot welding, it is mandatory to select the best tool material. The other properties which are considered for the tool material selection include wear resistance, low coefficient of thermal expansion, and no detrimental reaction with the parent metal [43,46].

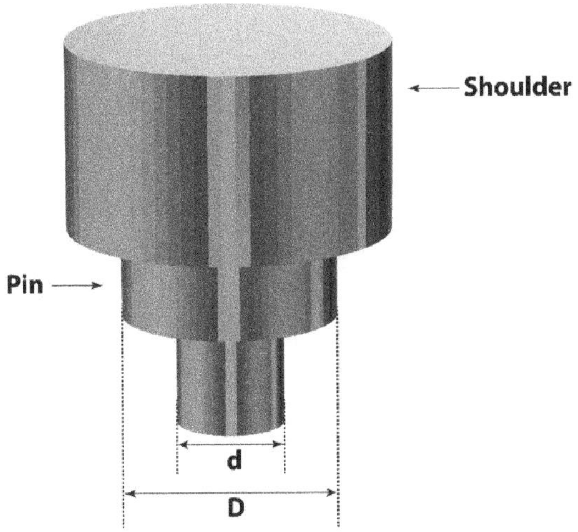

Figure 3. A schematic diagram of the FSSW tool.

2.2. Tool Steel

Steel tools are usually used for welding of aluminum or magnesium alloys and aluminum matrix composites (AMC) [45,47–55]. Tool steel can weld up to a thickness of 50 mm in materials like aluminum and magnesium alloys [46]. This type of tool material has also been used for both lap and butt welding of dissimilar alloys or materials [44,56–63]. The types of steel tools used for different alloys are given in Table 1.

Table 1. Different types of steel tools used for different materials of joining.

Materials for Joining	Thickness of the Joined Material	Tool Material
6061-T6 Al and AISI 1018 mild steel [17,21,64]	6 mm	H13
AA2011 and AA6063 alloy [13,65,66]	10 mm	HSS
AA5083-H111 Al alloy [67]		HCHCr
AA6082 and AA2024 [26,68]	4 mm	C40
Commercial grade Al alloys [69]	6 mm	SS310
AA5754 and C11000 copper [70,71]	3.175 mm	H13

The best tool material for welding Al-Mg and low-carbon steel is steel, and to avoid excessive tool wear, the softer plate is placed on top of the steel plate for lap welds. In butt joint configuration, cold-worked X155CrMoV12-1 steel tool has been used for joining 99.5% pure Cu with CuZn30 brass with the help of friction stir spot welding [63]. Moreover, for successful welding of Al 6061 + 20% Al_2O_3 aluminum matrix composite and Al 359 + 20% SiC aluminum matrix composite, oil-hardened (62 HRC) tool steel has been used as a tool material [48,51]. Some investigations have shown that while performing FSSW on AMCs at first, the tool wear gets a self-optimized shape, but later on, it gets much less conspicuous [48,51,72]. But the increase in total wear has been observed with the increase in rotational speed [48,51]. Different types of materials such as stainless steel (SS), high-carbon high-chromium steel (HCHCr), high-speed steel (HSS), H13, and C40 have been used as tool material for tool steel to follow joining methods of friction welding [44,73,74]. Figure 4 shows the schematic diagram and photographs of the FSSW tools made of H13 and HCHCR-D2 tool materials.

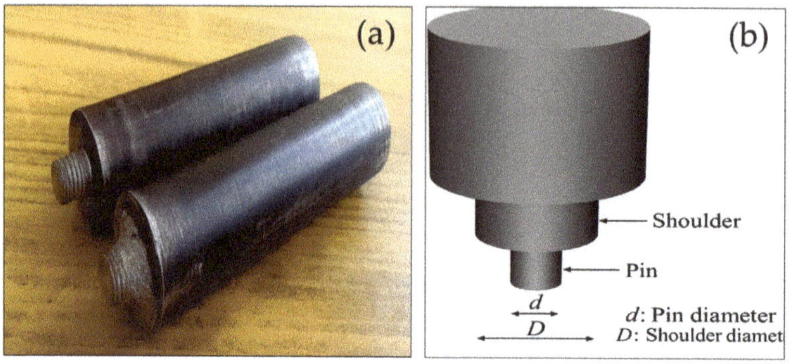

Figure 4. (**a**) Cylindrical tools made of H13 and HCHCR-D2; (**b**) schematic of the tool [75].

Among these types of steel tools, HCHCr (high-carbon high-chromium steel) has the highest wear resistance. Another type, H13 (H13 chromium hot-work steel is widely used in hot and cold work tooling applications) has 5% of chromium in its material composition [75]. This steel tool is combined with red hardness along with shock and abrasion resistance. High-chromium stainless steel (HCSS) is specialized with high corrosion-resisting capacity. On the other hand, HSS, which is a high alloy grade tool, usually has high wear resistance capability. The quality of the steel tool relies on the heat treatment process they go through [44]. Table 2 depicts the chemical composition of a few steel tool materials.

Table 2. Composition of tool materials [75–78].

Steel Type	%C	%Ni	%Mn	%Cr	%Si	%Mo	%W	%Cu	%V
H13	0.45	0.30	0.50	5.25	1.20	1.75	-	0.25	1.20
HCHCr-D2	1.5		0.45	12.00	0.30	0.9	-		1.00
C40	0.37–0.44	0.4	0.5–0.8	0.4	0.4	0.1	-	-	-
HSS	0.87		0.26	3.99		4.61	5.83		1.76
SS310	0.25	19.0–22.0	2.00	24.0–26.0	1.50				

2.3. PCBN Tools

PCBN or polycrystalline cubic boron nitride and polycrystalline diamond (PCD) are mostly used as super abrasive tool materials. Small crystals of ultra-hard material such as diamond or CBN are both present in PCBN and PCD. For the formation of the matrix, those small crystals bond together with a second-phase material in a skeletal matrix to serve as a catalyst [79–82]. With the purpose of machining and turning super alloys, cast iron, and tool steel, this tool was initially developed. But now, because of its high mechanical and thermal activities, the PCBN tool is well accepted in friction welding [43]. Furthermore, PCBN is the most used tool for materials like steels and Ti-alloys because of its temperature stability, hardness, and high strength at an exalted temperature [81,83–92].

Studies show that boron nitride has two sorts of crystal structures. One of them is hexagonal, and the other one is cubic. Following a similar process of producing diamonds from graphite, the cubic form is prepared. For preparing this, the hexagonal version is subjected to high temperature and pressure. Even the carbon itself has lesser thermal and chemical stability than this cubic form, and only diamonds can compete with it in hardness. It has been reported that the chemical inertness of this phase can be compared with iron [93] and can go up to 1573 K [91,94]. Mechanical properties like high thermal conductivity are responsible for the development of the design of liquid-cooled tools as well as for not generating hot spots on the tool [95]. Without the requirement of any binder, the single-phase cubic boron nitride (CBN) can be achieved with the best mechanical properties. Commercially, it is possible to obtain such material for which hexagonal boron nitride should be sintered at high pressure (6–8 Gpa) and temperature (1773–2673 K) [91,96,97]. At ambient temperature, the fracture toughness with a grain size that is in the range of 2–12 μm showed a Young's Modulus and shear modulus of 900 GPa and 405 GPa, respectively [96].

A ductile-to-brittle transition can be seen when CBN mixes with the binders, where the transition temperature can be in the range of 1323–1423 K [98]. Studies show that, after running experiments on PCBN as a cutting tool for a hundred sheets of steel and super alloys, wear mechanisms like abrasion and diffusion can be observed. This study also allows the showing of PCBN with different types of CBN and binders, which are of two grades [99]. This study also appraised the different grades of PCBN based on a real cutting test which was only for the friction welding process. Another study shows the presence of a different wear mechanism by comparing the Tic-CBN tool with that in CBN [100]. This study also suggested that a more stable protective layer in higher temperatures can be achieved when the lower thermal conductivity of the TiC-CBN-based tool is compared with the CBN based tool. Apart from these, studies have been conducted on the mechanisms of the cutting tool wear to understand it in a better manner [101–103]. The experiment was run on a 45 m long steel alloy that shows that the obtained tool life of the PCBN tool was sufficient enough for the welding process [104]. Later on, another experiment was carried out, where the high strength low alloy, 65 of 6 mm (thickness), was welded with the help of the PCBN tool [105–108].

In the past few years, the design of the PCBN tool has changed considerably. In Figure 4, it can be seen that the tool only had a truncated cone probe with a smooth concave shoulder and without any other features in it [75]; after that, the PCBN tool was upgraded with a newly added feature (truncated cone), as shown in Figure 5a [106]. The probe was modified with a stepped spiral thread (Figure 5b). Even after that, the tool design was again modified, providing a convex shoulder with scrolls added to it (Figure 5c). Because of these added features, it was possible to avoid the defects caused by the tool, and the features also improved the productivity of the tool. The need for the tilting can be omitted in the case of the threaded tools.

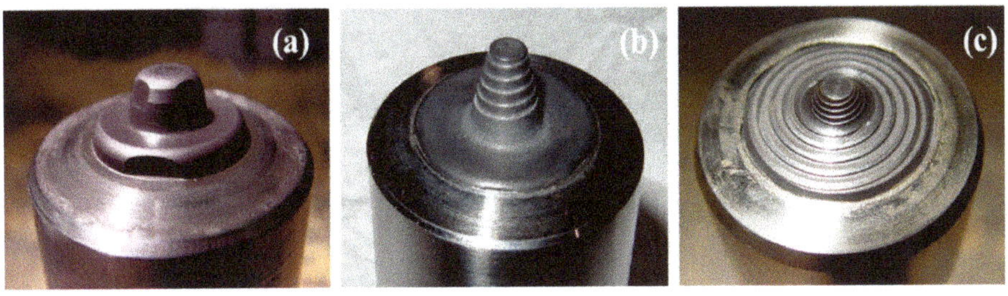

Figure 5. Evolution of the design of PCBN tool: (**a**) early featureless design, (**b**) step spiral probe, and (**c**) convex scrolled shoulder step spiral probe [106].

Another new type of PCBN tool material has been developed where WRe works as the binder or the catalyst. At the cost of tool wear resistance. This new type of PCBN-WRe can offer better fracture toughness than the regular PCBN tool material [106]. There is an enhanced control over the temperature of the tool as well as the whole friction welding process due to the development which has been carried out over the past few years with the new grades of PCBN by improving the wear and fracture resistance of both PCBN and PCBN-WRe tool material [95].

In a nutshell, it can be observed that a variety of the tool materials are being used depending upon the specific application. Common tool materials include various steel types like H13, HCHCr, HSS, and SS310 which are selected based on wear resistance, thermal stability, and compatibility with work material. Steel tools are effective for aluminum and magnesium alloys. For harder materials, PCBN (polycrystalline cubic boron nitride) tools are preferred due to their high hardness, temperature stability, and wear resistance. As a general rule, the tool material should be harder than the work material. The harder the tool material is, the tougher the machining of the tool would be in a proper shape. As the shape of the tool pin and shoulder can influence the weld quality, a careful consideration should be made in deciding the tool materials so that there will not be any issue in the fabrication of a specific and intricate tool design.

2.4. Temperature Distribution

The temperature profile of friction stir spot welding of aluminum alloys is shown in Figures 6 and 7, where the maximum temperature generated during the FSSW of different aluminum alloys at different parameters is revealed. The peak temperature values at different dwell times are shown along the ordinate and dwell time along the abscissa where the type of metal was indicated at the top of the bar.

Figure 6 shows the compiled data of maximum temperature generated at different dwell times, whereas Figure 7 shows the variation of temperature at different speeds. Arul et al. [107] have used Al 6111-T4 sheets with a thickness of 1.3 mm and 1.7 mm as a welding sample and welded by using an H13 tool steel of 10 mm tool diameter. It can be observed that at a dwell time of 2.2 s, the maximum temperature rise was about 573 K at 1500 RPM, and at the same dwell time, at 3000 RPM, the temperature reached up to 688 K. The welding was performed in the Kawasaki SFW system, and the temperature was recorded by using a thermocouple, which was 2 mm away from the shoulder of the tool [107].

Pandey et al. [109] achieved the highest temperature of 598 K at a dwell time of 3.6 s. Aluminum 6061 series of 1 mm thickness was welded with an H-13 tool. The pin length of the tool was 1.4 mm with a diameter of 4 mm at 1500 RPM. A FE or finite element model was developed to compare the experimental data with the predicted data.

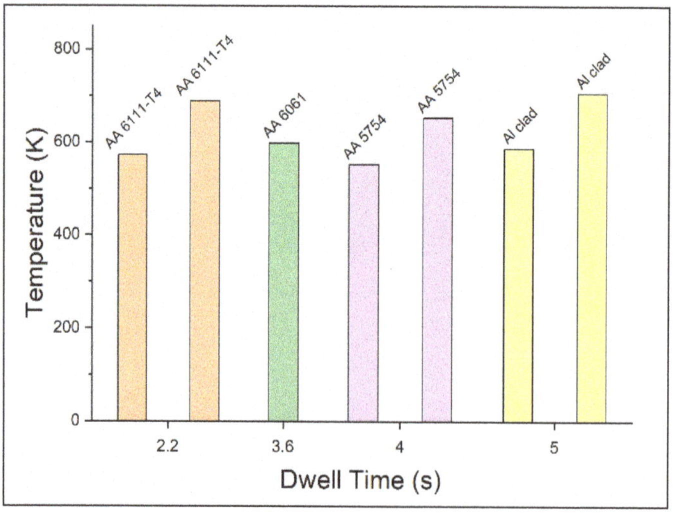

Figure 6. Temperature variation at different dwell times [108,109].

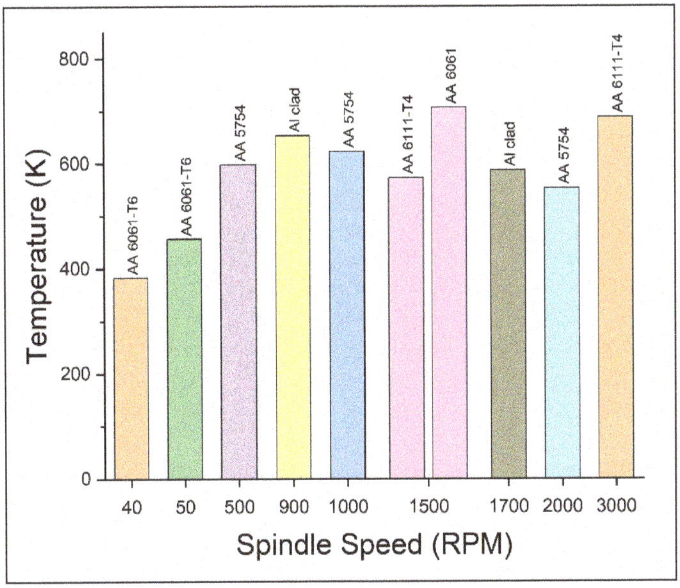

Figure 7. Temperature at different plunging speeds [71,110].

Aluminum-5754 sheets of 1.6 mm thickness were welded with a circular pin tool with a diameter of 5 mm by Pathak et al. [71]. The lowest peak temperature recorded at 4 s was 280 °C with an RPM of 500, and the temperature recorded at the same dwell time but at 2000 RPM was 553 K. The study also notes that, if the tool was replaced with a taped pin tool, the peak temperature increases. The circular pin happened to produce more heat than the tapered one, in the specimen, and near the sheet–tool interface, an asymmetrical temperature profile has been observed with both circular and tapered pins. Moreover, with the increase in rotational speed and dwell time, the peak temperature has also been observed to be increasing. The maximum and minimum temperature recorded for a 5 s dwell time was 706 K and 587 K at 1700 RPM and 900 RPM, respectively. Li et al. [110] used Alclad 2A12-T4 aluminum sheets of 2 mm thickness, and the experiment was conducted by a self-developed RFSSW machine that had a 5.3 mm diameter. By using a K-type thermocouple, the thermal cycles were determined, and the readings were obtained 6 mm away from the weld center. The variation in the highest peak temperature and lowest peak temperature takes place due to the variation in the RPM of the given parameters.

Also in a friction welding process, incipient melting may occur, since the shear strength of the plasticized metals tends to decrease significantly if the temperature reaches the solidus temperature. Thus, no further temperature rise takes place as a drop-in heat efficiency occurs in the process. The maximum temperature generated at different spindle speeds is shown in Figure 7. It was observed that the temperature changes with a change in spindle speed. Then, some fluctuation can also be observed until the temperature reaches its highest and then with the increasing RPM, the value of the temperature becomes more identical. From the plot, it can be concluded that the welding peak temperature remains between 523 K and 723 K. The lowest temperature was achieved at low spindle speed, i.e., 40 RPM and 50 RPM because of the low strain rate of the material.

Along with the works on the Aluminum 5XXX and 6XXX series, authors have contributed significantly in the joining of Al 7XXX series, which is popular for its high strength, toughness, and corrosion resistance and it finds the majority of application in aerospace and transportation applications. Both numerical [111] and experimental works have been reported to investigate the effect of the thermal history on the final joint quality (microstructure and mechanical performance) in Al 7XX series welded through the FSSW refill FSSW process. Fully coupled thermal analysis was performed to predict the thermal history for Al 7075-T6 sheets welded by the refill FSSW process. The results were validated with the experimental thermal profile. The model showed a steep thermal gradient near the Stir Zone.

Gerlich et al. [112] reported peak temperatures reaching 527 °C during a traditional FSSW process at 3000 rpm. These high temperatures facilitated dynamic recrystallization, producing fine-grained equiaxed microstructures in the SZ. However, the study also found that high tool speeds could lead to transient local melting and tool slippage, which compromise weld quality. Shen et al. [113] observed melted films in the SZ with maximum temperatures of approximately 470.9 °C located 2.6 mm from the weld center. These elevated temperatures correlated with increased nugget thickness and lap shear strength, indicating a direct relationship between thermal input and joint integrity. However, excessive heat could degrade surrounding material properties. The thermo-mechanical modeling by Janga et al. [114] validated numerical predictions of temperature cycles using thermocouple measurements. Their DEFORM-3D model accurately captured the thermal history (Figure 8), strain, and material flow, highlighting the localized nature of heat generation and its critical role in defining the size and quality of the SZ.

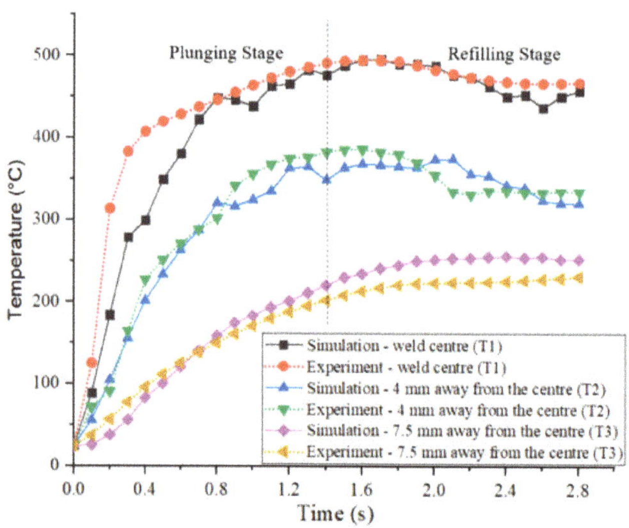

Figure 8. Comparison of experimental and numerical temperature results at different locations [114].

The temperature profile's impact on weld microstructure was also shown in the work reported by Kubit et al. [115], where micrographs of refill FSSW welds revealed the higher thermal input refined grain structures, but could also lead to softening in the heat-affected zone (HAZ), impacting fatigue strength. It has been observed that the combination of plunge depth and welding time governs the thermal activity [116]. Proper calibration ensures sufficient temperature to plastically deform the material without exceeding its melting point, maintaining the solid-state nature of the process.

Overall, from these studies, it can be understood that peak temperature during the welding process depends on various factors like tool type, dwell time, spindle speed, and tool geometry. High temperatures are essential for the softening of the base material, and on the other hard excessive temperature can cause local melting and generation of residual stresses. Therefore, controlling temperature distribution is very much essential in the

FSSW process. Both experimental and numerical analyses emphasize the need for precise process parameter optimization to ensure high-quality joints with favorable mechanical and metallurgical properties.

3. Mechanical Behavior

The highest load-bearing capacity at different parametric combinations is shown in Figure 9. The highest load-bearing capacity was achieved by Sun et al. [117] for a probe-less tool. The maximum load-bearing capacity recorded by the FSSW joint of commercial 6061-T6 series of aluminum alloy plate of 1 mm thickness was recorded at 5400 N with a displacement of 1.5 mm.

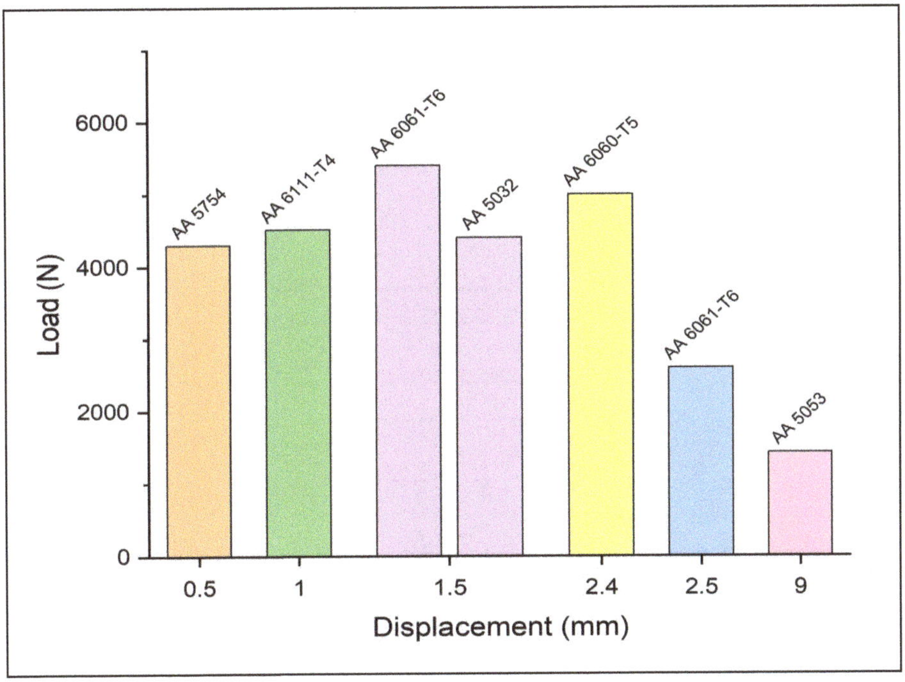

Figure 9. Displacement of the joint at different tensile shear loads [71,110,117].

Pathak et al. [71] analyzed the lap shear load of FSSW on Aluminum-5754 sheets of 1.6 mm thickness on different dwell times, tool plunge depth, and rotational speeds. The parameters used in that experiment are a plunge depth of 2.4 mm, a dwell time of 4 s constant, and a rotational speed of 2000 RPM with a circular pin that has a diameter of 5 mm. The highest load-bearing capacity or the peak load was obtained as 4.3 KN which creates a displacement of 0.5 mm in the specimen [71].

Farmanbar et al. [118] carried out an experiment on AA5052 al sheets of 1 mm thickness, and the type of welding performed was protrusion friction stir spot welding. The protrusion had a diameter of 11 mm and a height of 0.4 mm above the desk. The tool used for this welding was the H13 steel tool which was a pinless cylindrical tool with a 14 mm diameter. Welding was performed with a rotational speed of 800 RPM and with various dwelling times. It was observed that the highest tensile shear load obtained was 4400 N with a displacement of 1.5 mm [118].

The lowest tensile shear load of 1420 N with a displacement of around 9 mm was recorded by Badarinarayan et al. [119] at an RPM of 1500 with a plunge speed of

20 mm/min, dwell time of 2 s, and a plunge depth of 0.2 mm. For the study, annealed 5083AA sheets with two different thicknesses of 1.64 and 1.24 mm were used. A cylindrical pin was used to weld with a shoulder diameter of 12 mm, and a pin length of 1.6 mm. For AA6060-T5 alloys, the tensile shear load was 5000 N with a displacement of 2.4 mm before the rapture. The welding tool used was X210CR 12 with shoulder and a pin diameter of 14 mm and 5 mm, respectively. The tool also had a length of about 3.95 mm. The average failure load or load-bearing capacity was found to be 5 KN with a feed rate of 16 mm/min, along with a rotational speed of 1000 RPM [120]. FSSW of 6061-T6 al alloy of 2 mm thickness extends up to 2.5 mm at a load of 2600 N [32]. The H13 pinless tool was used for welding with a plunge depth of 1.1 mm, 1400 RPM, 6 s dwell time, and 18.6 mm/min feed rate. The study also shows that the weld samples which were made with the same welding parameters but using a tool with the probe have lesser yielded strength [32]. And the rest of the two values are 4512 with a displacement of 1.0 mm.

Residual stress [25,121] is always a matter of concern for any welding process. For the FSSW process, the thermal gradient and the mixing parameters influence the residual stress behavior. The research conducted by Sarfaraz et al. [116] and Boucherit et al. [122] shows the residual stress distribution in the weld center and around (Figure 10) for Al-Cu dissimilar FSSW joints. The highest amount of residual stress (RS) can be seen around the HAZ. The maximum value of RS was around 60 MPa. The longitudinal RS was reduced near the spot weld center. The SZ also noticed a decreased value of the RS, which may be due to the finer grains in this region. Though very less efforts have been found in the investigation of the residual stresses in the FSSW process, the presence of the residual stresses in the FSSW joint is much less compared to the fusion welding process [123]. The estimation of the RS could be an area for understanding the FSSP process in a better way.

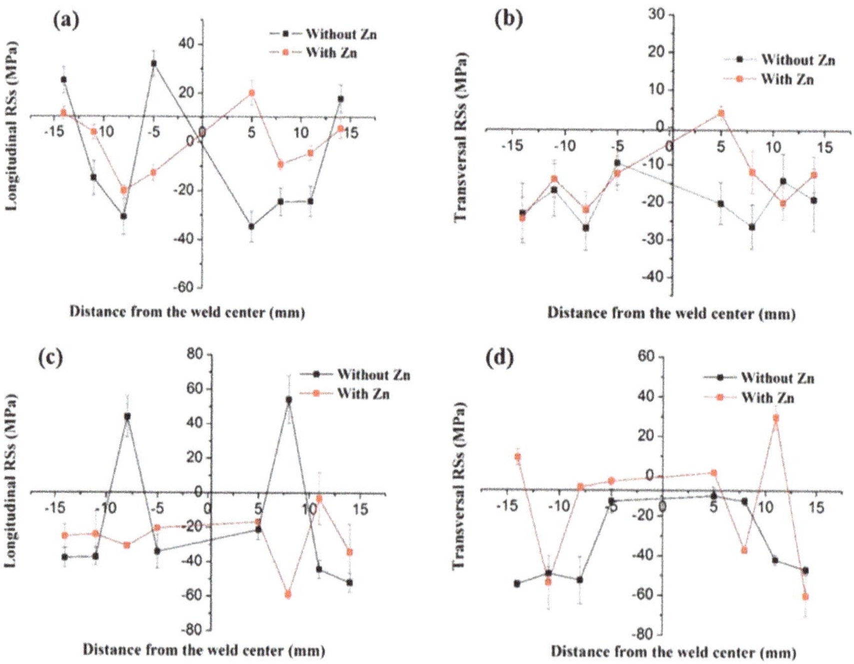

Figure 10. Residual stress distribution in Al–Cu dissimilar FSSW joint (**a**,**c**) Longitudinal, (**b**,**d**) Transverse for two different locations [122].

Recent advancements in the application of FSSW have extended beyond sheet joining and are now being explored for structural aluminum members such as beams, columns, and stiffened panels [124]. These developments are particularly significant in automotive and rail car body frames, where lightweight [121] and high-load-bearing components are essential. For example, FSSW has been applied to the fabrication of aluminum hollow extrusions and stiffeners used in crash-relevant substructures, enabling enhanced joint fatigue performance and crash energy absorption compared to traditional joining methods [125]. Aerospace-grade aluminum alloys (e.g., 7075) are being considered for structural ribs and bulkhead components using FSSW due to the lower residual stresses and improved dimensional stability offered by the process [124]. These applications demonstrate FSSW's scalability and reinforce the need to evaluate its design feasibility and mechanical reliability under large-scale loading scenarios [124–126].

4. Microhardness

The graph provided below represents the hardness profile of different aluminum alloys at different process parameters by various authors. Figure 11 shows the range of microhardness values at different spindle speeds ranging from 40 RPM to 3000 RPM, which gives a brief idea of the range of hardness values of friction stir spot welded joints of Al alloy. The compilation of highest and lowest hardness numbers by various authors was plotted for different spindle speeds and the type of the welded material was mentioned along with the bar in the graph.

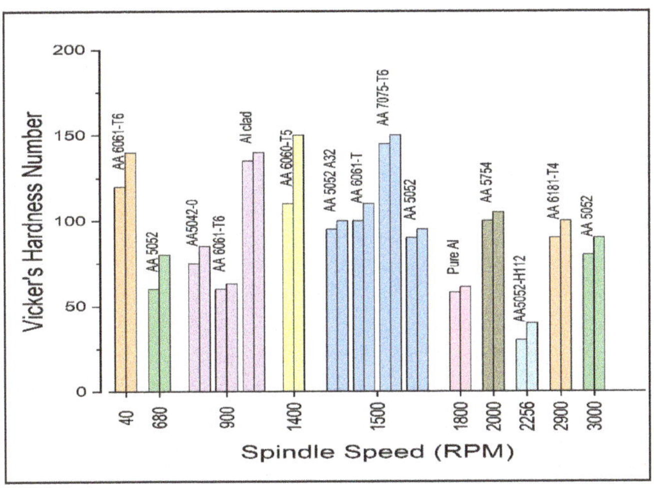

Figure 11. Microhardness analysis of different aluminum alloys [117–120].

It can be observed from Figure 11 that the value of microhardness of FSSW of aluminum alloy ranges between 55 HV to 160 HV depending on the input parameters. The highest microhardness in an aluminum series of 6060-T5 was achieved by Mohamed et al. [120] with a 14 mm shoulder diameter, 5 mm pin diameter, and 3.95 mm length at rotational speed from 1000 to 2000 RPM. The hardness value of the Stir Zone (SZ) and thermo-mechanical-affected zone (TMAZ) was higher than the base material and the HV in the SZ showed a peak value in the range of 110–150 with a rotational speed of 1400 RPM. The highest hardness value of 150 HV was recorded in the welding joint welded at a rotational speed of 1400 RPM. Sun et al. [117] used commercial 6061-T6 Al alloy plates with a thickness of 1 mm for that experimental process. Two types of tools were used for the analysis; one of the rotating tools had a diameter of 12 mm along with a probe diameter of 4 mm, and the length was 1 mm. The other rotating tool was flat-shouldered, i.e., probe less tool with a diameter of 12 mm. The hardness value noted in the range of 120–140 HV was achieved in the joints welded at a spindle speed of 40 RPM and welding time of 30 s. The hardness values changed slightly if probe-less tools were used for welding

Another experiment was carried out on AA5042-O al with 1.5 mm thickness by Tiera [127]. The welding was carried out by using the FSSW-Refill prototype machine which had an 18 mm diameter clamping ring, a 9 mm diameter threaded sleeve, and a 5.2 mm diameter grooved pin. At a tool speed of 900 RPM and 6.07 KN applied load, the hardness value was found to be 75–85 HV. Aluminum 6061-T6 rolled sheets of 2 mm thickness were chosen by Venukumar et al. [128] to analyze the welding properties. The tool which was used for the experiment had a shoulder diameter of 18 mm and a pin diameter of 5 mm. It was observed that the hardness was in the range of 60–63 HV with a rotational speed of 900 RPM. The hardness value of alclad 2A12-T4 aluminum alloy had a thickness of 2 mm and range of 135–140 with a rotational speed of 900 RPM in the SZ [110]. This alloy also showcases a hardness profile of W-shaped. The HV value was at its lowest (116 HV) near the edges of the Stir Zone. But then again, the hardness value increases in the SZ.

From the graph, it can be observed that four values of hardness were discussed for different Al alloys at 1500 RPM. These values are a compilation of experiments carried out by different authors, which are being discussed. Freeney et al. [129] carried an experiment with a 5052 H32 Al alloy of 1 mm and 1.6 mm thickness. The tool with a shoulder diameter of 12 mm, and a 1.77 mm long conical pin with a root diameter of 4.5 mm and a tip diameter of 3 mm was used for welding. At a rotational speed of 1500 RPM, the value of the hardness was in the range of 90–95 HV, and the hardness value decreased at a rotational speed of 3000 RPM in the range of 80–90 HV. The authors found that the factor affecting the HV values was dwelling time, but the only changes in the RPM do not make significant differences in HV values. It was also noted that at a speed of 680 RPM, the hardness value ranges between 60 and 80 HV. Aluminum alloy 6061 of 1 mm thickness was used by Pandey et al. [109] to conduct the study and found that the hardness was in the range of 100–110 near the edges of the Stir Zone at a spindle speed of 1500 RPM. The edges of the Stir Zone experience high material flow and due to the compression of more material at those points, the hardness values are usually higher. Another study by Badarinarayan et al. [119] was conducted by using Annealed 5083 Al sheets with two different thicknesses at a spindle speed value of 95–100 HV with an RPM of 1500. This alloy had a higher hardness value in the SZ than the BM and with a distance of around 2 mm from the periphery of the weld keyhole, the HV values were lower and quite stable at the same time. Shen et al. [130] experimented with a 7075-T6 aluminum alloy plate of 2.0 mm thickness. The hardness values of the same experiment range between 145 and 150 in the SZ area at 1500 RPM. The HV distribution was W-shaped and the lowest hardness values were obtained at the boundaries of the HAZ and TMAZ. The factor affecting the HV values was dwelling time, but the only change in spindle speed does not make significant changes in HV values. Freeney et al. [129] selected strained hardened 5052 H32 aluminum alloy to conduct their study. The next bar represents the HV value of the same study which ranges from 90–95 at the boundaries of the SZ, at a spindle speed of 1500 RPM.

The HV value of SZ of commercially pure aluminum was found to be in the range of 58–61 HV at a constant speed of 1800 RPM [9]. The HV distribution had a W-shaped appearance and the hardness values were lower in the HAZ. Also, the hardness value was found to be maximum in the SZ due to the formation of the fine grains. Pathak et al. [71] studied Aluminum-5754 sheets with a thickness of 1.6 mm for FSSW and found that the hardness value was in the range of 100–105 HV with a rotational speed of 2000 RPM when a distance of 2 mm from the keyhole was considered. Due to the fine dynamic recrystallization of the grain structure, the hardness values were higher in the nugget zone. Aluminum alloy 5052-H112 sheets of 1 mm thickness were used by Zhang et.al. [131] for analyzing the effect of parameters on the welding properties. The tool with a shoulder diameter, root, and tip of the pin of 10, 4.5, and 3 mm, respectively, was used for FSSW, respectively. The pin was about 1.8 mm in length and the concave face of the shoulder was angled at 4°. With a spindle speed of 2256 RPM and a dwell time of 5 s, the highest hardness value ranges between 30 and 40 HV in the SZ. The HV values had a W-shaped appearance and the hardness was found to be lowest in the HAZ and lower in the BM. One of the highest HV values (30–40) only near SZ was achieved at 2256 RPM; the spindle speed and dwell time influenced hardness values.

At an RPM of 2900, the hardness value was in the range of 80–100 HV. The experimental work was carried out by Rosendo et al. [132] using 1.7 mm thick sheets of AA6181-T4 aluminum alloy. The base metal has an average hardness value of 80 HV but it decreases in the HAZ. The SZ goes through dynamic recrystallization which leads to solubilization of the precipitates. Hence, this process was considered one of the main reasons behind the higher value of hardness in the SZ.

5. Microstructure

The weld zone in FSSW is composed of several regions, each exhibiting distinct microstructural characteristics. The Stir Zone (SZ), also known as the nugget zone, undergoes significant plastic deformation and dynamic recrystallization, resulting in a fine and equiaxed grain structure due to the intense stirring action of the tool. Adjacent to the SZ is the Thermo-Mechanically Affected Zone (TMAZ), which experiences both thermal and mechanical effects. However, the degree of plastic deformation here is less intense compared to the SZ, leading to elongated and deformed grains from the combined effects of heat and mechanical stirring. The Heat-Affected Zone (HAZ) is subjected to thermal cycles without undergoing plastic deformation. Its grain structure remains similar to the base material but may show signs of thermal exposure, such as grain growth or phase changes. Lastly, the Base Material (BM) represents the unaffected parent material, retaining its original microstructure. Figure 12A shows the microstructure of the Stir Zone (Stir Nugget), TMAZ, and HAZ [133].

The Stir Zone (SZ) typically features fine and equiaxed grains resulting from dynamic recrystallization. In contrast, the Thermo-Mechanically Affected Zone (TMAZ) exhibits elongated grains aligned with the material flow direction. The Heat-Affected Zone (HAZ) shows minimal changes in grain structure compared to the Base Material (BM). In precipitation-hardened aluminum alloys, thermal cycles can alter the distribution and morphology of precipitates. The SZ often displays a more uniform distribution of fine precipitates due to recrystallization and subsequent re-precipitation. Second-phase particles, such as intermetallic compounds, can be broken up and redistributed within the SZ. The SZ may also develop a unique crystallographic texture due to severe plastic deformation. The TMAZ and HAZ retain some of the original texture of the BM, though modified by thermal and mechanical effects. Intense plastic deformation and dynamic recrystallization in the Stir Zone (SZ) result in significant grain refinement. The stirring action also helps homogenize the distribution of alloying elements and second-phase particles. Additionally, thermal cycles and plastic deformation can introduce residual stresses, potentially affecting the mechanical properties of the weld. Understanding the microstructural evolution during FSSW of aluminum alloys is essential for optimizing welding parameters and enhancing the performance of the welded joints.

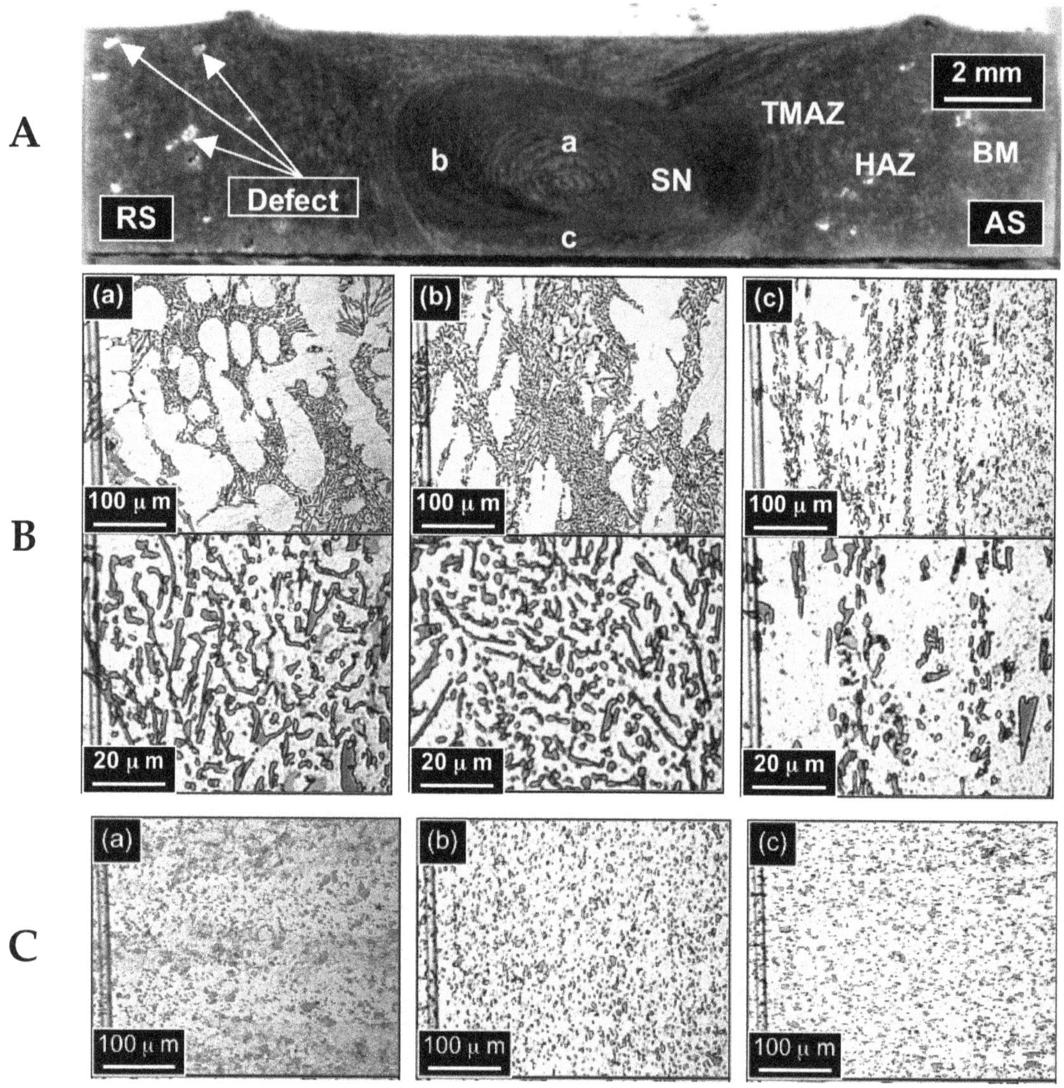

Figure 12. (**A**) Cross section of the FSW joint; (**B**) micrographs of (**a**) BM, (**b**) HAZ, (**c**) TMAZ; (**C**) micrograph of the (**a**–**c**) marked regions of the Stir Nugget Zone [133].

The microstructure of the parent material in the FSW process remains relatively unchanged in the Heat-Affected Zone (HAZ) with minimal effect on grain structure. Dynamic recrystallization occurs in the DXZ with subgrain growth absorbing dislocations into boundaries. Precipitates in the base alloy coarsen and increase in the HAZ after FSW. Al3Zr dispersoids remain unchanged in HAZ and TMAZ regions but act as nucleation sites in the DXZ where as Al7Cu2Fe phase changes shape in the DXZ due to FSW tool stirring [1]. A macrostructural investigation revealed two distinct FSpW weld regions: the Stir Zone (SZ) and the Thermo-Mechanically Affected Zone (TMAZ), as shown in Figure 13. Analysis of the cross sections found consistent geometric and metallurgical patterns in all connections investigated, including hooking, partial bonding, and bonding ligament. The hook is created by the upward bending of the sheet due to tool penetration, forming an upside-down V shape controlled by energy input. Partial bonding is a weaker transition area between sheets, while the bonding ligament represents strong adhesion resistant to separation [132].

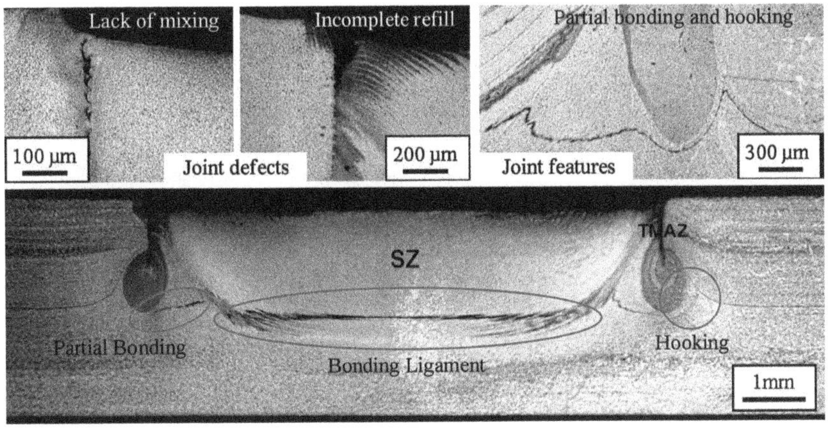

Figure 13. Macrographs of a typical FSpW connection cross section showing the weld zones, geometric/metallurgical features, and weld defects [132].

The RFSSW process produces a more homogeneous and refined microstructure in the Stir Zone (SZ), leading to different hardness values along the welded joint. The SZ also exhibits higher localized corrosion resistance and thicker passive film compared to the Base Material (BM) and Heat-Affected Zone (HAZ), making RFSSW a promising option for welding dissimilar joints [34]. Figure 14 shows the difference of grain size in the Stir Zone (SZ) and Thermo-Mechanically Affected Zone (TMAZ) at various tool rotational speeds. As the tool rotational speed increases from 900 to 1700 RPM, the change in average grain sizes occurs in the SZ. The elevated temperature provides more energy for grain growth, leading to grain coarsening in the recrystallized grains in the SZ. Similarly, in the TMAZ, grains also become coarser at higher rotational speeds for the same reason [110].

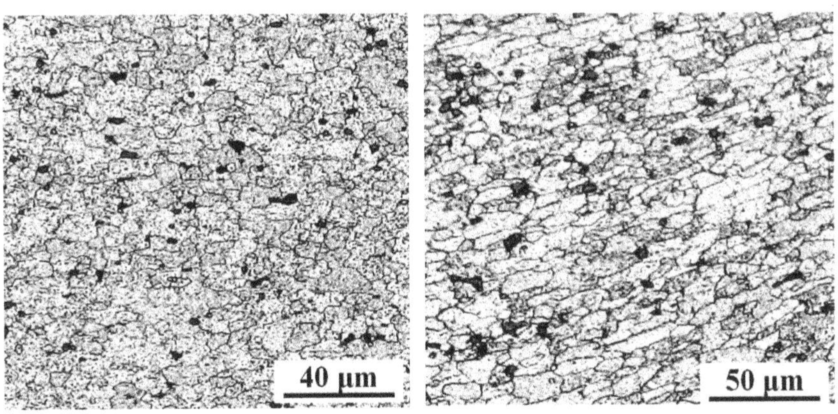

(a) (b)

Figure 14. Grain structures in joints at different tool rotational speeds [110]. (**a**) SZ-900 rpm, (**b**) TMAZ-900 rpm.

Authors have tried to analyze micro fracture surfaces of tensile-tested specimens using SEM to understand failure patterns. Figure 15 shows two micro graphs of welded samples with and without heat treatment. Fractured surfaces showed dimples of various sizes and shapes, indicating ductile fracture. Voids typically form before necking in ductile materials, but early necking can lead to more prominent void formation. Heat-treated specimens showed shallow and small dimples, while welded joints had deep and elongated dimples. A larger population of fine dimples indicated a more ductile behavior before failure. Similar findings were reported for fatigue-tested FSW joints [49].

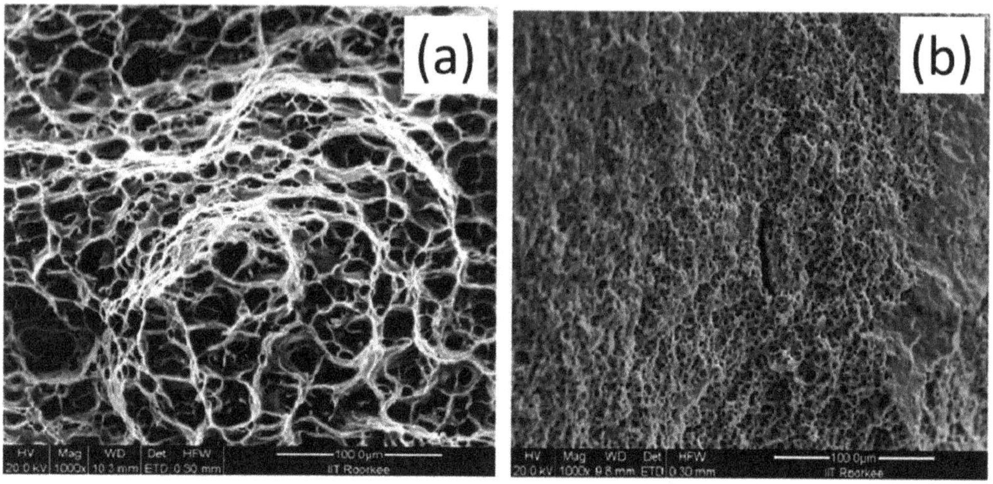

Figure 15. Fracture surface of tensile specimen [49]: (**a**) as welded at a welding speed of 8 mm/min, (**b**) heat treated and at a welding speed of 8 mm/min.

The formation of intermetallics is a matter of concern in most of the friction stir processes as the intermetallic compounds are generally brittle in nature, hence weakening the joint and it can produce further fracture to the joint [134]. Intermetallics can be formed more in the liquid–solid interface than the solid–solid interface [135].

Friction stir welding can be beneficial in some cases where it can limit the formation of the intermetallics. Authors have reported that dissimilar Al-Mg joints formed by fusion welds have more tendency to form the intermetallics than the FSW joints. Various types of IMCs can be observed for the dissimilar Al-Mg: β-Al3Mg2 and γ-Al12Mg17, etc. There is definite effect of the process parameters on the formation of the IMCs. Lv et al. [135] reported that tool offset can influence the overall thickness of the IMC layer. The utilization of the ultrasonic energy disjointed the IMCs and in turn enhanced the strength. With higher ultrasonic energy the complete removal of the IMCs is also possible. Baghdadi et al. [136] reported that the mechanical property is mainly affected by the IMCs which can be controlled by controlling the process parameters.

Among the popular methods, energy dispersive spectroscopy, X-Ray diffraction (Figure 16), etc., can be used to investigate the presence of the IMCs. As reported by Medhi at al. [25], the inter-diffusion performance of Al and Mg was responsible for the formation the IMCs (Al12Mg17 and Al3Mg2). TEM images as shown in Figure 17 also confirm the presence of the IMC in the Stir Zone at various tool rotational speeds. The investigation by Kim et al. [137] shows an in-depth analysis of the Al-Cu joint welded by the refill FSSP. It was reported that definite inter-metallic formation was found during the FSSW process (Figure 16). Intermetallic lumps composed of a variety of the inter-metallic compounds of the aluminum, viz., Al4Cu9, Al2Cu, Al6Mn, and Al2CuMg was observed at various locations at the joint interface.

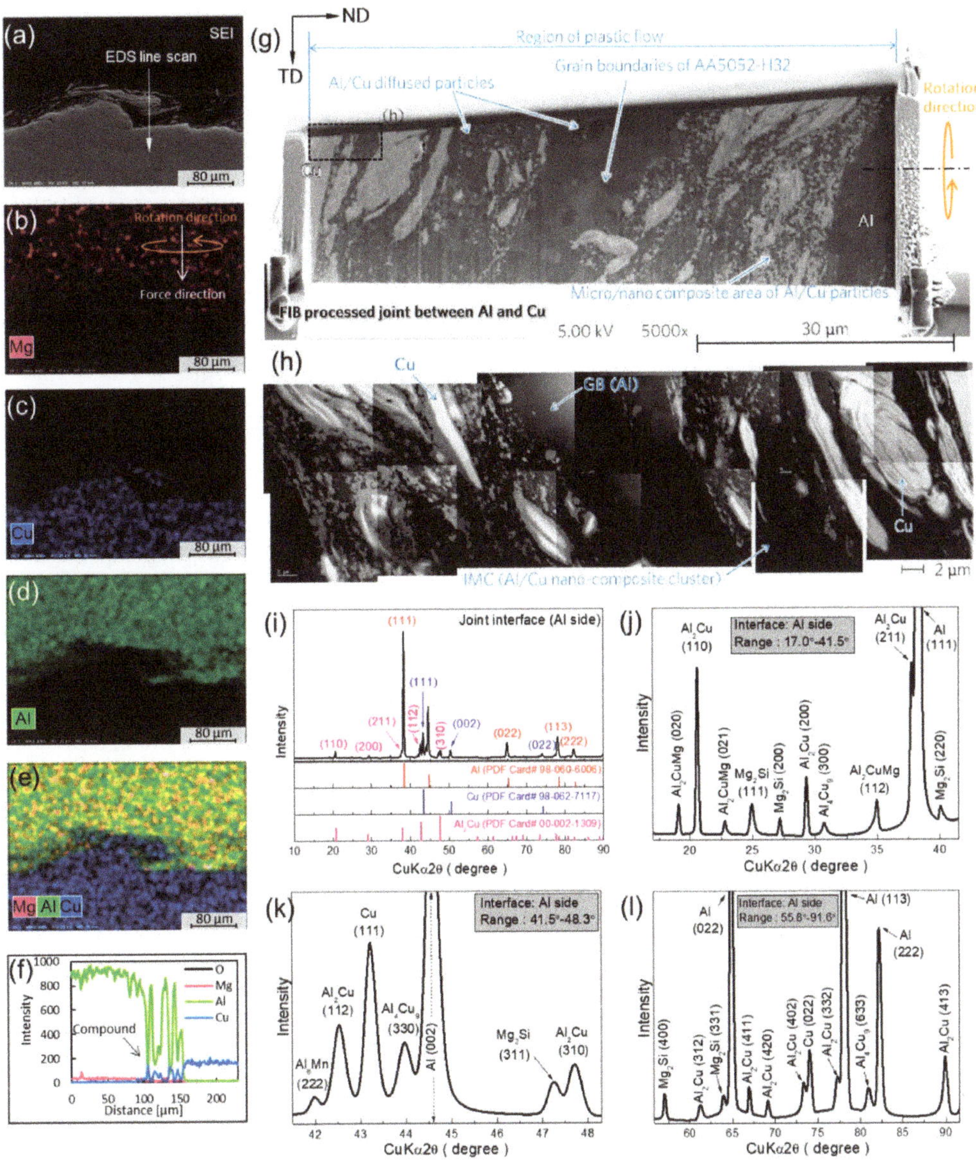

Figure 16. EDS images of Al-Cu joint: (**a**) joint interface (**b**–**e**), mapping of Mg, Cu, Al, and Mg-Cu-Al, (**f**) line scan of Al-Cu joint interface, (**g**) magnified joint interface, (**h**) various locations of the joint interface using HAADF-STEM, (**i**–**l**) XRD of the joint interface at various angles [121].

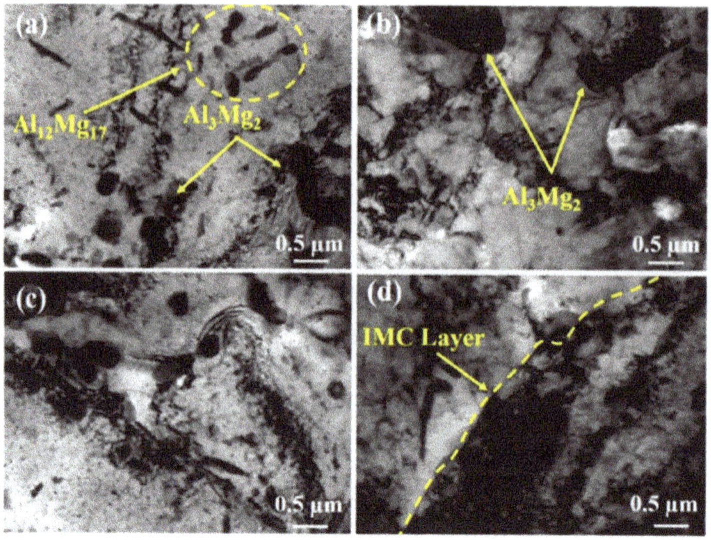

Figure 17. TEM images of Stir Zone (AZ91D and Al-7Si) at different tool rotational speeds (**a**) 700, (**b**) 800, (**c**) 900, and (**d**) 1000 rpm [25].

The intermetallic formation plays a very significant role for deciding the final mechanical behavior of the joint. As the joint strength of the friction stir spot welding process depends on both the mechanical locking due to mixing and the partial change in the microstructure, the inter-metallic formation can significantly influence the joint strength and fatigue behavior. Therefore, the process parameters should be controlled precisely to decide the final microstructural behavior of the weld joint for a sound joint, otherwise it may lead to premature failure of the joint.

6. Design Rules and Code Comparisons for FSSW of Aluminum Alloys

Although friction stir spot welding (FSSW) is widely researched, its inclusion in formal design codes is still in a developmental stage. However, some progress has been made in incorporating spot welding practices, including those applicable to aluminum alloys, in various regional standards such as the Eurocode 9 [138], American Welding Society (AWS) D17.3/D17.3M [139], AS/NZS 1665 [140], CSA S157 [141] and AISC Manual [142]. The Table 3 below provides a comparative summary of these standards with respect to aluminum spot welding, particularly highlighting provisions relevant to FSSW or translatable from resistance spot welding (RSW).

Table 3. Comparative summary of design codes for aluminum spot welding.

Code/Standard	Region	Design Focus	Relevant Provisions for FSSW	Advantages	Limitations
Eurocode 9 (EN 1999-1-1)	Europe	Structural design of aluminum structures	Allows mechanical fastening and fusion welding; FSSW not explicitly included	Comprehensive structural safety checks	Lacks FSSW-specific guidance; conservative assumptions for weld strength
AWS D17.3/D17.3M	USA	Friction stir welding of aluminum in aerospace	Covers friction stir welding, thus indirectly applicable to FSSW	Aerospace-focused, high reliability	Limited FSSW-specific joint design parameters

Table 3. *Cont.*

Code/Standard	Region	Design Focus	Relevant Provisions for FSSW	Advantages	Limitations
AS/NZS 1665	Australia/New Zealand	Welding of aluminum structures	Primarily deals with fusion welding; spot weld strength limits provided	Some design guidance can be extrapolated	Does not consider solid-state welding processes like FSSW
AISC Manual	USA	Structural steel/aluminum joints	Not directly applicable but useful for understanding allowable stress design	Well-established engineering principles	No mention of FSSW or aluminum-specific solid-state welding
CSA S157	Canada	Strength of aluminum joints	Offers general aluminum weld joint design guidance	Includes provisions for performance-based assessment	No separate treatment of friction stir-based spot welding

Currently, most design codes do not directly incorporate provisions for FSSW of aluminum alloys. Instead, design engineers and researchers often rely on RSW design principles as a baseline, adjusted for the improved mechanical behavior and reliability of FSSW joints [121,143,144]. AWS D17.3 [139] is the only available design which talks about friction welding techniques, making it the most relevant framework currently available, especially for aerospace applications.

The lack of harmonized and FSSW-specific design equations presents a major limitation, particularly when adopting FSSW in load-bearing or fatigue-sensitive structures. Variability in experimental data, differences in alloy behavior, and tool configuration challenges complicate the development of universal equations. To bridge this gap, future efforts should aim to establish standardized design protocols for FSSW based on comprehensive experimental validation, fatigue life assessment, and numerical simulation [136]. These could then feed into code revisions or the creation of dedicated FSSW appendices within existing welding standards.

7. Summary

In the past few years, rampant growth has been maintained for the investigation of FSSW for aluminum alloys. A huge number of promising results have been obtained which include the wide application of FSSW for various industries. This is why the FSSW process plays an important role to reduce pollutant emissions, and moreover, this process requires less temperature than the melting point. Moreover, this method is also used for the overlap joining of aluminum alloys which are not weldable by traditional techniques. Though due to limited scope, this study could not cover a few of the areas like the modeling and simulation works and microstructural study of the FSSW joints, but rather a multifold analysis of the recent studies on FSSW has been performed. To maximize the application of FSSW with minimum defects, various experiments were carried out by worldwide researchers, which concludes the following points:

- It can be seen that a wide range of tool materials are being used; H13 is the most widely used, other popular ones are HSS, C40, CBN, etc.
- From the study, it can be concluded that the peak welding temperatures vary from 280 °C to 450 °C and there is an increase in temperature with an increase in RPM; however, it depends upon the aluminum grade.
- Most of the researchers have performed the Vicker's hardness study for all the similar Al FSSW welding ranges of 50–160 HV, where HV at the Stir Zone of the upper plate has the highest hardness value.

- For achieving the highest load-bearing capacity, the dwell time should be kept minimum and the load-bearing capacity can be slightly improved using a pinless tool.
- Numerous researchers have conducted tensile testing of the FSSW joints, Al 6XXX series seems to show the maximum load-bearing capacity.
- From the studies, it is evident that the thermal history has a critical influence on the final mechanical behavior.
- Studies have been conducted to analyze the design standards, and it was found that currently there is no defined standard for the FSSW process, at present only AWS has a specific standard for the friction stir welding.

8. Future Directions

The current review has highlighted several critical aspects and recent developments in friction stir spot welding (FSSW) of aluminum alloys, particularly emphasizing tool design, temperature control, mechanical behavior, microstructural evolution, and process outcomes. While FSSW has shown clear advantages over conventional welding techniques, several research gaps and opportunities for advancement remain.

One of the key findings of this review is the pivotal role of tool geometry and material in controlling heat generation and material flow. Future research should focus on the design and optimization of non-traditional tool geometries, including multi-featured and adaptive tools, to enhance weld quality and reduce cycle time. Additionally, while tool steels and PCBN tools have been extensively studied, the development of cost-effective, high-durability alternatives, especially for dissimilar or high-strength aluminum alloys, remains underexplored.

A scope can be the comparative study with the other variants of the spot welding processes (RSW, LFW, or hybrid processes) with the merits and demerits of the solid-state process. There is a need for real-time process monitoring and closed-loop control systems to ensure repeatability and consistency. Advanced sensing techniques (e.g., thermography, acoustic emission, and force feedback) coupled with AI-based predictive models could enable adaptive process control tailored to material and joint conditions.

Residual stress evolution and its interaction with microstructure, particularly in multi-pass or hybrid processes, require deeper investigation through coupled thermo-mechanical simulations validated by experimental methods such as neutron diffraction and digital image correlation. Furthermore, long-term performance studies, including fatigue, corrosion resistance, and fracture behavior under service-like conditions, are essential to validate FSSW joints for critical applications in aerospace and automotive structures.

Standardization of design rules specific to FSSW by reviewing and harmonizing provisions across Eurocodes, AWS, and AS/NZS standards is crucial for broader structural implementation. Comparative analysis and experimental validation of design equations can provide a foundation for code development and improve industry acceptance.

In a nutshell, the future of FSSW research lies in a multidisciplinary approach, integrating materials science, mechanical engineering, process modeling, and machine learning. The insights from this review provide a roadmap for developing a more robust, intelligent, and industrially scalable FSSW process for aluminum alloys.

Author Contributions: Conceptualization, B.H. and B.L.; methodology, M.J.B., K.S., B.H., A.K.M. and H.J.; validation, A.K.M., B.L. and H.J.; formal analysis, Y.N., A.K.M. and H.J.; investigation, K.S.; resources, K.S.; data curation, M.J.B. and A.K.M.; writing—original draft preparation, M.J.B., Y.N., B.H., A.K.M., B.L. and H.J.; writing—review and editing, Y.N., B.H., A.K.M. and H.J.; visualization, Y.N., B.H., A.K.M., B.L. and H.J.; supervision, A.K.M.; project administration, B.H. and A.K.M.; funding acquisition, B.H. and B.L. All authors have read and agreed to the published version of the manuscript.

Funding: This work was supported and funded by the Deanship of Scientific Research at Imam Mohammad Ibn Saud Islamic University (IMSIU) (grant number IMSIU-DDRSP2502).

Data Availability Statement: No new data were created or analyzed in this study. Data sharing is not applicable to this article.

Acknowledgments: The authors are thankful to the National Institute of Technical Teachers' Training & Research (NITTTR), Kolkata for continuous support.

Conflicts of Interest: The authors declare that they have no conflicts of interest.

References

1. Su, J.Q.; Nelson, T.W.; Mishra, R.; Mahoney, M. Microstructural investigation of friction stir welded 7050-T651 aluminium. *Acta Mater.* **2003**, *51*, 713–729. [CrossRef]
2. Mubiayi, M.P.; Akinlabi, E.T.; Makhatha, M.E. Current state of friction stir spot welding between aluminium and copper. *Mater. Today Proc.* **2018**, *5*, 18633–18640. [CrossRef]
3. Çam, G.; Mistikoglu, S. Recent developments in friction stir welding of al-Alloys. *J. Mater. Eng. Perform.* **2014**, *23*, 1936–1953. [CrossRef]
4. Lin, Y.C.; Liu, J.J.; Chen, J.N. Material flow tracking for various tool geometries during the friction stir spot welding process. *J. Mater. Eng. Perform.* **2013**, *22*, 3674–3683. [CrossRef]
5. Shen, Z.; Ding, Y.; Gerlich, A.P. Advances in friction stir spot welding. *Crit. Rev. Solid State Mater. Sci.* **2019**, *45*, 457–534. [CrossRef]
6. Hsu, T.I.; Wu, L.T.; Tsai, M.H. Resistance and friction stir spot welding of dual-phase (DP 780)—A comparative study. *Int. J. Adv. Manuf. Technol.* **2018**, *97*, 2293–2299. [CrossRef]
7. Yang, X.W.; Fu, T.; Li, W.Y. Friction stir spot welding: A review on joint macro- and microstructure, property, and process modelling. *Adv. Mater. Sci. Eng.* **2014**, *2014*, 697170. [CrossRef]
8. Andalib, H.; Farahani, M.; Enami, M. Study on the new friction stir spot weld joint reinforcement technique on 5754 aluminum alloy. *Proc. Inst. Mech. Eng. Part C J. Mech. Eng. Sci* **2018**, *232*, 2976–2986. [CrossRef]
9. Venukumar, S.; Muthukumaran, S.; Swaroop, Y. Microstructure and mechanical properties of refilled friction stir spot welding of commercial pure aluminium. *Mater. Sci. Forum* **2013**, *765*, 776–780. [CrossRef]
10. Ahmed, M.M.Z.; EI-Sayed Seleman, M.M.; Fydrych, D.; Çam, G. Friction Stir Welding of Aluminum in the Aerospace Industry: The Current Progress and State-of-the-Art Review. *Materials* **2023**, *16*, 2971. [CrossRef]
11. Sengupta, K.; Chowdhury, I.; Banerjee, A.; Mondal, A.K.; Bose, D. Analysis of suitability of WC tool for joining Inconel 601 alloy by electric assisted friction stir welding. *Mater. Today Proc.* **2022**, *60*, 2093–2098. [CrossRef]
12. Sengupta, K.; Mondal, A.K.; Bose, D.; Chowdhury, I. Experimental Investigation on Metallurgical Properties on Effect of Dual Hybridization by Ultrasonic and Electric Energy in Friction Stir Welding for Joining Inconel 625 to Stainless Steel 304. *Trans. Indian Inst. Met.* **2024**, *77*, 941–953. [CrossRef]
13. Chowdhury, I.D.; Sengupta, K.; Singh, D.K.; Maji, K.K.; Roy, S.; Ghosal, S.; Mondal, A.K. Investigation of mechanical properties of dissimilar joint of 6063 aluminium and C26000 copper alloy by friction stir welding. *Mater. Today Proc.* **2021**, *44*, 4039–4047. [CrossRef]
14. Sengupta, K.; Singh, D.K.; Mondal, A.K.; Bose, D.; Patra, D.; Dhar, A. Characterization of tool wear in similar and dissimilar joints of MS and SS using EAFSW. *Mater. Today Proc.* **2021**, *44*, 3967–3975. [CrossRef]
15. Shen, Z.; Ding, Y.; Guo, W.; Hou, W.; Liu, X.; Chen, H.; Liu, F.; Li, W.; Gerlich, A. Refill Friction Stir Spot Welding Al Alloy to Copper via Pure Metallurgical Joining Mechanism. *Chin. J. Mech. Eng.* **2021**, *34*, 75. [CrossRef]
16. Myśliwiec, P.; Kubit, A.; Szawara, P. Optimization of 2024-T3 Aluminum Alloy Friction Stir Welding Using Random Forest, XGBoost, and MLP Machine Learning Techniques. *Materials* **2024**, *17*, 1452. [CrossRef]
17. Mehdi, H.; Singh, J.; Mouria, P.K.; Mishra, S.; Prakash, C. Effect of Tool Pin Profile on Mechanical and Metallurgical Properties of the Friction Stir Welded Joint of AA6061 and AZ91D. *J. Mater. Eng. Perform.* **2025**, *34*, 7976–7990. [CrossRef]
18. Mehdi, H.; Singh, B.; Salah, A.N.; Dubey, M.K.; Mishra, S.; Kumar, S. Effect of PWHT on metallurgical and mechanical characterization of dissimilar welded joint of P91 and P92 steels. *J. Adhes. Sci. Technol.* **2024**, *38*, 1395–1412. [CrossRef]
19. Reyaz, M.S.B.; Sinha, A.N.; Mehdi, H.; Murtaza, Q. Effect of Pulsed TIG Welding Parameters on the Microstructural Evolution and Mechanical Properties of Dissimilar AA6061-T6 and AA7075-T6 Weldments. *Arab. J. Sci. Eng.* **2024**, *49*, 10891–10911. [CrossRef]
20. Kumar, J.; Kumar, G.; Mehdi, H.; Kumar, M. Optimization of FSW parameters on mechanical properties of different aluminum alloys of AA6082 and AA7050 by response surface methodology. *Int. J. Interact. Des. Manuf.* **2024**, *18*, 1359–1371. [CrossRef]
21. Mehdi, H.; Batra, L.; Singh, A.P.; Malla, C. Multi-response optimization of FSW process parameters of dissimilar aluminum alloys of AA2014 and AA6061 by response surface methodology (RSM). *Int. J. Interact. Des. Manuf.* **2024**, *18*, 1507–1522. [CrossRef]

22. Jain, S.; Mishra, R.S.; Mehdi, H.; Gupta, R.; Dubey, A.K. Optimization of processing variables of friction stir welded dissimilar composite joints of AA6061 and AA7075 using response surface methodology. *J. Adhes. Sci. Technol.* **2024**, *38*, 949–968. [CrossRef]
23. Msomi, V.; Mabuwa, S.; Ngonda, T.; Mehdi, H.; Saxena, K.K. Microstructure and mechanical properties of tungsten inert gas welded dissimilar aluminum alloys joint subjected to normal multi-pass friction stir process. *Proc. Inst. Mech. Eng. Part E J. Process Mech. Eng.* **2023**, *235*, 2531–2546. [CrossRef]
24. Mehdi, H.; Jain, S.; Salah, A.N.; Singh, A.P.; Chawla, S. Effect of friction stir welding parameters on microstructure and mechanical properties of the dissimilar alloys of AZ91D and AA7075. *Proc. Inst. Mech. Eng. Part E J. Process Mech. Eng.* **2023**, *239*. [CrossRef]
25. Mehdi, H.; Jain, S.; Msomi, V.; Mabuwa, S.; Malla, C. Effect of Intermetallic Compounds on Mechanical and Microstructural Properties of Dissimilar Alloys Al-7Si/AZ91D. *J. Mater. Eng. Perform.* **2023**, *33*, 4781–4793. [CrossRef]
26. Jain, S.; Mishra, R.S.; Mehdi, H. Influence of SiC Microparticles and Multi-Pass FSW on Weld Quality of the AA6082 and AA5083 Dissimilar Joints. *Silicon* **2023**, *15*, 6185–6197. [CrossRef]
27. Boldsaikhan, E.; Fukada, S.; Fujimoto, M.; Kamimuki, K.; Okada, H. Refill friction stir spot welding of surface-treated aerospace aluminum alloys with faying-surface sealant. *J. Manuf. Process.* **2019**, *42*, 113–120. [CrossRef]
28. Yuan, W. Friction Stir Spot Welding of Aluminum Alloys. Master's Thesis, Missouri University of Science and Technology, Rolla, MO, USA, 2008. Available online: https://scholarsmine.mst.edu/masters_theses/5429/ (accessed on 11 July 2025).
29. Kah, P.; Rajan, R.; Martikainen, J.; Suoranta, R. Investigation of weld defects in friction-stir welding and fusion welding of aluminium alloys. *Int. J. Mech. Mater. Eng.* **2015**, *10*, 26. [CrossRef]
30. Yang, Q.; Mironov, S.; Sato, Y.S.; Okamoto, K. Material flow during friction stir spot welding. *Mater. Sci. Eng. A* **2010**, *527*, 4389–4398. [CrossRef]
31. Bakavos, D.; Chen, Y.; Babout, L.; Prangnell, P. Material interactions in a novel pinless tool approach to friction stir spot welding thin aluminum sheet. *Metall. Mater. Trans. A Phys. Metall. Mater. Sci.* **2011**, *42*, 1266–1282. [CrossRef]
32. Yazdi, S.R.; Beidokhti, B.; Haddad-Sabzevar, M. Pinless tool for FSSW of AA 6061-T6 aluminum alloy. *J. Mater. Process. Technol.* **2019**, *267*, 44–51. [CrossRef]
33. Garg, A.; Bhattacharya, A. Strength and failure analysis of similar and dissimilar friction stir spot welds: Influence of different tools and pin geometries. *Mater. Des.* **2017**, *127*, 272–286. [CrossRef]
34. Vacchi, G.S.; Silva, R.; Plaine, A.H.; Suhuddin, U.F.H.; Alcântara, N.G.; Sordi, V.L.; Rovere, C.A.D. Refill friction stir spot welded AA5754-H22/Ti-6Al-4V joints: Microstructural characterization and electrochemical corrosion behavior of aluminum surfaces. *Mater. Today Commun.* **2020**, *22*, 100759. [CrossRef]
35. Shin, H.S.; De Leon, M. Weldability assessment of friction stir spot welded lightweight alloys using pin and pinless tools. *Sci. Technol. Weld. Join.* **2016**, *21*, 99–105. [CrossRef]
36. Cox, C.D.; Gibson, B.T.; Delapp, D.R.; Strauss, A.M.; Cook, G.E. A method for double-sided friction stir spot welding. *J. Manuf. Process.* **2014**, *16*, 241–247. [CrossRef]
37. Cox, C.D.; Gibson, B.T.; Strauss, A.M.; Cook, G.E. Energy input during friction stir spot welding. *J. Manuf. Process.* **2014**, *16*, 479–484. [CrossRef]
38. Hirasawa, S.; Badarinarayan, H.; Okamoto, K.; Tomimura, T.; Kawanami, T.; Hirano, S. Analysis of Temperature and Plastic Flow during Friction Stir Spot Welding Using Particle Method. *J. Therm. Sci. Technol.* **2009**, *4*, 260–271. [CrossRef]
39. Membala, S.B.; Sutresman, O.S.; Arsyad, H.; Syahid, M.; Widyianto, A. Metallurgical Characterization of Micro Friction Stir Spot Welding (μFSSW) Parameters on Similar Materials AA1100. *J. Phys. Conf. Ser.* **2024**, *2739*, 012043. [CrossRef]
40. Safitri, L.A.; Fakhrurrozi, I.F.; Amat, M.A.; Baskoro, A.S.; Rupajati, P.; Kiswanto, G. Effect of Tool Profile on Mechanical Properties, Temperature, and RPM Using Micro Friction Stir Spot Welding (mFSSW) on Aluminum Alloy AA1100. *J. Phys. Conf. Ser.* **2024**, *2739*, 012024. [CrossRef]
41. Li, W.; Li, J.; Zhang, Z.; Gao, D.; Wang, W.; Dong, C. Improving mechanical properties of pinless friction stir spot welded joints by eliminating hook defect. *Mater. Des.* **2014**, *62*, 247–254. [CrossRef]
42. Rai, R.; De, A.; Bhadeshia, H.K.D.H.; DebRoy, T. Review: Friction stir welding tools. *Sci. Technol. Weld. Join.* **2011**, *16*, 325–342. [CrossRef]
43. Zhang, Y.N.; Cao, X.; Larose, S.; Wanjara, P. Review of tools for friction stir welding and processing. *Can. Metall. Q.* **2012**, *51*, 250–261. [CrossRef]
44. Chiteka, K. Friction Stir Welding/Processing Tool Materials and Selection. *Int. J. Eng. Res. Technol.* **2013**, *2*, 8–18.
45. Mishra, R.S.; Ma, Z.Y. Friction stir welding and processing. *Mater. Sci. Eng. R Rep.* **2005**, *50*, 1–78. [CrossRef]
46. Meilinger, A.; Torok, I. The Importance of Friction Stir Welding Tool. *Prod. Process. Syst.* **2013**, *6*, 25–34.
47. Prado, R.A.; Murr, L.E.; Shindo, D.J.; Soto, K.F. Tool wear in the friction-stir welding of aluminum alloy 6061+20% Al_2O_3: A preliminary study. *Scr. Mater.* **2001**, *45*, 75–80. [CrossRef]
48. Prado, R.A.; Murr, L.E.; Soto, K.F.; McClure, J.C. Self-optimization in tool wear for friction-stir welding of Al 6061+20% Al_2O_3 MMC. *Mater. Sci. Eng. A* **2003**, *349*, 156–165. [CrossRef]

49. Singh, R.K.R.; Sharma, C.; Dwivedi, D.K.; Mehta, N.K.; Kumar, P. The microstructure and mechanical properties of friction stir welded Al-Zn-Mg alloy in as welded and heat treated conditions. *Mater. Des.* **2011**, *32*, 682–687. [CrossRef]
50. Nami, H.; Adgi, H.; Sharifitabar, M.; Shamabadi, H. Microstructure and mechanical properties of friction stir welded Al/Mg2Si metal matrix cast composite. *Mater. Des.* **2011**, *32*, 976–983. [CrossRef]
51. Fernandez, G.J.; Murr, L.E. Characterization of tool wear and weld optimization in the friction-stir welding of cast aluminum 359+20% SiC metal-matrix composite. *Mater. Charact.* **2004**, *52*, 65–75. [CrossRef]
52. Lee, W.B.; Yeon, Y.M.; Jung, S.B. Joint properties of friction stir welded AZ31B—H24 magnesium alloy. *Mater. Sci. Technol.* **2003**, *19*, 785–790. [CrossRef]
53. Rodriguez, N.A.; Almanza, E.; Alvarez, C.J.; Murr, L.E. Study of friction stir welded A319 and A413 aluminum casting alloys. *J. Mater. Sci.* **2005**, *40*, 4307–4312. [CrossRef]
54. Rajakumar, S.; Muralidharan, C.; Balasubramanian, V. Influence of friction stir welding process and tool parameters on strength properties of AA7075-T6 aluminium alloy joints. *Mater. Des.* **2011**, *32*, 535–549. [CrossRef]
55. Thomas, W.M.; Nicholas, E.D.; Watts, E.R.; Staines, D.G. Friction based welding technology for aluminium. *Mater. Sci. Forum* **2002**, *396–402*, 1543–1548. [CrossRef]
56. Lee, C.Y.; Choi, D.H.; Yeon, Y.M.; Jung, S.B. Dissimilar friction stir spot welding of low carbon steel and Al-Mg alloy by formation of IMCs. *Sci. Technol. Weld. Join.* **2009**, *14*, 216–220. [CrossRef]
57. Elrefaey, A.; Gouda, M.; Takahashi, M.; Ikeuchi, K. Characterization of aluminum/steel lap joint by friction stir welding. *J. Mater. Eng. Perform.* **2005**, *14*, 10–17. [CrossRef]
58. Chen, T. Process parameters study on FSW joint of dissimilar metals for aluminum-steel. *J. Mater. Sci.* **2009**, *44*, 2573–2580. [CrossRef]
59. Chen, T.P.; Lin, W.B. Optimal FSW process parameters for interface and welded zone toughness of dissimilar aluminium-steel joint. *Sci. Technol. Weld. Join.* **2010**, *15*, 279–285. [CrossRef]
60. DebRoy, T.; Bhadeshia, H.K.D.H. Friction stir welding of dissimilar alloys—A perspective. *Sci. Technol. Weld. Join.* **2010**, *15*, 266–270. [CrossRef]
61. Dressler, U.; Biallas, G.; Alfaro Mercado, U. Friction stir welding of titanium alloy TiAl6V4 to aluminium alloy AA2024-T3. *Mater. Sci. Eng. A* **2009**, *526*, 113–117. [CrossRef]
62. Miles, M.P.; Feng, Z.; Kohkonen, K.; Weickum, B.; Steel, R.; Lev, L. Spot joining of AA 5754 and high strength steel sheets by consumable bit. *Sci. Technol. Weld. Join.* **2010**, *15*, 325–330. [CrossRef]
63. Meran, C.; Kovan, V. Microstructures and mechanical properties of friction stir welded dissimilar copper/brass joints. *Mater. Und Werkst.* **2008**, *39*, 521–530. [CrossRef]
64. Jiang, W.H.; Kovacevic, R. Feasibility study of friction stir welding of 6061-T6 aluminium alloy with AISI 1018 steel. *Proc. Inst. Mech. Eng. Part B J. Eng. Manuf.* **2004**, *218*, 1323–1331. [CrossRef]
65. Karthikeyan, L.; Puviyarasan, M.; Sharath Kumar, S.; Balamugundan, B. Experimental studies on friction stir welding of AA2011 and AA6063 aluminium alloys. *Int. J. Adv. Eng. Technol.* **2012**, *3*, 144–145.
66. Hariharan, V.; Elumalai, P.V.; Nambiraj, M.; Jayakar, J.; Parthasarathy, M.; Venkata Kamesh, V. Experimental investigation of friction stir welding on aluminium AA6063. *Mater. Today Proc.* **2021**, *47*, 6830–6834. [CrossRef]
67. Palanivel, R.; Koshy Mathews, P. Prediction and optimization of process parameter of friction stir welded AA5083-H111 aluminum alloy using response surface methodology. *J. Cent. South Univ. Technol. (English Ed.)* **2012**, *19*, 1–8. [CrossRef]
68. Cavaliere, P.; De Santis, A.; Panella, F.; Squillace, A. Effect of welding parameters on mechanical and microstructural properties of dissimilar AA6082-AA2024 joints produced by friction stir welding. *Mater. Des.* **2009**, *30*, 609–616. [CrossRef]
69. Parida, B.; Mohapatra, M.M.; Biswas, P.; Mandal, N.R. Study of Mechanical and Micro-structural Properties of Friction Stir Welded Al-Alloy. *Int. J. Emerg. Technol. Adv. Eng.* **2012**, *2*, 307–312.
70. Akinlabi, E.T.; Reddy, R.D.; Akinlabi, S.A. Microstructural characterizations of dissimilar friction stir welds. *Lect. Notes Eng. Comput. Sci.* **2012**, *3*, 1779–1783.
71. Pathak, N.; Bandyopadhyay, K.; Sarangi, M.; Panda, S.K. Microstructure and mechanical performance of friction stir spot-welded aluminum-5754 sheets. *J. Mater. Eng. Perform.* **2013**, *22*, 131–144. [CrossRef]
72. Shindo, D.J.; Rivera, A.R.; Murr, L.E. Shape optimization for tool wear in the friction-stir welding of cast Al359-20% SiC MMC. *J. Mater. Sci.* **2002**, *37*, 4999–5005. [CrossRef]
73. Feizollahi, V.; Yousefi, M.; Elahifar, A.; Pourmirza, B.; Hasab, M.G.; Moghadam, A.H. Effect of shoulder diameter, tool rotation speed, and arrangement of plates on mechanical and metallurgical properties of dissimilar aluminum 2024-T3 and 7075-T6 friction stir spot welding (FSSW). *Eng. Fail. Anal.* **2024**, *163*, 108548. [CrossRef]
74. Tobin, D.; O'Shaughnessy, S.; Trimble, D. Characterisation of force and torque with auxiliary heating during friction stir spot welding of AA2024-T3. *Results Mater.* **2024**, *21*, 100535. [CrossRef]
75. Manohar, M.V.N.S.; Mahadevan, K.; Aravindan, A. Performance of H13 and HCHCR Tool Materials on Friction Stir Welded 6070Aluminium Alloy Plates Subjected to Different Temper Conditions. *Int. J. Recent Technol. Eng.* **2019**, *8*, 5261–5266. [CrossRef]

76. Singh, K.; Khatirkar, R.K.; Sapate, S.G. Microstructure evolution and abrasive wear behavior of D2 steel. *Wear* **2015**, *328–329*, 206–216. [CrossRef]
77. Zhou, X.F.; Fang, F.; Li, F.; Jiang, J.Q. Morphology and microstructure of M2C carbide formed at different cooling rates in AISI M2 high speed steel. *J. Mater. Sci.* **2011**, *46*, 1196–1202. [CrossRef]
78. Vander Voort, G.F.; Lucas, G.M. *Metallography and Microstructures of Stainless Steels and Maraging Steels*; ASM Handbook: Materials Park, OH, USA, 2019. [CrossRef]
79. Tanaka, T.; Hirata, T.; Shinomiya, N.; Shirakawa, N. Analysis of material flow in the sheet forming of friction-stir welds on alloys of mild steel and aluminum. *J. Mater. Process. Technol.* **2015**, *226*, 115–124. [CrossRef]
80. Ragu Nathan, S.; Balasubramanian, V.; Malarvizhi, S.; Rao, A.G. An investigation on metallurgical characteristics of tungsten based tool materials used in friction stir welding of naval grade high strength low alloy steels. *Int. J. Refract. Met. Hard Mater.* **2016**, *56*, 18–26. [CrossRef]
81. Sato, Y.S.; Yamanoi, H.; Kokawa, H.; Furuhara, T. Microstructural evolution of ultrahigh carbon steel during friction stir welding. *Scr. Mater.* **2007**, *57*, 557–560. [CrossRef]
82. Ghosh, M.; Kumar, K.; Mishra, R.S. Analysis of microstructural evolution during friction stir welding of ultrahigh-strength steel. *Scr. Mater.* **2010**, *63*, 851–854. [CrossRef]
83. Park, S.H.C.; Sato, Y.S.; Kokawa, H.; Okamoto, K.; Hirano, S.; Inagaki, M. Rapid formation of the sigma phase in 304 stainless steel during friction stir welding. *Scr. Mater.* **2003**, *49*, 1175–1180. [CrossRef]
84. Park, S.H.C.; Sato, Y.S.; Kokawa, H.; Okamoto, K.; Hirano, S.; Inagaki, M. Microstructural characterisation of stir zone containing residual ferrite in friction stir welded 304 austenitic stainless steel. *Sci. Technol. Weld. Join.* **2005**, *10*, 550–556. [CrossRef]
85. Miles, M.P.; Nelson, T.W.; Steel, R.; Olsen, E.; Gallagher, M. Effect of friction stir welding conditions on properties and microstructures of high strength automotive steel. *Sci. Technol. Weld. Join.* **2009**, *14*, 228–232. [CrossRef]
86. Zhang, Y.; Sato, Y.S.; Kokawa, H.; Park, S.H.C.; Hirano, S. Stir zone microstructure of commercial purity titanium friction stir welded using PCBN tool. *Mater. Sci. Eng. A* **2008**, *488*, 25–30. [CrossRef]
87. Sato, Y.S.; Nelson, T.W.; Sterling, C.J.; Steel, R.J.; Pettersson, C.O. Microstructure and mechanical properties of friction stir welded SAF 2507 super duplex stainless steel. *Mater. Sci. Eng. A* **2005**, *397*, 376–384. [CrossRef]
88. Santella, M.; Hovanski, Y.; Frederick, A.; Grant, G.; Dahl, M. Friction stir spot welding of DP780 carbon steel. *Sci. Technol. Weld. Join.* **2010**, *15*, 271–278. [CrossRef]
89. Park, S.H.C.; Sato, Y.S.; Kokawa, H.; Okamoto, K.; Hirano, S.; Inagaki, M. Boride formation induced by PCBN tool wear in friction-stir-welded stainless steels. *Metall. Mater. Trans. A Phys. Metall. Mater. Sci.* **2009**, *40*, 625–636. [CrossRef]
90. Hovanski, Y.; Santella, M.L.; Grant, G.J. Friction stir spot welding of hot-stamped boron steel. *Scr. Mater.* **2007**, *57*, 873–876. [CrossRef]
91. Sumiya, H.; Uesaka, S.; Satoh, S. Mechanical properties of high purity polycrystalline CBN synthesized by direct conversion sintering method. *J. Mater. Sci.* **2000**, *35*, 1181–1186. [CrossRef]
92. Sato, Y.S.; Harayama, N.; Kokawa, H.; Inoue, H.; Tadokoro, Y.; Tsuge, S. Evaluation of microstructure and properties in friction stir welded superaustenitic stainless steel. *Sci. Technol. Weld. Join.* **2009**, *14*, 202–209. [CrossRef]
93. Agarwala, B.K.; Singh, B.P.; Singhal, S.K. Synthesis and characterization of polycrystalline sintered compacts of cubic boron nitride. *J. Mater. Sci.* **1986**, *21*, 1765–1768. [CrossRef]
94. Giménez, S.; Van der Biest, O.; Vleugels, J. The role of chemical wear in machining iron based materials by PCD and PCBN super-hard tool materials. *Diam. Relat. Mater.* **2007**, *16*, 435–445. [CrossRef]
95. Collier, M.; Steel, R.; Nelson, T.; Sorensen, C.; Packer, S. Grade development of polycrystalline cubic boron nitride for friction stir processing of ferrous alloys. *Mater. Sci. Forum* **2003**, *426–432*, 3011–3016. [CrossRef]
96. D'Evelyn, M.P.; Taniguchi, T. Elastic properties of translucent polycrystalline cubic boron nitride as characterized by the dynamic resonance method. *Diam. Relat. Mater.* **1999**, *8*, 1522–1526. [CrossRef]
97. Casanova, C.A.M.; Balzaretti, N.M.; Voronin, G.; Da Jornada, J.A.H. Experimental study of plastic deformation during sintering of cubic boron nitride compacts. *Diam. Relat. Mater.* **1999**, *8*, 1451–1454. [CrossRef]
98. Harris, T.K.; Brookes, E.J.; Taylor, C.J. The effect of temperature on the hardness of polycrystalline cubic boron nitride cutting tool materials. *Int. J. Refract. Met. Hard Mater.* **2004**, *22*, 105–110. [CrossRef]
99. König, W.; Neises, A. Wear mechanisms of ultrahard, non-metallic cutting materials. *Wear* **1993**, *162–164*, 12–21. [CrossRef]
100. Hooper, R.M.; Shakib, J.I.; Brookes, C.A. Microstructure and wear of TiC-cubic BN tools. *Mater. Sci. Eng.* **1988**, *105–106*, 429–433. [CrossRef]
101. Nabhani, F. Wear mechanisms of ultra-hard cutting tools materials. *J. Mater. Process. Technol.* **2001**, *115*, 402–412. [CrossRef]
102. Poulachon, G.; Moisan, A.; Jawahir, I.S. Tool-wear mechanisms in hard turning with polycrystalline cubic boron nitride tools. *Wear* **2001**, *250–251*, 576–586. [CrossRef]
103. Zimmermann, M.; Lahres, M.; Viens, D.V.; Laube, B.L. Investigations of the wear of cubic boron nitride cutting tools using Auger electron spectroscopy and X-ray analysis by EPMA. *Wear* **1997**, *209*, 241–246. [CrossRef]

104. Nelson, T.W.; Sorensen, C.D.; Mayfield, D.W. Friction Stir Welds of HSLA steel panels for shipyard applications. In Proceedings of the 8th International Symposium on Friction Stir Welding, Hamburg, Germany, 18–20 May 2010; The Welding Institute: Hamburg, Germany.
105. Mahoney, M.; Nelson, T.; Sorenson, C.; Packer, S. Friction stir welding of ferrous alloys: Current status. *Mater. Sci. Forum* **2010**, *638–642*, 41–46. [CrossRef]
106. Liu, F.C.; Hovanski, Y.; Miles, M.P.; Sorensen, C.D.; Nelson, T.W. A review of friction stir welding of steels: Tool, material flow, microstructure, and properties. *J. Mater. Sci. Technol.* **2018**, *34*, 39–57. [CrossRef]
107. Borah, M.; Saha, N.; Rajbongshi, S.K.; Dharam, B. Friction stir spot welding process: An innovative approach for transforming from engineering design to production. *Int. J. Interact. Des. Manuf. (IJIDeM)* **2023**, *17*, 2259–2270. [CrossRef]
108. Arul, S.G.; Miller, S.F.; Kruger, G.H.; Pan, T.Y.; Mallick, P.K.; Shih, A.J. Experimental study of joint performance in spot friction welding of 6111-T4 aluminium alloy. *Sci. Technol. Weld. Join.* **2008**, *13*, 629–637. [CrossRef]
109. Pandey, A.K.; Mahapatra, S.S. Investigation of weld zone obtained by friction stir spot welding (FSSW) of aluminium-6061 alloy. *Mater. Today Proc.* **2019**, *18*, 4491–4500. [CrossRef]
110. Li, G.; Zhou, L.; Luo, L.; Wu, X.; Guo, N. Microstructural evolution and mechanical properties of refill friction stir spot welded alclad 2A12-T4 aluminum alloy. *J. Mater. Res. Technol.* **2019**, *8*, 4115–4129. [CrossRef]
111. Kubit, A.; Trzepiecinski, T. A fully coupled thermo-mechanical numerical modelling of the refill friction stir spot welding process in Alclad 7075-T6 aluminium alloy sheets. *Archiv. Civ. Mech. Eng.* **2020**, *20*, 117. [CrossRef]
112. Gerlich, A.; Avramovic-Cingara, G.; North, T.H. Stir zone microstructure and strain rate during Al 7075-T6 friction stir spot welding. *Metall. Mater. Trans. A* **2006**, *37*, 2773–2786. [CrossRef]
113. Shen, Z.; Chen, Y.; Hou, J.S.C.; Yang, X.; Gerlich, A.P. Influence of processing parameters on microstructure and mechanical performance of refill friction stir spot welded 7075-T6 aluminium alloy. *Sci. Technol. Weld. Join.* **2015**, *20*, 48–57. [CrossRef]
114. Janga, V.S.R.; Awang, M.; Yamin, M.F.; Suhuddin, U.F.H.; Klusemann, B.; Santos, J.F.d. Experimental and Numerical Analysis of Refill Friction Stir Spot Welding of Thin AA7075-T6 Sheets. *Materials* **2021**, *14*, 7485. [CrossRef]
115. Kubit, A.; Kluz, R.; Trzepieciński, T.; Wydrzyński, D.; Bochnowski, W. Analysis of the mechanical properties and of micrographs of refill friction stir spot welded 7075-T6 aluminium sheets. *Archiv. Civ. Mech. Eng.* **2018**, *18*, 235–244. [CrossRef]
116. Sarfaraz, Z.; Awan, Y.R.; Saeed, H.A.; Khan, R.; Wieczorowski, M.; Din, N.A. Residual Stress in Friction Stir Welding of Dissimilar Aluminum Alloys: A Parametric Study. *Materials* **2025**, *18*, 316. [CrossRef]
117. Sun, Y.; Morisada, Y.; Fujii, H.; Tsuji, N. Ultrafine grained structure and improved mechanical properties of low temperature friction stir spot welded 6061-T6 Al alloys. *Mater. Charact.* **2018**, *135*, 124–133. [CrossRef]
118. Farmanbar, N.; Mousavizade, S.M.; Ezatpour, H.R. Achieving special mechanical properties with considering dwell time of AA5052 sheets welded by a simple novel friction stir spot welding. *Mar. Struct.* **2019**, *65*, 197–214. [CrossRef]
119. Badarinarayan, H.; Yang, Q.; Zhu, S. Effect of tool geometry on static strength of friction stir spot-welded aluminum alloy. *Int. J. Mach. Tools Manuf.* **2009**, *49*, 142–148. [CrossRef]
120. Merzoug, M.; Mazari, M.; Berrahal, L.; Imad, A. Parametric studies of the process of friction spot stir welding of aluminium 6060-T5 alloys. *Mater. Des.* **2010**, *31*, 3023–3028. [CrossRef]
121. Subramanian, S.; Natarajan, E.; Khalfallah, A.; Muthukutti, G.P.; Beygi, R.; Louhichi, B.; Sengottuvel, R.; Ang, C.K. Current Trends and Emerging Strategies in Friction Stir Spot Welding for Lightweight Structures: Innovations in Tool Design, Robotics, and Composite Reinforcement—A Review. *Crystals* **2025**, *15*, 556. [CrossRef]
122. Boucherit, A.; Abdi, S.; Aissani, M.; Medhi, B.; Abib, K.; Badji, R. Weldability, microstructure, and residual stress in Al/Cu and Cu/Al friction stir spot weld joints with Zn interlayer. *Int. J. Adv. Manuf. Technol.* **2020**, *111*, 1553–1569. [CrossRef]
123. Balamurugan, M.; Gopi, S.; Dhanesh, G.M. Influence of tool pin profiles on the filler added friction stir spot welded dissimilar aluminium alloy joints. *Mater. Res. Express* **2021**, *8*, 096531. [CrossRef]
124. Yousefi, A.; Serjouei, A.; Hedayati, R.; Bodaghi, M. Fatigue Modeling and Numerical Analysis of Re-Filling Probe Hole of Friction Stir Spot Welded Joints in Aluminum Alloys. *Materials* **2021**, *14*, 2171. [CrossRef]
125. Kubit, A.; Święch, Ł.; Trzepieciński, T.; Faes, K. Experimental Analysis of the Post-Buckling Behaviour of Compressed Stiffened Panel with Refill Friction Stir Spot Welded and Riveted Stringers. *Adv. Sci. Technol. Res. J.* **2022**, *16*, 159–167. [CrossRef]
126. Kubit, A.; Drabczyk, M.; Trzepiecinski, T.; Bochnowski, W.; Kaščák, L'.; Slota, J. Fatigue Life Assessment of Refill Friction Stir Spot Welded Alclad 7075-T6 Aluminium Alloy Joints. *Metals* **2020**, *10*, 633. [CrossRef]
127. Tier, M.D.; Rosendo, T.S.; dos Santo, J.F.; Huber, N.; Mazzaferro, J.A.; Mazzaferro, C.P.; Strohaecker, T.R. The influence of refill FSSW parameters on the microstructure and shear strength of 5042 aluminium welds. *J. Mater. Process. Technol.* **2013**, *213*, 997–1005. [CrossRef]
128. Venukumar, S.; Baby, B.; Muthukumaran, S.; Kailas, S.V. Microstructural and Mechanical Properties of Walking Friction Stir Spot Welded AA 6061-T6 Sheets. *Procedia Mater. Sci.* **2014**, *6*, 656–665. [CrossRef]
129. Freeney, T.A.; Sharma, S.R.; Mishra, R.S. Effect of welding parameters on properties of 5052 Al friction stir spot welds. *SAE Tech. Pap.* **2006**, *1*, 0969. [CrossRef]

130. Shen, Z.; Yang, X.; Zhang, Z.; Cui, L.; Li, T. Microstructure and failure mechanisms of refill friction stir spot welded 7075-T6 aluminum alloy joints. *Mater. Des.* **2013**, *44*, 476–486. [CrossRef]
131. Zhang, Z.; Yang, X.; Zhang, J.; Zhou, G.; Xu, X.; Zou, B. Effect of welding parameters on microstructure and mechanical properties of friction stir spot welded 5052 aluminum alloy. *Mater. Des.* **2011**, *32*, 4461–4470. [CrossRef]
132. Rosendo, T.; Parra, B.; Tier, M.A.D.; da Silva, A.A.M.; dos Santos, J.F.; Strohaecker, T.R.; Alcântara, N.G. Mechanical and microstructural investigation of friction spot welded AA6181-T4 aluminium alloy. *Mater. Des.* **2011**, *32*, 1094–1100. [CrossRef]
133. Nandan, R.; DebRoy, T.; Bhadeshia, H.K.D.H. Recent advances in friction-stir welding—Process, weldment structure and properties. *Prog. Mater. Sci.* **2008**, *53*, 980–1023. [CrossRef]
134. Mofid, M.A.; Abdollah-zadeh, A.; Gür, C.H. Investigating the formation of intermetallic compounds during friction stir welding of magnesium alloy to aluminum alloy in air and under liquid nitrogen. *Int. J. Adv. Manuf. Technol.* **2014**, *71*, 1493–1499. [CrossRef]
135. Lv, X.Q.; Wu, C.S.; Padhy, G.K. Diminishing intermetallic compound layer in ultrasonic vibration enhanced friction stir welding of aluminum alloy to magnesium alloy. *Mater. Lett.* **2017**, *203*, 81–84. [CrossRef]
136. Baghdadi, A.H.; Sajuri, Z.; Selamat, N.F.M.; Omar, M.Z.; Miyashita, Y.; Kokabi, A.H. Effect of intermetallic compounds on the fracture behavior of dissimilar friction stir welding joints of Mg and Al alloys. *Int. J. Miner. Metall. Mater.* **2019**, *26*, 1285–1298. [CrossRef]
137. Kim, D.; Baek, S.; Nishijima, M.; Lee, H.-C.; Geng, P.; Ma, N.; Zhang, Z.; Park, H.; Chen, C.; Lee, S.-J.; et al. Toward defect-less and minimized work-hardening loss implementation of Al alloy/high-purity Cu dissimilar lap joints by refill friction stir spot welding for battery tab-to-busbar applications. *Mater. Sci. Eng. A* **2024**, *892*, 146089. [CrossRef]
138. *EN 1999-1-1*; Eurocode 9: Design of Aluminium Structures—Part 1-1: General Structural Rules. European Committee for Standardization (CEN): Brussels, Belgium, 2007.
139. *AWS D17.3/D17.3M:2016*; Specification for Friction Stir Welding of Aluminum Alloys for Aerospace Applications. AWS: Miami, FL, USA, 2016.
140. *AS/NZS 1665:2004*; Welding of Aluminium Structures. Standards Australia/Standards New Zealand: Sydney, Australia; Wellington, New Zealand, 2004.
141. *CSA S157-17*; Strength Design in Aluminium. Canadian Standards Association (CSA Group): Mississauga, ON, Canada, 2017.
142. *AISC 360-22*; Specification for Structural Steel Buildings. American Institute of Steel Construction (AISC): Chicago, IL, USA, 2022.
143. Draper, J.; Fritsche, S.; de Traglia Amancio-Filho, S.; Galloway, A.; Toumpis, A. A numerical modelling approach to predict material flow and defect formation in refill friction stir spot welded joints. *Int. J. Adv. Manuf. Technol.* **2025**, *139*, 2907–2921. [CrossRef]
144. Mondal, A.K.; Biswas, P.; Bag, S. Influence of tacking sequence on residual stress and distortion of single sided fillet submerged arc welded joint. *J. Marine. Sci. Appl.* **2015**, *14*, 250–260. [CrossRef]

crystals

MDPI

Article

Effect of Y on the Microstructures and Heat-Resistant Property of ZL109 Alloys

Xiangdu Qin [1], Anmin Li [1,2,3,*] and Xiang Li [1]

1 School of Resources, Environment and Materials, Guangxi University, Nanning 530004, China; 15078266480@163.com (X.Q.); 15177617346@163.com (X.L.)
2 State Key Laboratory of Featured Metal Materials and Life-Cycle Safety for Composite Structures, Guangxi University, Nanning 530004, China
3 MOE Key Laboratory of New Processing Technology for Nonferrous Metals and Materials, Guangxi University, Nanning 530004, China
* Correspondence: lianmin@gxu.edu.cn

Abstract: The effects of rare earth Y on the microstructures and tensile properties of ZL109 alloys were studied through metallographic observation, scanning electron microanalysis, X-ray diffraction, and tensile experiments, and the existence form and mechanism of Y were analyzed. The results show that the grain size of the ZL109 alloy is obviously reduced and that the strength of the ZL109 alloy is significantly increased after adding the Y element. When the Y content is increased to 0.2 wt.%, the tensile properties of the ZL109 alloy at room temperature and 350 °C are better than those without a rare earth addition, and the comprehensive tensile properties are better. This is due to the addition of the Y element; α-Al dendrites are obviously refined, and there is a tendency for them to change into fine isometric crystals. The size of the eutectic Si decreases, and its shape is modified. The morphology and size of high temperature enhanced phases, such as Al_3CuNi, are optimized. The heat-resistant enhanced phase Al_2Si_2Y is formed after the addition of rare earth Y. However, with the addition of the Y element, the Al_2Si_2Y phase increases, and coarsening results in the decrease in alloy strength.

Keywords: ZL109; Y; heat-resistant property

1. Introduction

Due to their excellent specific strength and low density, aluminum alloys are gradually replacing traditional steel materials in automotive parts and other applications [1]. However, components made of aluminum alloys tend to significantly decrease their high temperature mechanical properties due to the growth of their grains and precipitates at high temperatures, thus making them unusable in high temperature environments [2]. The development of cast heat-resistant aluminum alloys that can be used safely for a long time in the temperature range of 300 to 400 °C has become an international research hotspot in recent years [3].

A cast heat-resistant aluminum alloy refers to an aluminum alloy prepared by casting, served at 150 to 350 °C, with sufficient oxidation resistance, and under the long-term action of heat-force coupling, with certain creep resistance and mechanical properties [4]. Cast heat-resistant aluminum alloys can be roughly divided into an Al-Si series, an Al-Cu series, and an Al-Mg series of cast heat-resistant aluminum alloys. The Al-Si cast heat-resistant aluminum alloy is the most used and widely used cast heat-resistant aluminum alloy at present, and its output and application amount account for more than 90% of the total cast heat-resistant aluminum alloys [5]. Due to the presence of Si elements, A1-Si alloys have good air tightness, fluidity, small casting shrinkage, and other casting properties, and they can be widely used in the pistons and cylinders of automotive engines [6]. However, the plastic toughness of the Al-Si alloy is still relatively poor, and the mass fraction of Si in industrial production is generally between 4 and 22 wt.%. The heat resistance strengthening phase is often formed by adding Cu, Ni, Mg, Mn and other elements to improve the mechanical properties of the alloy [7]. Al-Si cast heat-resistant aluminum alloys can be divided into Al-Si-Mg, Al-Si-Cu, and Al-Si-Cu-Mg-Ni [8]. The strength of the Al-Si-Mg alloy is improved mainly by the Mg_2Si strengthening phase, but the properties of the Al-Si-Mg alloy decrease rapidly due to the obvious coarsening during the Mg2Si phase at high temperature [9]. The Al-Si-Cu alloy relies on the θ-Al_2Cu phase and the Q-$Al_5Cu_2Mg_8Si_6$ phase to play a strengthening role, but the high temperature stability of the θ-Al_2Cu phase and the Q-$Al_5Cu_2Mg_8Si_6$ phase will also decline sharply when the temperature exceeds 200 °C and 250 °C [10]. Studies have shown that after adding Ni elements to the Al-Si-Cu-Mg alloy, Ni-rich strengthening phases, such as Al_3CuNi, Al_3Ni, and Al_7Cu_4Ni, will be formed in the alloy, which can significantly improve the high temperature performance of the alloy above 300 °C [11]. Therefore, the research of casting an Al-Si-Cu-Mg-Ni heat-resistant aluminum alloy that can meet the more severe high temperature environment will make the goal of achieving lightweight automobiles using aluminum instead of steel one step closer. Among these alloys, the Al-Si-Cu-Mg-Ni series ZL109 heat-resistant aluminum alloy, independently developed by China, is still the most widely used piston alloy material in China. Selection of the ZL109 alloy as the matrix alloy that is to be used to cast a heat-resistant aluminum alloy with higher performance is the objective of this study.

The addition of Ni, Ag [12], Mn [13], Zr [14], and other transition metal elements and rare earth elements such as Sc [15], La [16], and Y [17] for the basis of casting the Al-Si-Cu-Mg aluminum alloy can reduce the layer fault energy of the alloy system and form a high temperature stable heat-resistant phase and a nanometer coherent precipitated phase. Pinning dislocations and blocking grain boundary movement at high temperatures are effective ways to improve heat resistance. Abhishek et al. [12] studied the effects of trace amounts of Ag and Sn elements on the precipitates and mechanical properties of the 7075 alloy and showed that adding a small amount of Ag and Sn to the 7075 alloy could significantly reduce the average grain size of the alloy. Ag and Sn reduce the formation of coarse η precipitates by trapping quenched vacancies and reducing the concentration of free vacancies. Ag and Sn accelerate the kinetic rate of the precipitated phase, and all phase transitions are controlled by the dynamics of Ag and Sn. After adding Ag and Sn, the strength and ductility of the 7075 alloy are significantly improved. Jun et al. [14] studied the effect of Zr on the microstructure and mechanical properties of the Al-Si-Cu-Mg alloy at intermediate temperature (100–200 °C). It was found that Zr can effectively inhibit the nucleation and grain growth of the alloy and refine the grain. The size of the eutectic Si and Fe-rich phase of the alloy decreased significantly. The highest intermediate temperature thermal exposure performance (150 °C/20 h) was obtained by adding 0.2% Zr, and the tensile strength and elongation were 208.1 MPa and 11.4%, respectively, which were significantly better than those without Zr. Rare earth elements have good chemical activity and can form intermetallic compounds with other elements in the alloy. It Zr promotes heterogeneous nucleation, which allows the thinning of α-Al dendrites [18]. The microstructure optimization of the Al-Si alloy can be divided into three parts, namely, eutectic silicon refining, primary silicon refining, and α-Al refining. Li [19] systematically studied the influences of the Y element on the morphology and size of α-Al, primary Si, and eutectic Si in the multi-component eutectic Al-13Si-3Cu-2Ni-0.5Mg alloy. When the addition of rare earth Y is 0.8%, the coarse α-Al dendrites in the alloy are refined into uniform cellular or equiaxial crystals, and the secondary dendrite arm spacing decreases from 12.7 μm to 4.5 μm—a decrease of 64.6%. The coarse lamellar or long acicular eutectic Si is transformed into a fine fibrous structure with many branches. The tensile properties of an as-cast Al-13Si-3Cu-2Ni-0.5Mg alloy at room temperature and at high temperature are also significantly improved. Yu et al. [20] explored the effect of Y on the microstructure and mechanical properties of the ZL107 alloy at room temperature. When the content of Y was equal to or greater than 0.1 wt.%, the eutectic Si was completely modified, and the shape of the eutectic Si became short rod or granular. With the further increase in the Y content, the shape of the eutectic Si does not change. After adding 0.1 wt.% Y, the tensile strength increased from 194.3 MPa to 203.4 MPa, and the elongation increased from 2.3% to 2.6%. The tensile strength and elongation decreased with the increase in the Y content. In addition, some studies have shown that rare earth Y and the aluminum form rare earth phase Al_3Y, which prevents grain growth and grain boundary slip at high temperature, plays a role in grain boundary strengthening, and can improve the high temperature strength of aluminum alloys [21].

At present, there are many studies on the modification and refinement of rare earth Y in casting Al-Si alloys, and most of them focus on the mechanical properties at room temperature. However, the effect of rare earth Y on the strengthening phase in the Al-Si-Cu-Mg-Ni alloy and its mechanism still need further experimental investigation. The study of high temperature properties in this paper has certain novelty. The aims of this study are to optimize a cast heat-resistant aluminum alloy on the basis of the existing ZL109 alloy, so that the optimized alloy has better room temperature and high temperature properties than the ZL109 alloy, and to analyze the existence form of rare earth Y in the Al-Si-Cu-Mg-Ni alloy and the strengthening mechanism of rare earth Y on the ZL109 alloy.

2. Materials and Methods

In the experiment, the ZL109 piston alloy was used as the base alloy, and different contents of the Y element were added to form 5 control groups. The nominal chemical compositions and the measured actual chemical compositions of the alloys are shown in Tables 1 and 2, respectively. The raw materials used in the experiment are industrial high-pure aluminum blocks (99.90%), Al-50Si, Al-50Cu, Al-20Mg, Al-20Ni and Al-10Y intermediate alloys produced by Haichuan Metal Materials Co., Ltd. (Suzhou, China).

Table 1. The nominal chemical compositions of ZL109-xY alloys (wt.%).

Alloy	Si	Cu	Mg	Ni	Y	Al
Y-free	12	1.2	1.2	1.2	0	Bal.
2Y	12	1.2	1.2	1.2	0.2	Bal.
4Y	12	1.2	1.2	1.2	0.4	Bal.
6Y	12	1.2	1.2	1.2	0.6	Bal.
8Y	12	1.2	1.2	1.2	0.8	Bal.

Table 2. The actual compositions of ZL109-xY alloys (wt.%).

Alloy	Si	Cu	Mg	Ni	Y	Al
Y-free	12.13	1.18	1.22	1.21	0	Bal.
2Y	12.25	1.21	1.19	1.22	0.21	Bal.
4Y	11.98	1.23	1.23	1.18	0.37	Bal.
6Y	11.93	1.17	1.17	1.23	0.59	Bal.
8Y	12.07	1.23	1.18	1.18	0.78	Bal.

First, the pure aluminum ingot and graphite crucible together were put into a well resistance furnace. The furnace was heated to about 800 °C, and then the required intermediate alloys were added in turn. The covering agent was added, and the melt was left to keep warm for 10 min; then, the melt was stirred and slagged. When the furnace temperature dropped to about 680–700 °C, the Mg patch was quickly pressed into the bottom of the solution, held for 5 min, and refined; then, the power was adjusted again to make the furnace temperature rise to 710 °C for mechanical stirring for 2 min. After being degassed by C_2Cl_6, the melt was cast into a preheated steel mold at around 730 °C. The heat treatment process used for the samples was a T6 heat treatment: after the samples were held at 505 °C for 6 h, they were quenched with hot water at 70–90 °C, and then the cooled samples were held at 180 °C for 6 h. Subsequently, the samples were taken out for air cooling. After that, the alloy samples were etched with Keller's reagent (1% HF + 1.5% HCl + 2.5% HNO_3 + 95% H_2O). An X-ray diffractometer (Rigaku D/MAX 2500V, Rigaku Corporation, Akishima, Japan), a metallographic microscope (BH200M, Sunyu Instrument Corporation, Ningbo, Chian), a field emission scanning electron microscope (Sigma 300, ZEISS, Oberkochen, Germany), and an EDS (S3400N, Hitachi Corporation, Tokyo, Japan, accelerating voltage, 20 KV) were used to characterize the microstructures and phase analysis of the samples. Elemental analysis of the samples was performed using an X-ray fluorescence element analyzer (S8 TIGER, BRUKER, Berlin, Germany). The tensile properties of the series of alloys at room temperature were performed on a universal testing machine (AGS-X 100KN, Shimadzu, Shanghai, China) with a standard drawing distance of 25 mm and a drawing rate of 0.5 mm/min. The high temperature mechanical properties were performed on a testing machine (Kappa, ZWICK, Ulm, Germany) at a tensile rate of 0.5 mm/min.

3. Results and Discussion

3.1. Microstructure of Alloys

Figure 1 shows the XRD results of the ZL109 alloys with different Y content. It can be found that the alloys consist of α-Al, a eutectic Si phase, an Al_3Ni phase, an Al_3CuNi phase, a Mg_2Si phase, and a small amount of Al_2Cu. However, the $Al_5Cu_2Mg_8Si_6$, Al_7Cu_4Ni, and other possible second phases were not found. No new diffraction peak appeared after the Y element was added, and no phase containing Y was found. The XRD results were correlated with the content of each component phase. Because some alloying elements are solidly dissolved in an Al matrix and the alloying phase content is small, the corresponding diffraction peak is not found in the back of the diffraction pattern. The existence form of rare earth Y in the ZL109 aluminum alloy and the type and variation of the strengthening phase in the alloy need to be further analyzed based on the SEM–EDS results.

As can be found from Figure 2, the alloy structure is primarily composed of primary Al and eutectic Si. According to the results of the metallographic microstructures and phase analysis, the most abundant light-colored phase is the α-Al matrix with continuous distribution. The black phase is mostly a spherical and rod-like eutectic Si phase, and only contains a small amount of the irregular bulk primary Si phase. A small amount of light gray intermetallic compounds is also observed. After the Y element is added to the matrix alloy, the α-Al structure is further refined and tends to change to fine isometric crystals. With the increase in the Y content, the eutectic Si becomes more rounded and changes from a short rod-like to a spherical shape, indicating that Y has a significant modification effect on the eutectic Si, which is consistent with the research results of Li [22] et al. The Si phase and the intermetallic compounds are mainly concentrated in the dendrites, and small amounts are distributed in the crystals. In view of the metamorphic mechanism of the eutectic Si, some researchers have proposed that the ideal ratio of the atomic radius of the metamorphic element to the atomic radius of silicon is about 1.65 [23]. The r_Y/r_{Si} value of 1.55 is close to the ideal value; therefore, the Y element has an obvious metamorphic effect on the eutectic Si, and the size of the eutectic Si is obviously refined.

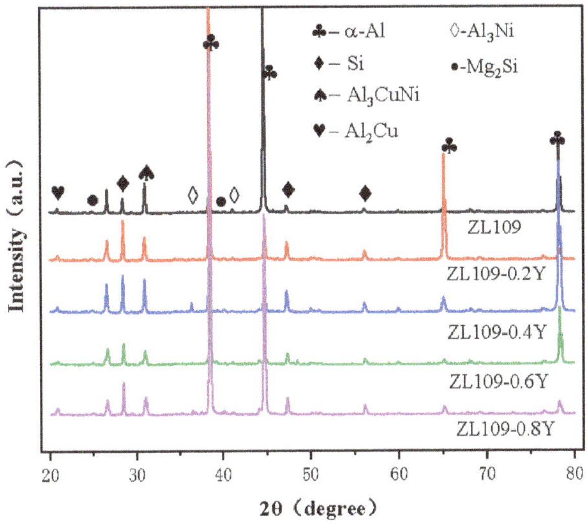

Figure 1. X-ray diffraction patterns of ZL109-xY alloys.

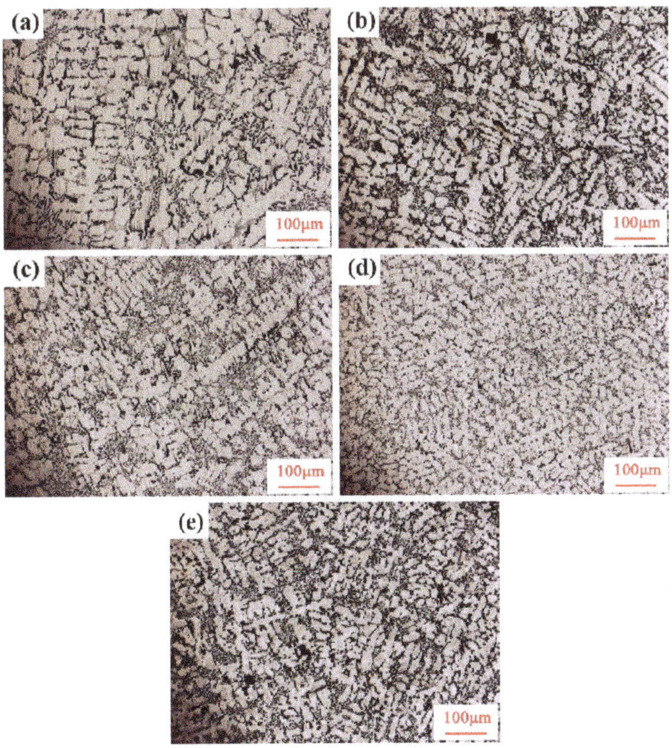

Figure 2. Effect of different Y additions on microstructures of ZL109-xY alloys: (**a**) Y-free; (**b**) 0.2Y; (**c**) 0.4Y; (**d**) 0.6Y; and (**e**) 0.8Y.

In order to explore the metamorphism of the Y element on the eutectic Si, the Image J 2024 software was used to measure and analyze the size and form factor of the eutectic Si alloy particles, and the results are shown in Figure 3. The size of the eutectic Si decreases with the increase in the Y content. When the Y content increases to 0.6 wt.%, the size of the eutectic Si is basically unchanged by adding the Y element. When the Y content is 0.8 wt.%, the minimum average size is 3.58 μm. Form factor, also known as roundness [24], is the aspect ratio of the eutectic Si particles, and the test results are basically consistent with the size change rule. When the Y content is 0.6 wt.%, the form factor is closest to 1, and the form factor increases slightly with the addition of the Y element.

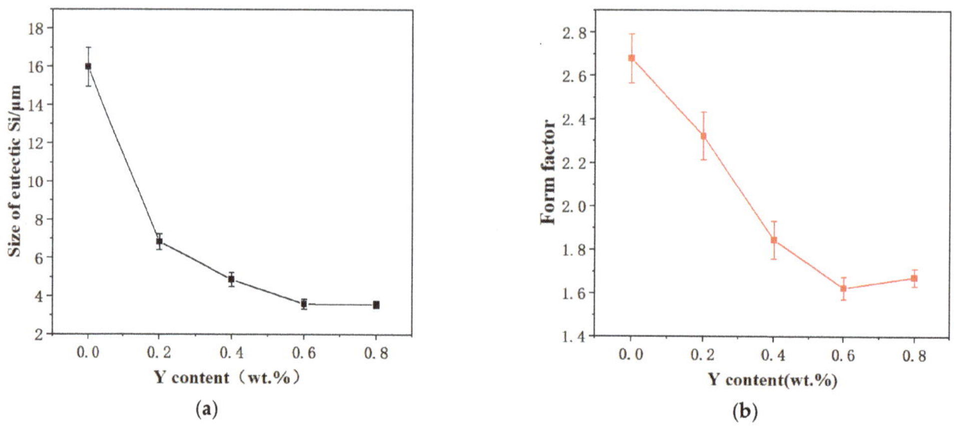

Figure 3. (**a**) Eutectic Si size of ZL109-xY alloys; (**b**) Form factor of eutectic Si in ZL109-xY alloys.

Figure 4 shows the SEM morphology of the ZL109-0Y alloy. According to the EDS results (as shown in Figure 6), it was determined that the strip structure is the Al_3Ni phase, the network or Chinese character is the Al_3CuNi phase, and the black microstructure is the Mg_2Si phase. However, the Al_2Cu may not have been detected due to its small content and relatively small size.

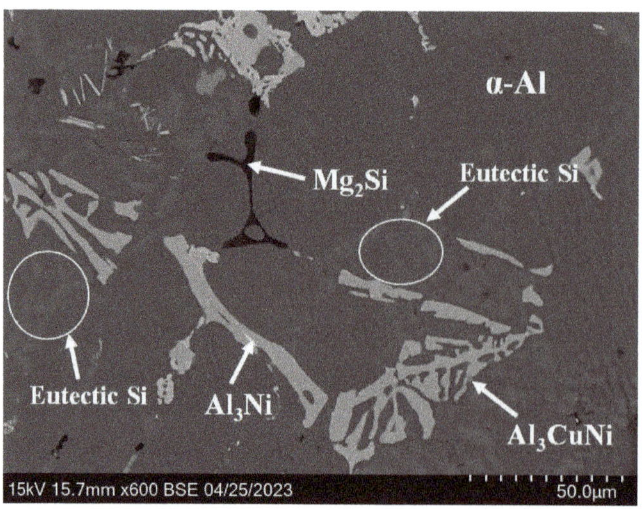

Figure 4. SEM microstructure of ZL109-0Y alloy.

The microstructures of the ZL109 alloys with different Y content are shown in Figure 5. It can be found from the figure that after the Y element is added, the brighter Ni-rich phase in the alloy is striped and radial, and the size is significantly reduced. The average size of the Ni-rich phase in the ZL109-0Y, ZL109-0.2Y, ZL109-0.4Y, ZL109-0.6Y, and ZL109-0.8Y alloys is 28.61 μm, 12.69 μm, 14.80 μm, 15.56 μm, and 17.79 μm, respectively. The Ni-rich phase decreased significantly after the Y element was added. Moreover, according to the surface scanning energy spectrum results of the ZL109-0.6Y alloy (as shown in Figure 6), the Y element is enriched in the Ni-rich phase and the eutectic Si phase, and is less distributed in the Al matrix. With the increase in the Y content, the thinning effect of the Y element is weakened, and the coarsening phenomenon of the Ni-rich phase is obvious.

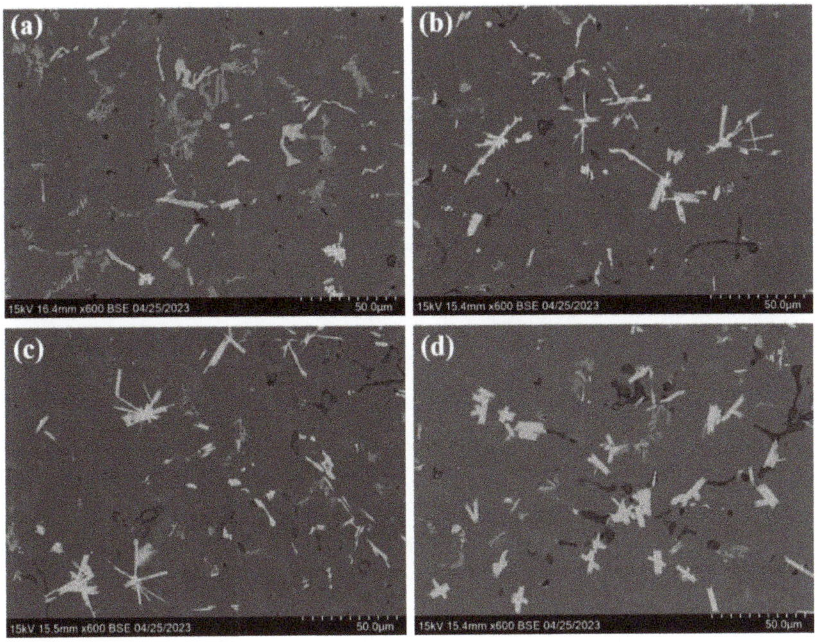

Figure 5. SEM microstructures of ZL109-xY alloys: (**a**) 0.2Y; (**b**) 0.4Y; (**c**) 0.6Y; and (**d**) 0.8Y.

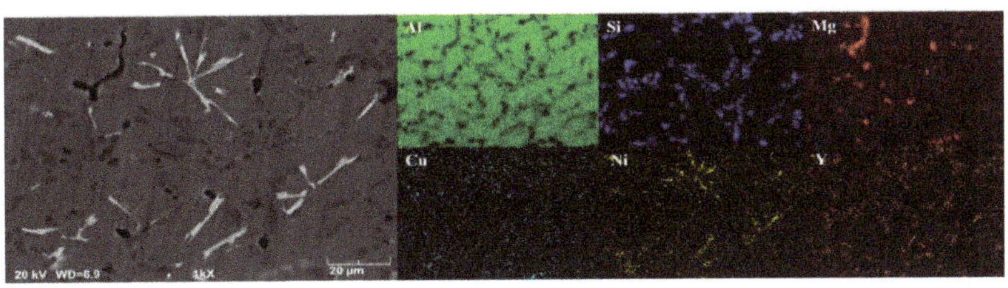

Figure 6. SEM microstructure of ZL109-0.4Y alloy.

In order to further determine the existence form of the Y element in the alloy, the microstructures of the ZL109-xY (x = 0.4, 0.8) alloys were analyzed. The SEM topography and point analysis results are shown in Figure 7 and Table 3, respectively. According to the results, the Ni-rich phase is mainly the Al_3CuNi phase, and the Y element tends to form the Al, Si, Ni, Y multi-component phase; that is, there are two closely combined phases—an Al_3Ni phase and a yttrium-rich phase—that are difficult to distinguish. Figure 7b shows that when the Y content is 0.8 wt.%, the morphology of the Al_3CuNi phase changes significantly, and these phases always attach to the compound phase. The energy spectrum results of Spot 4 and Spot 5 in Figure 7b show that the atom ratio of silicon yttrium is close to 2:1, no Cu is detected, and a small amount of Mg is contained—which is basically consistent with the energy spectrum results of the Al_2Si_2Y phase reported by Wan [25] et al. Therefore, it is speculated that the existing Y-rich phase is the Al_2Si_2Y phase. The existence of the Al_3Ni phase is confirmed by the XRD phase analysis; therefore, it is further proved that the Al_3Ni phase and the Al_2Si_2Y phase coexist And that Al_2Si_2Y can inhibit dislocation and grain boundary movement and improve the mechanical properties of the aluminum alloy [25].

Figure 7. SEM microstructures of ZL109-xY alloys: (**a**) 0.4Y and (**b**) 0.8Y.

Table 3. EDS analysis is of phases at locations depicted in Figure 2 (wt.%).

Spot	Al	Cu	Ni	Si	Mg	Y	Phase
1	45.681	23.222	27.369	2.010	0.869	0.264	Al_3CuNi
2	41.491	1.395	20.179	15.538	1.079	17.468	Al_3Ni, Al_2Si_2Y
3	70.728	13.289	14.543	0.000	1.440	0.000	Al_3CuNi
4	52.130	0.505	8.710	24.521	2.338	11.796	Al_3Ni, Al_2Si_3Y
5	50.528	0.424	7.787	26.346	2.112	12.802	Al_2Si_2Y

3.2. *Tensile Properties of Alloys*

The relationship between yield strength and grain size of metallic materials can be expressed by the Hall–Petch equation (as shown in Equation (1) [26]). In the formula, σ_s is the yield stress of the material, σ_0 represents the constant of the initial stress of dislocation motion, k represents the strengthening coefficient, and d represents the average grain diameter.

$$\sigma_s = \sigma_0 + kd^{-\frac{1}{2}} \tag{1}$$

Figure 8 and Table 4 show the tensile strength, yield strength, and elongation of the ZL109-xY alloys at room temperature. The relevant data of tensile test performance of the ZL109-xY alloys at room temperature are shown in Table 4. By comparing the ZL109-xY alloys, it can be found that with the increase in the Y content, the tensile strength of the ZL109-xY alloys at room temperature first increases and then decreases. When the Y content is increased to 0.2 wt.%, the tensile strength of the alloys is increased from 273.73 MPa to 313.08 MPa by 14.4%, and the comprehensive tensile property is better. However, with the addition of the Y element, the tensile strength and plasticity of the alloys decrease.

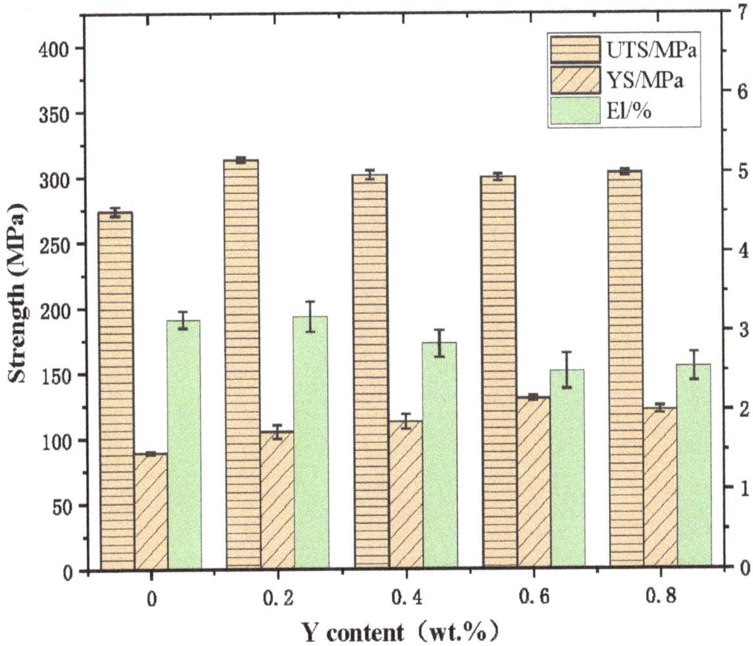

Figure 8. Tensile strength, yield strength, and elongation of ZL09-xY alloys with different Y content at room temperature.

Table 4. Relevant tensile test data of ZL109-xY alloys at room temperature.

Y (wt.%)	Tensile Strength (MPa)	Yield Strength (MPa)	Elongation (%)
0	273.73	89.14	3.15
0.2	313.08	105	3.18
0.4	301.43	112.56	2.84
0.6	299.52	130.3	2.49
0.8	302.91	121.63	2.54

Under the combined action of the Y element and the strengthening phase, the ZL109-0.2Y alloy has good mechanical properties at room temperature. However, according to relevant studies and the analysis of the Griffith crack theory, the thick and slender eutectic Si particles are more prone to fracture. With the increase in the Y content, Si particle size decreases, and the critical stress of crack instability propagation increases [27]. Smooth spherical eutectic Si greatly reduces stress concentration, and the strength and plasticity of the alloy should be significantly improved. However, with the addition of the Y element, the tensile strength and plasticity of the alloy decrease. Griffith's formula [28] is shown in Equation (2): σ_c is the critical stress for the unstable propagation of a crack in the Si phase; E is the elastic modulus of the Si phase; γ_s is the specific surface energy of the Si phase crack; a is the half-length of the Si phase crack.

$$\sigma_c = \left(\frac{2E\gamma_s}{\pi a}\right)^{\frac{1}{2}} \quad (2)$$

In order to investigate the reason for the plasticity decline, the fracture of the tensile specimen at room temperature was further analyzed. Figure 9 shows the tensile fracture

SEM morphologies of the ZL109 alloys with different the Y content at room temperature. The fractures are mainly composed of a cleavage plane—a large number of short and curved tearing edges and quasi-cleavage facets that are characterized by cleavage and quasi-cleavage—and are typical brittle fractures [28–30], consistent with the result of low elongation and poor plasticity after fracture. At the same time, it can be seen that there are a large number of micro-cracks in the fracture and that the quasi-cleavage fracture is a discontinuous fracture, with each hidden small crack growing constantly; this connection often has a large plastic deformation, thus forming tearing edges and quasi-cleavage facets [28–30].

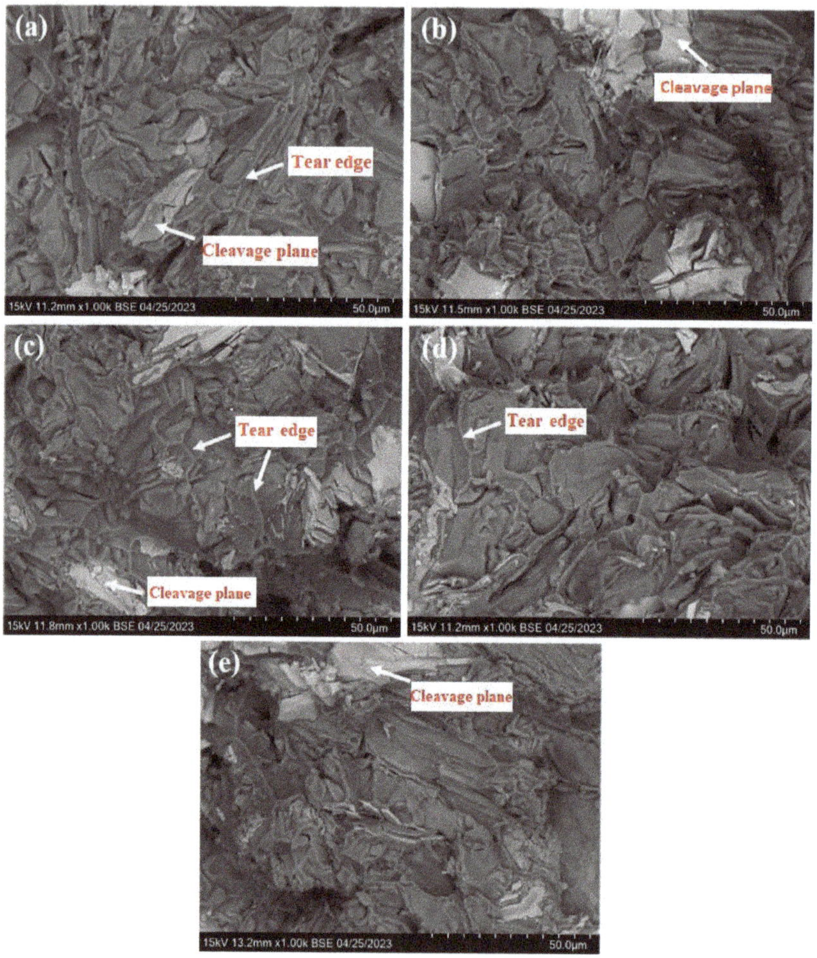

Figure 9. Tensile fracture morphologies of ZL109-xY alloys at room temperature: (**a**) 0Y; (**b**) 0.2Y; (**c**) 0.4Y; (**d**) 0.6Y; and (**e**) 0.8Y.

In order to further explore the causes of plastic decline and fracture sources of the alloy, energy spectrum analysis was also carried out on the tensile fracture structure of the ZL109-0.6Y alloy at room temperature, and the results are shown in Figure 10. There are micro-cracks in one convex structure and two sections formed by the quasi-cleavage fracture, which belong to the fracture source. At the same time, according to the point analysis results, the fracture source is the Al_3Ni phase and the Al_2Si_2Y phase. The results show that with the increase in the Y content, the Al_2Si_2Y phase content and coarsening are more likely to crack, reducing the mechanical properties of the alloy.

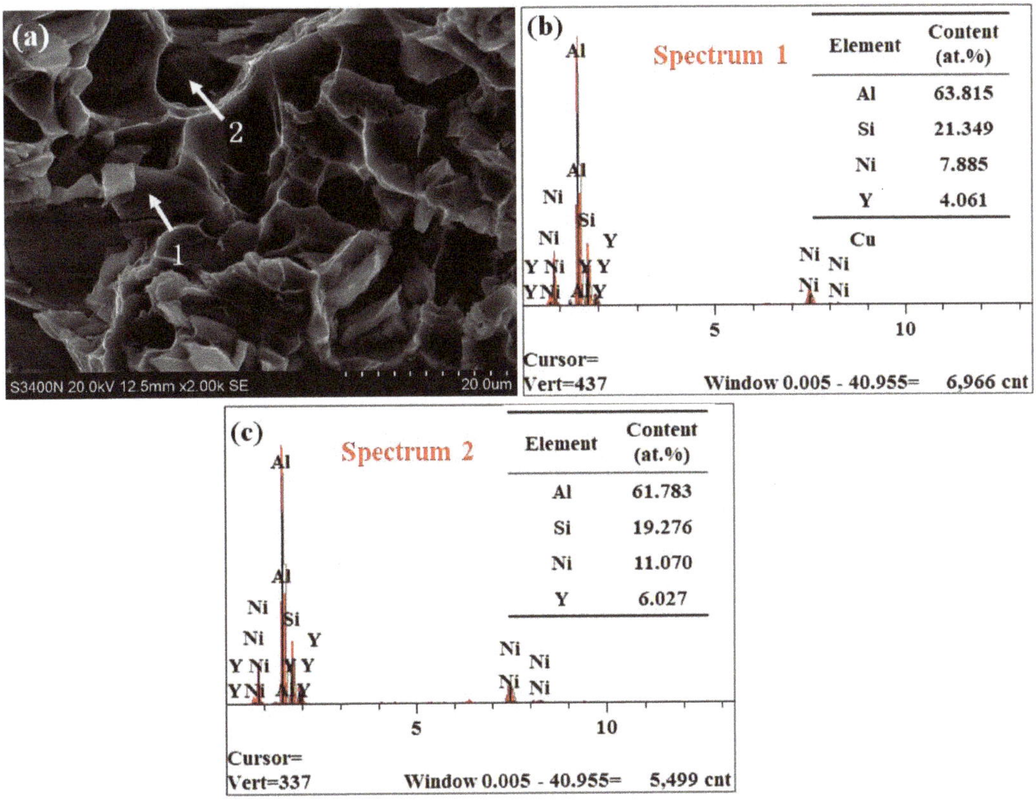

Figure 10. (**a**) Tensile fracture morphology of ZL109-0.6Y alloy; (**b**) EDS data at point 1; (**c**) EDS data at point 2.

3.3. Test Results and Analysis of Tensile Properties of Alloys at High Temperature

The tensile test results of the ZL109-xY alloys at high temperature (350 °C) are shown in Figure 11 and Table 5. The results show that the tensile strength at high temperature increases first and then decreases with the increase in the Y content. The high temperature tensile strength and yield strength of the ZL109-0.2y alloy are the highest, which are increased by 17.4% and 18.9%, respectively, compared with the ZL109 alloy. With the increase in the Y content, the elongation of the alloy shows an increasing trend. The results show that the second phase strengthening is the main strengthening mechanism under high temperature conditions, and the alloy strength is related to the shape, size and distribution of the second phase thermal stability [31]. Combined with the above results, it can be seen that when the Y content is 0.2%, the size of the Ni-rich phase is the smallest and the distribution is uniform, so the ZL109-0.2Y alloy has the best tensile strength. The addition of rare earth Y led to the coarsening of the Ni-rich phase and the Y-rich phase, which led to the decrease in the high temperature strength of the alloy. At the high temperature of 350 °C, the grain boundary strength decreases, the hindering effect on dislocation decreases, and the grain boundary slip is easier to carry out [32]. Combined with the above results, it can be seen that the size of the eutectic silicon and α-Al decreases with the increase in the Y content. This forms more grain boundaries, making the alloy more plastic; therefore, the elongation increases. This result is consistent with the experimental results of Yang [32].

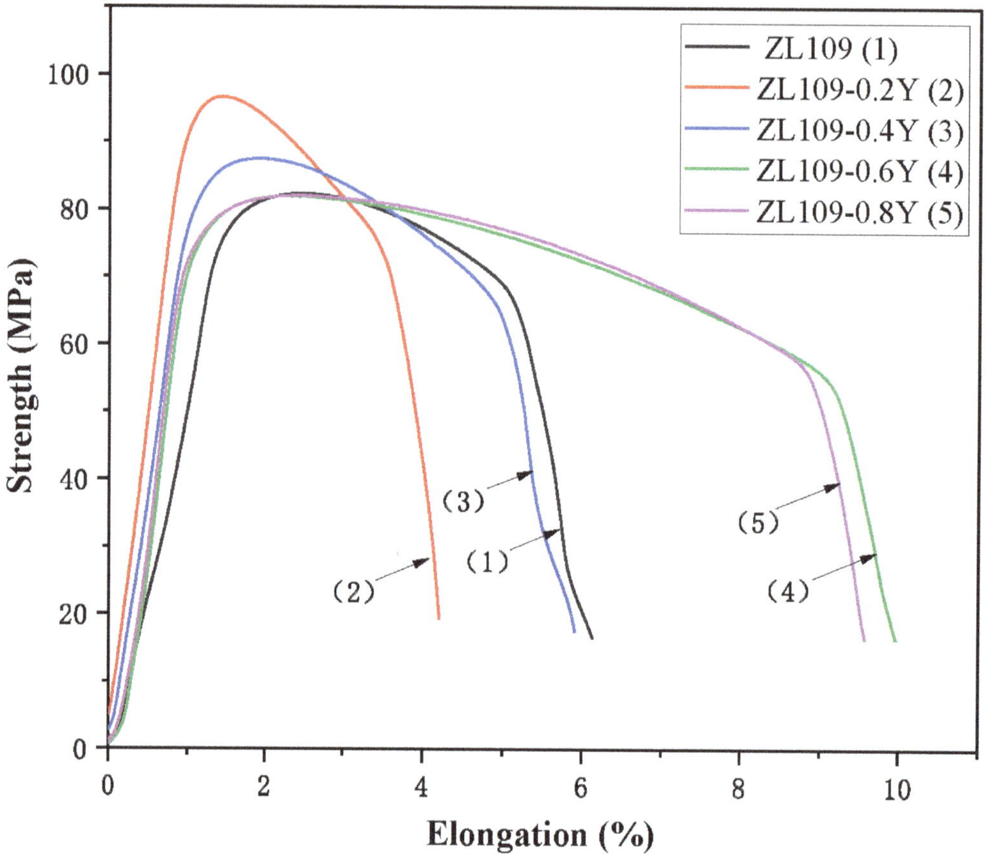

Figure 11. Tensile properties of ZL109-xY alloys at high temperature (350 °C).

Table 5. Relevant tensile test data of ZL109-xY alloys at high temperature (350 °C).

Y (wt.%)	Tensile Strength (MPa)	Yield Strength (MPa)	Elongation (%)
0	82.3	78.71	5.21
0.2	96.65	93.62	3.66
0.4	87.48	81.34	5.14
0.6	81.94	75.77	8.76
0.8	82	77.1	8.62

Figure 12 shows the SEM morphology of tensile fractures of the ZL109 alloys with different Y content at room temperature. Compared with the tensile fracture at room temperature, the tensile fracture at high temperature mainly has two kinds of micromorphologies: equiaxial dimple and tear edge, and there are a few cleavage planes, which belong to the mixed ductile and brittle fracture. Figure 12b shows the tensile fracture morphology of the alloy with a Y content of 0.2 wt.% at high temperature. The number of isometric dimples is relatively small, and there are more tearing edges in the fracture. Therefore, the ZL109-0.2Y alloy shows the worst plasticity. With the increase in the Y content, the number of isometric dimples increases, and the plasticity of the alloy increases gradually [28–30].

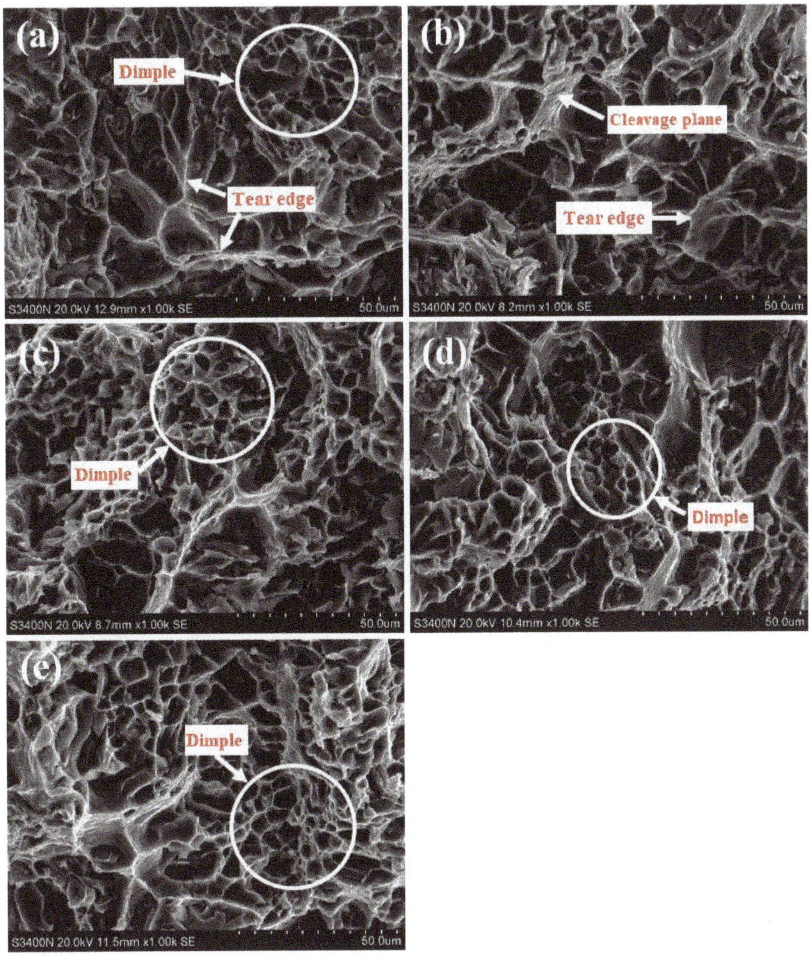

Figure 12. Tensile fracture morphologies of ZL109-xY alloys at high temperature (350 °C): (**a**) 0Y; (**b**) 0.2Y; (**c**) 0.4Y; (**d**) 0.6Y; and (**e**) 0.8Y.

4. Conclusions

In this paper, the microstructures and tensile properties of the ZL109-xY (x = 0.2, 0.4, 0.6, 0.8, wt.%) series alloys were analyzed; the existence form of rare earth Y in the ZL109 aluminum alloy was explored; and the influence of the Y element on the tensile properties of the alloy was analyzed, along with the strengthening mechanism of the alloy. The conclusion is as follows:

(1) The strengthening phases of the ZL109-xY alloy include the eutectic Si phase, the Al_3Ni phase, the Al_3CuNi phase, the Al_2Si_2Y phase, the Mg_2Si phase, and a small amount of the Al_2Cu phase; moreover, the Al_3Ni phase and the Al_2Si_2Y phase always exist simultaneously.

(2) The tensile strength of the ZL109-xY alloy at room temperature and at high temperature increases first and then decreases with the increase in the Y content. When the Y content is 0.2 wt.%, the tensile strength of the alloy at room temperature is 313.08 MPa,

which is 14.4% higher than that of the ZL109 alloy, and the comprehensive tensile property is better. The tensile strength and the yield strength of the ZL109-0.2Y alloy at high temperature are the highest, which are increased by 17.4% and 18.9%, respectively, compared with the ZL109 alloy. Therefore, the ZL109-0.2Y alloy has the better comprehensive mechanical properties.

(3) The strengthening phase Al_2Si_2Y is formed by adding the Y element to the ZL109 alloy, which can inhibit dislocation and grain boundary movement. The α-Al dendrites are obviously refined and tend to become fine isometric crystals. The size of the eutectic Si decreases, and its shape is modified. The morphology and size of high temperature strengthening phases, such as Al_3CuNi, are optimized. The tensile properties of the alloy are improved at both room temperature and high temperature. However, with the addition of the Y element, the Al_2Si_2Y phase increases, and coarsening results in decreasing the strength of the alloy.

Author Contributions: Conceptualization, X.Q. and A.L.; methodology, X.Q.; investigation, X.Q.; formal analysis, X.Q.; writing—original draft, X.Q. and X.L.; funding acquisition, A.L.; resources, A.L.; supervision, A.L.; writing—review and editing, A.L.; data curation, X.L. All authors have read and agreed to the published version of the manuscript.

Funding: This research was funded by the 2021 Special funds for central guidance of local scientific and technological development funds, grant number GuiKeZY21195030; the Guangxi Key Research and Development Program, grant number GuiKeAB22080015; and the Specific Research Project of Guangxi for Research Bases and Talents, grant number GuiKeAD21238010.

Data Availability Statement: The original contributions presented in the study are included in the article; further inquiries can be directed to the corresponding author.

Conflicts of Interest: The authors declare no conflicts of interest.

References

1. Zhang, H.W.; Liu, Y.; Fan, T.X. Research Progress and Prospect of Casting Heat-resistant Aluminum Alloy. *Mater. Rev.* **2022**, *36*, 1–17.
2. Li, Q.S.; Li, A.M.; Sui, Q.P. Microstructure and High Temperature Creep Properties of ZrB2/Al-12Si Composites Prepared by in-situ Reaction Method. *Mater. Rev.* **2019**, *12–19*, 1–19.
3. Shyam, A.; Bahl, S. Heat-resistant aluminium alloys. *Nat. Mater.* **2023**, *22*, 425. [CrossRef] [PubMed]
4. Manasijevic, S.; Radisa, R.; Markovic, S.; Acimovic-Pavlovic, Z.; Raic, K. Thermal Analysis and Microscopic Characterization of the Piston Alloy AlSi13Cu4Ni2Mg. *Intermetallics* **2011**, *19*, 486–492. [CrossRef]
5. Zhang, S.H. Progress of Application of Aluminum Alloy in automobile. *Automot. Ind. Res.* **2003**, *3*, 36–39.
6. Sui, Y.D.; Wang, Q.D. Application Research and Development of cast Heat-resistant Aluminum Alloy in Engine. *Mater. Rev.* **2015**, *29*, 14–19.
7. Bi, J.; Liu, L.; Zhang, D.S.; Wang, H.X.; Zheng, L.Q.; Dong, G.J. Research Progress of casting, Quick-setting and Additive Heat-resistant Aluminum Alloys. *Chin. J. Non-Ferr. Met.* **2022**, 1–41.
8. Ram, S.C.; Chattopadhyay, K.; Chakrabarty, I. Microstructures and High Temperature Mechanical Properties of A356-Mg2Si Functionally Graded Composites in as-cast and Artificially aged (T6) Conditions. *J. Alloys Compd.* **2019**, *805*, 454–470. [CrossRef]
9. Zuo, L.; Ye, B.; Feng, J.; Zhang, H.; Kong, X.; Jiang, H. Effect of ε-Al3Ni Phase on Mechanical Properties of Al-Si-Cu-Mg-Ni Alloys at Elevated Temperature. *Mater. Sci. Eng. A* **2020**, *772*, 138794. [CrossRef]
10. Yang, Y.; Li, Y.; Wu, W.; Zhao, D.; Liu, X. Effect of Existing Form of Alloying Elements on the Microhardness of Al-Si-Cu-Ni-Mg Piston Alloy. *Mater. Sci. Eng.* **2011**, *528*, 5723–5728. [CrossRef]
11. Liu, Y.; Jia, L.; Wang, W.; Jin, Z.; Zhang, H. Effects of Ni Content on Microstructure and Wear Behavior of Al-13Si-3Cu-1Mg-xNi-0.6Fe-0.6Mn Alloys. *Wear* **2022**, *500*, 204365. [CrossRef]
12. Ghosh, A.; Manojit, G.; Gyan, S. On the Role of Precipitates in Controlling Microstructure and Mechanical Properties of Ag and Sn Added 7075 Alloys During Artificial Ageing. *Mater. Sci. Eng.* **2018**, *738*, 399–411. [CrossRef]
13. Qian, Z.; Liu, X.F.; Zhao, D.G.; Zhang, G.H. Effects of Trace Mn Addition on the Elevated Temperature Tensile Strength and Microstructure of a Low-Iron Al–Si Piston Alloy. *Mater. Lett.* **2008**, *62*, 2146–2149. [CrossRef]

14. Li, J.; Peng, Y.; Guo, X.; Liu, J.; Li, L.; Zhang, N.; Cui, X.; Wang, X.; Yuan, Z.; Lu, K. On the Microstructures and Properties of a Zr-modified Al-Si-Cu-Mg alloy at Intermediate Temperature. *J. Alloys Compd.* **2025**, *1010*, 178328. [CrossRef]
15. Emadi, D.; Rao, A.P.; Mahfoud, M. Influence of Scandium on the Microstructure and Mechanical Properties of A319 Alloy. *Mater. Sci. Eng.* **2010**, *527*, 6123–6132. [CrossRef]
16. Li, X.; Li, A.M.; Qin, X.D.; Yang, H.L.; Cheng, P. Effect of La on the Microstructures and Mechanical Properties of Al-5.4Cu-0.7Mg-0.6Ag Alloys. *Materials* **2024**, *17*, 4141. [CrossRef] [PubMed]
17. Lu, B.; Li, A.M.; Rao, Y.; Wang, L.; Zuo, T.; Hu, Y. Effect of Rare Earth Y and Heat Treatment on Microstructure and Properties of 6016 Aluminum Alloy. *Mater. Rev.* **2002**, *36*, 139–146.
18. Li, Q.; Zhang, Y.; Lan, Y.; Pei, R.; Feng, X.; Xia, T.; Liu, D. Effect of Scandium Addition on Microstructure and Mechanical Properties of as-cast Al-5%Cu alloys. *Vacuum* **2020**, *177*, 109385. [CrossRef]
19. Li, Q.L.; Li, B.Q.; Zhu, Y.Q. Modification of multi-component Al-Si Castingpiston Alloys by Adding of Rare Earth Yttrium. *Microsc. Microanal.* **2019**, 1–15.
20. Yu, S.R.; Yan, Z.J.; Xiong, W.; Yan, L. Study on Microstructure and Mechanical Properties of ZL107 Alloy Added with Yttrium. *J. Rare Earths* **2013**, *31*, 198–203. [CrossRef]
21. Jia, J.Y.; Chen, Z.Q.; Hu, W.X. Effect of La and Y on Microstructure of Refined A356 Aluminum Alloy. *Rare Earth* **2020**, *41*, 41–46.
22. Li, B.; Wang, H.W.; Jie, J.C.; Wei, Z.J. Effects of Yttrium and Heat Treatment on the Microstructure and Tensile Properties of Al–7.5Si–0.5Mg Alloy. *Mater. Des.* **2011**, *32*, 1617–1622. [CrossRef]
23. Nogita, K.; Yasuda, H.; Yoshiya, M. The role of trace element segregation in the eutectic modification of hypoeutectic Al-Si alloys. *J. Alloys Compd.* **2010**, *489*, 415–420. [CrossRef]
24. Sui, Y.D.; Wang, Q.D.; Wang, G.L.; Liu, T. Effects of Sr Content on the Microstructure and Mechanical Properties of Cast Al–12Si–4Cu–2Ni–0.8 Mg Alloys. *J. Alloys Compd.* **2015**, *622*, 572–579. [CrossRef]
25. Wan, B.; Chen, W.; Liu, L. Effect of Trace Yttrium Addition on the Microstructure and Tensile Properties of Recycled Al–7Si–0.3Mg–1.0Fe Casting Alloys. *Mater. Sci. Eng.* **2016**, *666*, 165–175. [CrossRef]
26. Song, Z.X.; Li, Y.D.; Liu, W.J.; Yang, H.K.; Cao, Y.J.; Bi, G.L. Effect of La and Sc Co-Addition on the Mechanical Properties and Thermal Conductivity of As-Cast Al-4.8% Cu Alloys. *Metals* **2021**, *11*, 1866. [CrossRef]
27. Wei, F.F.; Lei, Y.S.; Yan, H.; Xu, X.; He, J. Microstructure and Mechanical Properties of A356 Alloy with Yttrium Addition Processed by Hot Extrusion. *J. Rare Earths* **2019**, *37*, 659–667. [CrossRef]
28. Xiao, D.H.; Song, M.; Chen, K.H.; Huang, B.Y. Effect of Rare Earth Yb Additionon Mechanical Properties of Al–5.3Cu–0.8Mg–0·6Ag Alloy. *Mater. Sci. Technol.* **2007**, *23*, 1156–1160. [CrossRef]
29. Srivastava, N.; Chaudhari, G.P.; Qian, M. Grain Refinement of Binary Al-Si, Al-Cu and Al-Ni Alloys by Ultrasonication. *J. Mater. Proces. Technol.* **2017**, *249*, 367–378. [CrossRef]
30. Kumar, R.; Villanova, J.; Lhuissier, P.; Salvo, L. In Situ Nanotomography Study of Creep Cavitiesin Al-3.6-Cu Alloy. *Acta Mater.* **2019**, *166*, 18–27. [CrossRef]
31. Feng, J.; Ye, B.; Zuo, L.; Wang, Q.; Wang, Q.; Jiang, H.; Ding, W. Bonding of Aluminum Alloys in Compound Casting. *Metall. Mater. Trans. A* **2017**, *48*, 4636–4644. [CrossRef]
32. Yang, Y.; Zhong, S.Y.; Chen, Z.; Wang, M.; Ma, N.; Wang, H. Effect of Cr Content and heat-treatment on the High Temperature Strength of Eutectic Al–Si Alloys. *J. Alloys Compd.* **2015**, *647*, 63–69. [CrossRef]

crystals

MDPI

Article

Effect of Yttrium on Iron-Rich Phases and Mechanical Properties of As-Cast Al-Fe Alloy with Low Si Concentration

Wenjie Wu, Wenxia Lai, Ziteng Cao, Chengdong Li * and Mei Zhao *

College of Materials Science and Engineering, Qingdao University of Science and Technology, Qingdao 266000, China; 2023061042@mails.qust.edu.cn (W.W.); 4024062018@mails.qust.edu.cn (W.L.); 4025062002@mails.qust.edu.cn (Z.C.)

* Correspondence: lichengdong@qust.edu.cn (C.L.); zhaomei@qust.edu.cn (M.Z.)

Abstract

In Al–Fe alloys, the mechanical properties are determined by the morphology of iron-rich phases. In this work, AA8176(Al-1Fe)-nY (n = 0, 0.3, 0.5, 0.7, and 0.9 wt.%) alloys were prepared by the cast method. The effects of yttrium (Y) addition on the microstructure and mechanical properties of AA8176 alloy were studied using various techniques including optical microscopy (OM), scanning electron microscopy (SEM), transmission electron microscopy (TEM), X-ray diffraction (XRD), cooling curve analysis and tensile tests. The results revealed that the optimal refinement effect was achieved when the amount of Y content was 0.5 wt.%. When the Y content increased from 0 to 0.5 wt.%, the coarse needle-like $Al_{13}Fe_4$ phases were gradually transformed into short rod-like morphology and some fine $Al_{10}Fe_2Y$ phases were formed around the $Al_{13}Fe_4$ phases. The average length of iron-rich phases was decreased from 10.01 μm to 2.65 μm. Additionally, as the Y content increased from 0 to 0.5 wt.%, the secondary dendrite arm spacing (SDAS) of AA8176 alloy was reduced from 31.33 μm to 20.24 μm. Furthermore, the mechanical properties of the AA8176 alloy were improved due to the modified microstructure. With the addition of 0.5 wt.% Y, the ultimate tensile strength, yield strength, elongation, and Vickers hardness were improved to 96.86 MPa, 57.21 MPa, 23.1%, and 30.55 HV, respectively, compared to 84.47 MPa, 50.71 MPa, 18.6%, and 27.28 HV for the unmodified AA8176 alloy. It is proposed that the growth of α-Al dendrite and $Al_{13}Fe_4$ phases were effectively inhibited by segregation of Y atoms around α-Al dendrite and $Al_{13}Fe_4$ phases during solidification. And the $Al_{10}Fe_2Y$ phases were formed by these Y atoms with Al and Fe elements. However, the formation of coarse $Al_{10}Fe_2Y$ phases was promoted by excessive Y content, resulting in a substantial degradation in mechanical properties.

Keywords: Al-Fe alloy; iron-rich phases; Y rare earth; microstructure; mechanical properties

1. Introduction

For Al-Fe-based aluminum (Al) alloys, excellent heat and wear resistance are achieved, and the high specific strength and low density of Al alloys are maintained [1–3]. The advantages of Al-Fe-based alloys are predominantly derived from various iron-rich phases (formed by Fe, Al, and other alloying elements such as Si, Mg, Mn) during solidification and heat treatment processes, which is attributed to the intentional addition of Fe [4,5]. Among these alloying elements, Si plays the most important role in determining the types of iron-rich phases. In Al-Fe alloy with high Si content, the iron-rich phases are mainly

 https://doi.org/10.3390/cryst16010028

composed of α-AlFeSi (Al_8Fe_2Si) and β-AlFeSi (Al_5FeSi). In contrast, in Al-Fe alloy with low Si content, the iron-rich phases are composed of $Al_{13}Fe_4$ (Al_3Fe), Al_6Fe, and Al_mFe phases [6].

Owing to the extremely low solubility of Fe in the Al matrix, a sharp increase in the size of iron-rich phases is achieved when the Fe content exceeds 0.1 wt.%. And their morphologies are gradually transformed from fine fibrous to large needle-like, flake-like, and plate-like structures [7,8]. These coarse iron-rich phases act as crack initiation sites, which lead to stress concentration during plastic deformation and thereby deteriorate the alloy's mechanical performance. Furthermore, the flow of melt is hindered by coarse iron-rich phases precipitated in the solidification, consequently inducing casting defects [9]. Therefore, it is essential to develop effective methods for modifying the morphology of iron-rich phases in Al-Fe Al alloys.

The 8000 series Al alloys, including AA8076, AA8079, AA8176 and other grades, belong to Al-Fe Al alloys with low Si concentration, which are mainly used in packaging, containers, cables and other fields. Currently, research on the refinement of iron-rich phases in such low-Si Al-Fe alloys remains insufficient.

Some strategies such as rapid solidification [10], mechanical alloying [11], applied magnetic field [12], and severe plastic deformation [13,14] have been employed to modify the iron-rich phases in Al alloys. However, for these methods, stringent equipment conditions and entail higher costs are required, which limit their applicability in industrial production.

In comparison to common refiners such as Mn, Ti, Ni, and Cr, the addition of rare earth elements, including La, Ce, Er, Sc, and Sm, is more attractive for Al casting alloys [15–25]. It has been demonstrated that the addition of rare earth elements significantly enhances alloy properties through the formation of thermally stable intermetallics, the promotion of grain refinement, and the modification of texture evolution. In contrast, non-rare-earth elements may not always match the performance improvements of RE elements [26,27]. Meanwhile, the melt is purified, and the content of gases and inclusions is reduced [28]. Among these rare earth elements, the widespread application of high-value rare earth elements such as Er, Sc, and Sm in large-scale Al alloy manufacturing is frequently hindered by their high costs. Additionally, in comparison to rare earth elements such as La and Ce, Y exhibits a stronger tendency to combine with Fe in Al-Fe alloys to form Al-Fe-Y ternary phases [29–31]. On the one hand, a portion of Fe atoms are consumed by the formation of the Al-Fe-Y ternary phases, thereby reducing the volume fraction of needle-like iron-rich phases. On the other hand, more favorable morphologies are typically exhibited by the Al-Fe-Y ternary phases, which results in an increase in the number of strengthening secondary phases.

The beneficial effect of Y addition on both the microstructure and the mechanical properties of Al alloys has been demonstrated by several studies [32,33]. Ding et al. [34] reported that a significant refinement effect on 6063 Al alloy was achieved by Y addition. The size of Mg_2Si particles was reduced by the addition of Y, and the β-AlFeSi phases were transformed into the α-AlFeSi phases. Furthermore, both tensile strength and elongation of the AA6063 alloy were improved by the combined addition of Y and the Al–Ti–B master alloy. Liu et al. [35] found that the microstructure of as-cast ADC12 Al alloy was significantly refined by Y addition, resulting in a notable decrease in the average size of the α-Al phase, as well as the coarse Si and iron-rich phases. And the mechanical properties of the ADC12 Al alloy were notably enhanced by the combination of Y addition and heat treatment. Wan et al. [36] reported on the effects of adding different contents of Y on the microstructure and tensile properties of recycled A356 cast alloy. When the Y content was 0.3 wt.%, the average length of the β-Fe phases was reduced from 78 μm to 20 μm and the finest α-Al phases were obtained. Meanwhile, the eutectic silicon was fully modified in morphology.

https://doi.org/10.3390/cryst16010028

While these studies have enhanced comprehension of Y modification for Al alloys, limited attention has been paid to the addition of Y in Al-Fe alloys with low Si concentration. Consequently, the underlying modification mechanism and the associated mechanical property response need to be further studied.

As described above, AA8176 was selected as the matrix material and Y was utilized as the refiner in this study to explore effective approaches for refining iron-rich phases in Al-Fe alloys with low Si concentration. Y was added to AA8176 alloys by the cast method. The influence of varying Y additions (0, 0.3, 0.5, 0.7, and 0.9 wt.%) on the evolution of the iron-rich phases, α-Al dendrite, and mechanical properties of as-cast AA8176 alloys was systematically studied. The aim of this study is to reveal the refinement mechanism of Y on the microstructure and the strengthening mechanism of mechanical properties of AA8176 Al alloys, and to propose a new strategy for achieving effective refinement of iron-rich phases in Al-Fe alloys with low Si concentration.

2. Materials and Methods

AA8176 alloy samples with varying Y additions (0, 0.3, 0.5, 0.7, and 0.9 wt.%) were fabricated using 99.7 wt.% pure Al ingots, Al-20 wt.% Fe master alloy, and Al-20 wt.% Y master alloy. The cast process was as follows: (1) 99.7 wt.% pure Al ingots, and Al-20 wt.% Fe master alloy were melted in furnace at 760 °C; (2) A suitable quantity of Al-20 wt.% Y master alloy was added at 730 °C and held for 15 min after addition; (3) The temperature of furnace was reduced to 720 °C for degassing and slag removal to obtain a purified melt; (4) The molten alloy was subsequently poured into a steel mold, thus producing the experimental cast samples for further analysis. Before the alloy was poured, the mold was placed in a muffle furnace for preheating, with the temperature set to 250 °C. The composition of the experimental alloys was determined using a direct reading spectrometer, as presented in Table 1.

Table 1. Chemical composition of Al alloy samples (wt.%).

Alloy	Fe	Y	Si	Al
AA8176	1.02	<0.01	<0.05	Bal.
03Y	1.09	0.29	<0.05	Bal.
05Y	1.04	0.52	<0.05	Bal.
07Y	0.98	0.68	<0.05	Bal.
09Y	1.01	0.91	<0.05	Bal.

To systematically analyze the effects of Y addition on the microstructural evolution and mechanical properties of AA8176 Al alloy, a comprehensive set of experimental procedures was established. The sizes of specimens for metallographic observation and tensile test are shown in Figure 1. Metallographic samples were taken from the prospective near-fracture zone of tensile specimens, with the sampling location illustrated in Figure 1. Before the metallographic observation, the samples were sequentially mechanically ground with SiC sandpapers of #180, #400, #800, #1200 and #2000, and subsequently followed by polishing with diamond paste to achieve a mirror-like surface finish. After that the samples were etched with a solution of Keller's reagent (2.5 mL HNO_3 + 1.0 mL HF + 1.5 mL HCl + 95 mL H_2O) for 10 s.

Metallographic observation was carried out using a PA53MET (Panthera, Chengdu, China) optical microscope. The SDAS of alloys was measured by the line intercept method using Image J 1.46r software. Phase analysis was performed with X-ray diffractometer using a D/MAX-2500/PC X-ray diffractometer (Rigaku, Tokyo, Japan) equipped with a Cu Kα radiation source. The diffraction patterns were recorded over a 2θ range of

https://doi.org/10.3390/cryst16010028

20° to 90° at a scanning speed of 5°/min. The microstructure and fracture morphology were examined using a Regulus8100 field emission scanning electron microscope (HITACHI, Tokyo, Japan). The $Al_{13}Fe_4$ and $Al_{10}Fe_2Y$ phases were further characterized by FEI Tecnai F20 transmission electron microscopy (FEI Company, Hillsboro, OR, USA). The samples for TEM observation were prepared using a TJ100-BE automatic twinjet electropolisher (LEBO, Jiangyin, China). The cooling curves of the alloy during solidification were measured using a THMA010K temperature recorder (TengHui, Yuyao, China) coupled with a K-type thermocouple inserted into the melt center to ensure reliable temperature monitoring throughout solidification. Tensile tests were conducted on an AI-700 tensile testing machine (Gotech, Taiwan, China), following GB/T 228.1-2021, at room temperature with a tensile speed of 5 mm·min^{-1}. As shown in Figure 1, the bone-shaped specimens were used for tensile property testing, with five tests conducted for each composition.

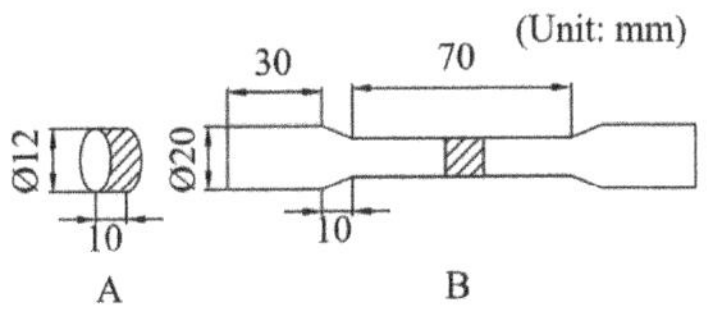

Figure 1. Schematic of (**A**) microstructure observation and (**B**) tensile test specimen.

3. Results

3.1. The Microstructure of AA8176-xY

3.1.1. Metallographic Observation

The microstructures of as-cast AA8176 Al alloys with varying Y contents were characterized using optical microscopy, as presented in Figure 2a–e. The microstructure of unmodified AA8176 alloy is predominantly composed of α-Al and eutectic structure [37]. As shown in Figures 2a and 3a, the eutectic iron-rich phases in the unmodified AA8176 alloys predominantly exhibited coarse needle-like morphology with an average length of 10.01 μm. These coarse needle-like iron-rich phases were located within interdendritic regions. And the stress concentration during plastic deformation is greatly promoted by these coarse needle-like iron-rich phases, leading to a decline in the mechanical properties.

With the Y content increasing from 0 wt.% to 0.5 wt.%, significant changes in the morphology of the iron-rich phases were observed. The average length of the iron-rich phases was decreased from 10.01 μm to 2.65 μm, transforming into short rod-like and particulate morphologies as shown in Figure 2b,c and Figure 3a. When the Y content was further increased to 0.7 wt.%, the average length of iron-rich phases exhibited an increasing trend as illustrated in Figure 2d. When the Y content was 0.9 wt.%, the average length of iron-rich phases was significantly increased to 7.87 μm, as shown in Figures 2e and 3a. Concurrently, a notable increase in the number of secondary phases within the alloy was observed.

To further analyze the effect of Y addition on the microstructure of AA8176 alloy, the SDAS of the α-Al matrix was measured for all alloys using the line intercept method in Image J 1.46r software. For the unmodified AA8176 alloy, the α-Al dendrites were relatively coarse, corresponding to an SDAS value of approximately 31.33 μm as shown in Figure 3b. With the Y addition to AA8176 alloys, the SDAS values of the AA8176 alloy with 0.3–0.9 wt.% Y were 24.03 μm, 20.24 μm, 20.45 μm, and 20.38 μm, respectively. It is confirmed by these results that the addition of Y promotes the refinement of both the iron-rich phases and the α-Al dendritic structure. Notably, no further enhancement in the refinement effect of Y on α-Al was observed when the Y content exceeded 0.5 wt.%.

https://doi.org/10.3390/cryst16010028

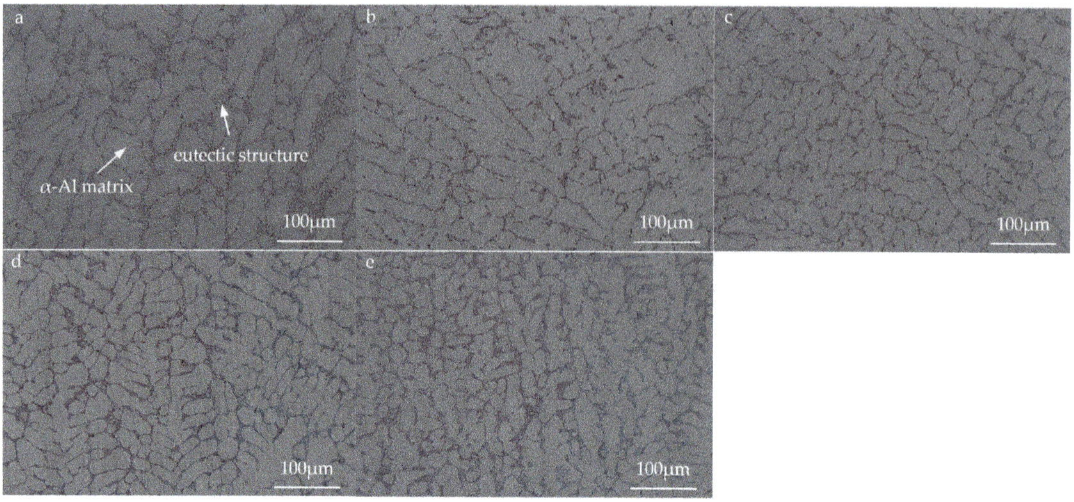

Figure 2. OM images of as-cast AA8176 alloys with different Y contents: (**a**) AA8176; (**b**) Alloy 03Y; (**c**) Alloy 05Y; (**d**) Alloy 07Y; (**e**) Alloy 09Y.

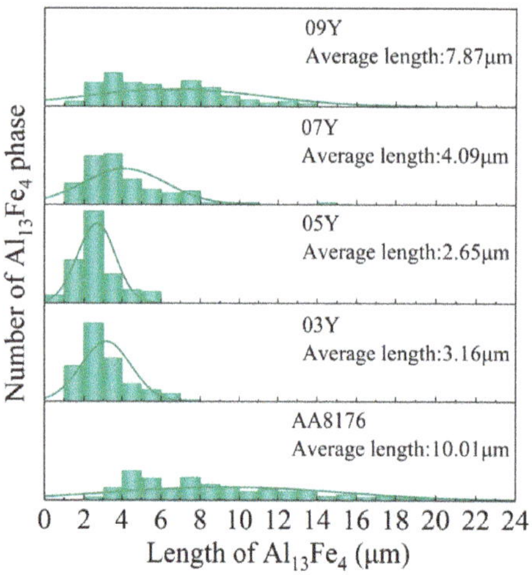

(**a**)

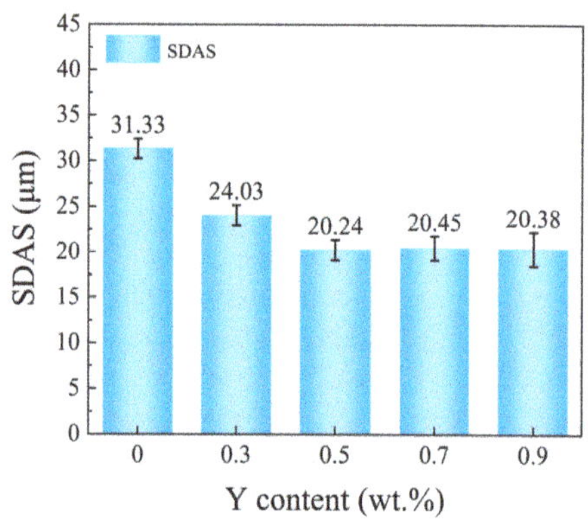

(**b**)

Figure 3. (**a**) Distribution of the length of iron-rich phases in as-cast AA8176 alloys with different Y contents; (**b**) SDAS of AA8176 alloys with different Y contents.

3.1.2. SEM and XRD Analysis

Figure 4 and Table 2 display SEM images and corresponding EDS analysis results for the AA8176 alloys with different Y contents. According to the X-ray diffraction patterns shown in Figure 5, except for α-Al matrix, the presence of the $Al_{13}Fe_4$ phases was detected in all AA8176 alloys with different Y contents, which is consistent with the reference code from the Joint Committee on Powder Diffraction Standards (JCPDS) Card No. 29-0042 reference code [38].

https://doi.org/10.3390/cryst16010028

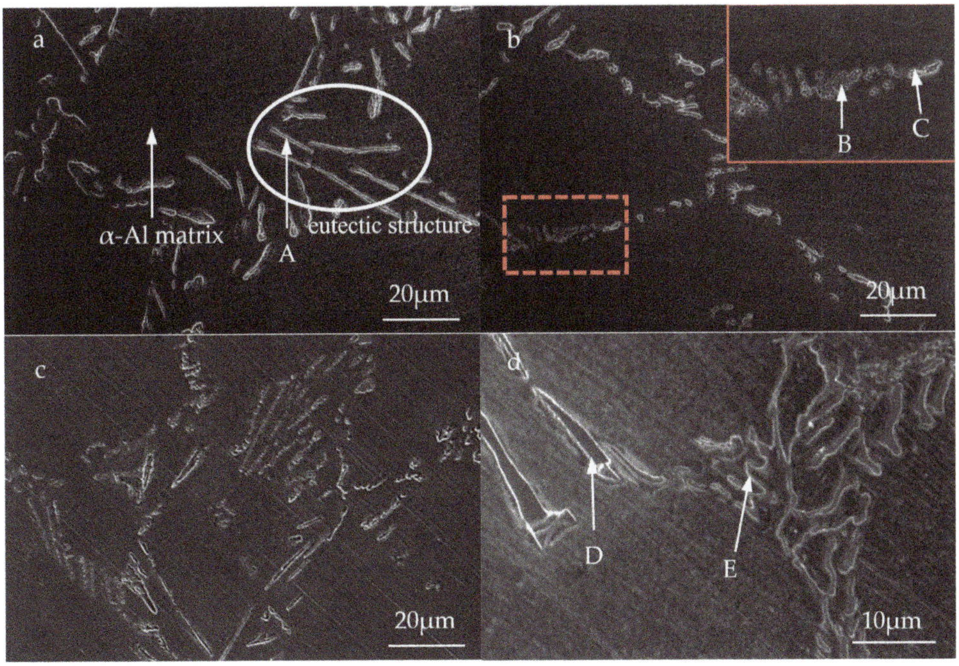

Figure 4. SEM images of as-cast AA8176 alloys with different Y contents: (**a**) AA8176; (**b**) Alloy 05Y; (**c**) Alloy 09Y. (**d**) Extremely coarse $Al_{13}Fe_4$ and $Al_{10}Fe_2Y$ phases in Alloy 09Y.

Table 2. Composition of Al alloy samples (at. %).

Point	Al K	Fe K	Y K	Possible Phase
A	74.44	25.56	—	$Al_{13}Fe_4$
B	89.90	6.49	3.61	$Al_{10}Fe_2Y$
C	78.12	21.88	—	$Al_{13}Fe_4$
D	76.82	23.72	—	$Al_{13}Fe_4$
E	84.21	9.70	6.09	$Al_{10}Fe_2Y$

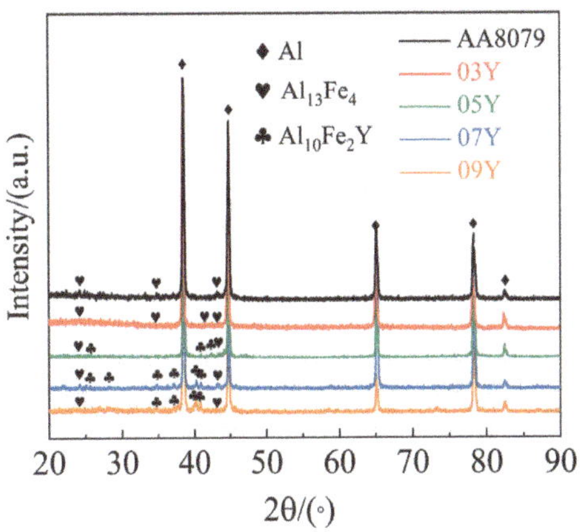

Figure 5. XRD patterns of the AA8176 alloys with different Y contents.

https://doi.org/10.3390/cryst16010028

As illustrated in Figure 4a, the microstructure of the as-cast AA8176 alloy was primarily composed of α-Al matrix and coarse needle-like $Al_{13}Fe_4$ phases distributed along the grain boundaries. When the Y content was increased to 0.5 wt.%, the iron-rich phases were refined into short rod-like and particulate morphologies instead of the coarse needle-like structures, as shown in Figure 4b. Furthermore, EDS results for points B and C (Table 2) indicated the presence of Al-Fe-Y ternary phases attached to $Al_{13}Fe_4$ phases in the Alloy 05Y. Statistical analysis was performed on the iron-rich phase sizes in Figure 4b. The results show that the average size of the $Al_{13}Fe_4$ phases is 2.41 μm, while that of the Al-Fe-Y ternary phases is 1.54 μm. In contrast, when the Y content exceeded 0.5 wt.%, the refinement efficiency of Y was diminished. As shown in Figure 4c, the average length of iron-rich phases in Alloy 09Y was increased, with an increase in both the quantity and size of the Al-Fe-Y ternary phases. The average size of the $Al_{13}Fe_4$ phases in Figure 4c is 6.75 μm, while that of the Al-Fe-Y ternary phases is 3.01 μm. Moreover, as illustrated in Figure 4d, a morphology characterized by the aggregation of extremely coarse $Al_{13}Fe_4$ and Al-Fe-Y ternary phases was observed in Alloy 09Y. And the average size of these coarse Al-Fe-Y ternary phases was increased to 5.82 μm. The formation of these Al-Fe-Y ternary phases around the coarse $Al_{13}Fe_4$ phases was confirmed by EDS results from points E and F (Table 2). And it was revealed by the comparison of the average sizes of the two iron-rich phases in alloys with different Y contents that the ternary Al-Fe-Y phases exhibit a finer average size than the $Al_{13}Fe_4$ phases.

According to the XRD patterns in Figure 5, the $Al_{10}Fe_2Y$ phases were detected in the Alloy 05Y, Alloy 07Y, and Alloy 09Y, which is consistent with the JCPDS Card No. 53-0532 reference code [39]. Furthermore, in the XRD pattern of Alloy 09Y, the peak intensity of $Al_{10}Fe_2Y$ phases was increased, while that of $Al_{13}Fe_4$ phases was decreased. This trend is mainly attributed to the consumption of Fe atoms by the formation of coarse $Al_{10}Fe_2Y$ phases, which results in a reduction of Fe available for $Al_{13}Fe_4$ phases precipitation. Based on the elemental ratios obtained from EDS and XRD patterns, the Al-Fe-Y ternary phases observed in the SEM images (Figure 4) are most likely to be the $Al_{10}Fe_2Y$ phases.

3.1.3. TEM Observation

Initial morphological characterization and EDS analysis were conducted on the secondary phases possibly containing Fe and Y elements in Alloy 05Y, as indicated by earlier SEM observations. The Y elemental mapping clearly reveals the enrichment of Y at the periphery of the $Al_{13}Fe_4$ phases, indicating the in situ formation of Al-Fe-Y ternary phases through its reaction with Al and Fe, as shown in Figure 6d.

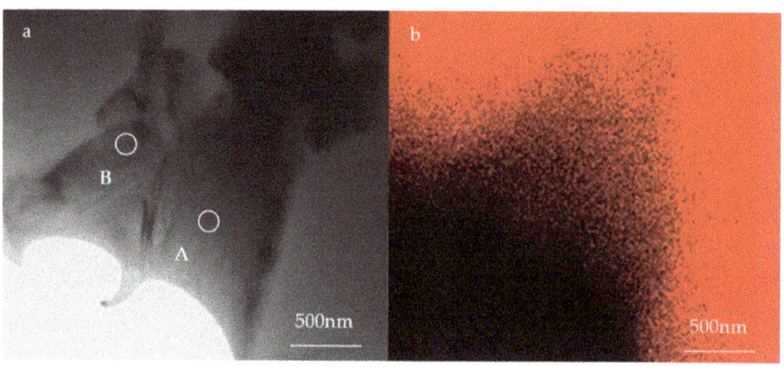

Figure 6. *Cont.*

https://doi.org/10.3390/cryst16010028

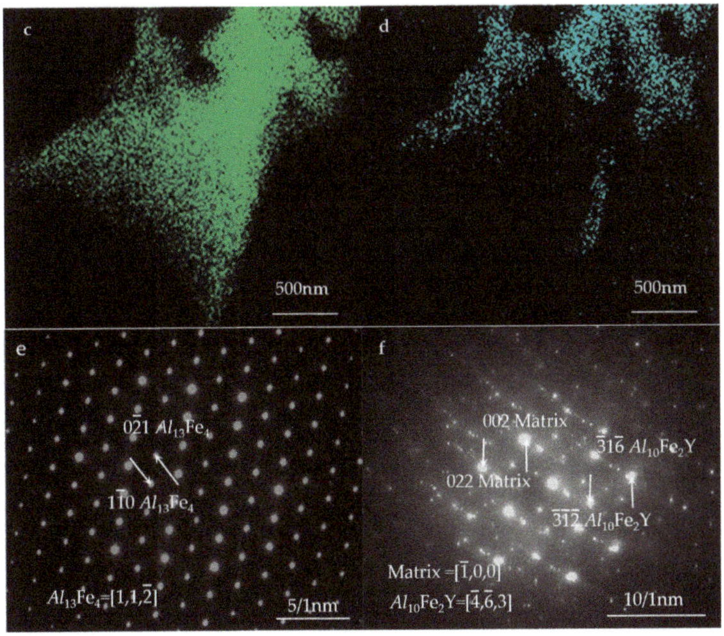

Figure 6. TEM characterization of the as-cast Alloy 05Y: (**a**): TEM bright-field image of Alloy; (**b–d**): EDS mapping results of $Al_{13}Fe_4$ phase and $Al_{10}Fe_2Y$ phase; (**e**,**f**): SAED patterns in the selected areas A and B, respectively.

To further identify the crystal structure of these phases, SAED patterns were collected from point A (corresponding to $Al_{13}Fe_4$ phases) and point B (corresponding to Al-Fe-Y ternary phases), as shown in Figure 6e,f, respectively. The SAED pattern acquired at point A can be indexed to the $Al_{13}Fe_4$ phases. Due to the greater sample thickness at point B, the SAED pattern obtained from this location contained contributions from more than one kind phase, which was confirmed to consist mainly of α-Al matrix and $Al_{10}Fe_2Y$ phase, as illustrated in Figure 6f.

3.1.4. Cooling Curves

The cooling curves and differential curves for AA8176 alloy, Alloy 05Y and Alloy 09Y are shown in Figure 7 and the characteristic temperatures of the eutectic reaction summarized in Table 3. T_{EN}, T_{EU}, and T_{EG} denote the eutectic nucleation temperature, eutectic minimum undercooling temperature and eutectic growth temperature, respectively. According to the Al-Fe alloy binary phase diagram, the eutectic temperature of Al-Fe alloys is approximately 655 °C [40], which is consistent with the characteristic values measured for the AA8176 alloy. With the addition of 0.5 wt.% Y, T_{EN} was decreased to 649.9 °C and T_{EG} was decreased to 643.6 °C, respectively, indicating the adsorption and hindering effects of Y during the nucleation and growth processes of $Al_{13}Fe_4$ eutectic phases [41]. Additionally, a new peak emerged in the cooling curve, with a nucleation temperature around 626.8 °C, corresponding to the nucleation and growth of the $Al_{10}Fe_2Y$ phases [42]. When the Y content was increased to 0.9 wt.%, only minor changes were detected on the values of both T_{EN} and T_{EG}, indicating the adsorption and hindering effects of Y on the eutectic structure were not further enhanced. In contrast, the T'_{EN} of Alloy 09Y was significantly increased, suggesting that the nucleation and growth of $Al_{10}Fe_2Y$ phases were greatly promoted by

https://doi.org/10.3390/cryst16010028

higher Y content. This also directly explains the presence of numerous coarse $Al_{10}Fe_2Y$ phases in Alloy 09Y.

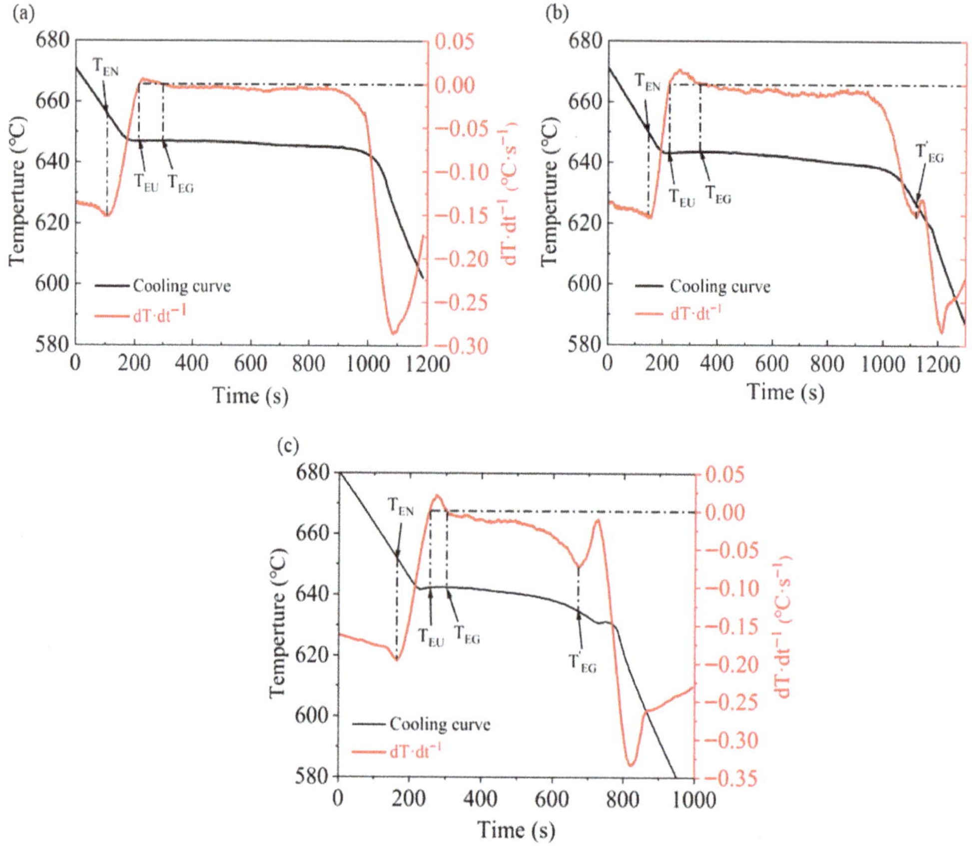

Figure 7. Cooling curves of (**a**): AA8176; (**b**): Alloy 05Y; (**c**): Alloy 09Y.

Table 3. Characteristic values of eutectic reaction.

Alloy	T_{EN} (°C)	T_{EU} (°C)	T_{EG} (°C)	T'_{EN} (°C)
AA8176	656.2	646.9	647.0	—
Alloy 05Y	649.9	643.1	643.5	626.8
Alloy 09Y	651.1	642.6	642.9	634.7

3.2. Mechanical Property

The mechanical properties of cast AA8176 alloys with different Y contents are presented in Figure 8 and Table 4. In unmodified AA8176 alloy, the ultimate tensile strength of the alloy was 84.47 MPa, the yield strength was 50.71 MPa, and the elongation was 18.6%. Optimal mechanical properties were achieved when the Y content was increased to 0.5 wt.%. The ultimate tensile strength was increased to 96.86 MPa, the yield strength was increased to 57.21 MPa and elongation was increased to 23.1%. However, further increase in Y content to 0.9 wt.% led to a reduction in ultimate tensile strength, yield strength and elongation. Therefore, it can be deduced that the ultimate tensile strength, yield strength and the elongation of the AA8176 alloys exhibited a trend of firstly increasing and then decreasing with rising Y content. As shown in Figure 8c and Table 4, the Vickers hardness

https://doi.org/10.3390/cryst16010028

of AA8176 alloy with varying Y content also followed a similar trend. When the Y content was increased from 0 to 0.5 wt.%, the Vickers hardness was increased from 27.28 HV to 30.55 HV. And the Vickers hardness was decreased to 28.74 HV with further Y addition to 0.9 wt.%.

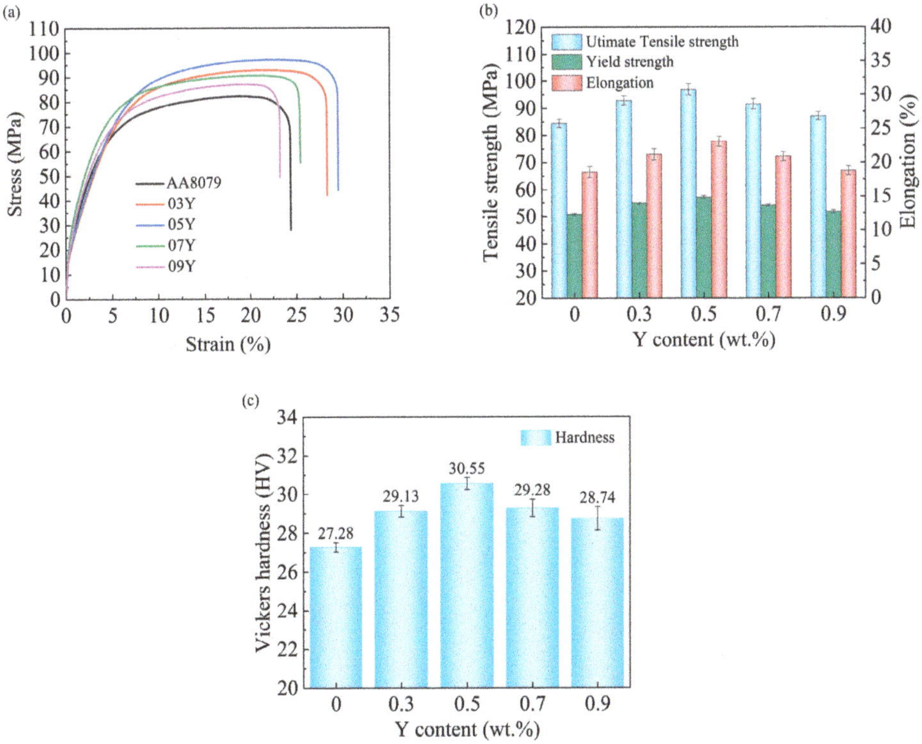

Figure 8. (**a**) Engineering stress–strain curves (**b**) Mechanical properties of AA8176 alloys with different Y contents (**c**) Vickers hardness of AA8176 alloys with different Y contents.

Table 4. Mechanical properties of AA8176 alloys with different Y contents.

Alloy	UTS (MPa)	YS (MPa)	Elongation (%)	Hardness (HV)
AA8176	84.47	50.71	18.6	27.28
Alloy 03Y	92.81	54.92	21.2	29.13
Alloy 05Y	96.86	57.21	23.1	30.55
Alloy 07Y	91.59	54.14	20.9	29.28
Alloy 09Y	87.10	51.78	18.8	28.74

Figure 9 presents the fracture surfaces of the as-cast AA8176 alloys with different Y contents. In the unmodified AA8176 alloy, shallow and irregularly sized dimples are observed with coarse iron-rich phase particles within the dimples (point A), as shown in Figure 9a. This is attributed to the presence of coarse, brittle iron-rich phases in AA8176 alloy. During deformation, stress concentration and rapid crack propagation, promoted by these coarse iron-rich phases, significantly inhibit the growth of dimples. Consequently, the non-uniform size and distribution of the iron-rich phases directly resulted in variability in dimple sizes. For Alloy 05Y, deeper and more uniformly distributed dimples were observed, with modified iron-rich phases observed within the dimples (point B), as shown in Figure 9c. The refined iron-rich phases mitigate stress concentration during deformation.

https://doi.org/10.3390/cryst16010028

This allows dimples to undergo more extensive growth, leading to the formation of deeper dimples. Moreover, the uniform size and distribution of these refined phases result in a corresponding uniformity in dimple size. However, when the Y content exceeded 0.5 wt.%, the weakening refinement effect of Y on the $Al_{13}Fe_4$ phases and formation of coarse $Al_{10}Fe_2Y$ phases resulted in a subsequent reduction in the ductility. Correspondingly, the fracture surfaces exhibited a tendency toward shallower dimples and an increased presence of coarse iron-rich phases (point C). Figure 9f shows the morphology of the tensile fracture side of Alloy 05Y. It reveals that obvious microcrack initiate at iron-rich phases and propagate along the interface between the iron-rich phases and the matrix. This behavior is consistent with the phenomenon observed in the AA8176 alloy shown in Supplementary Figure S1.

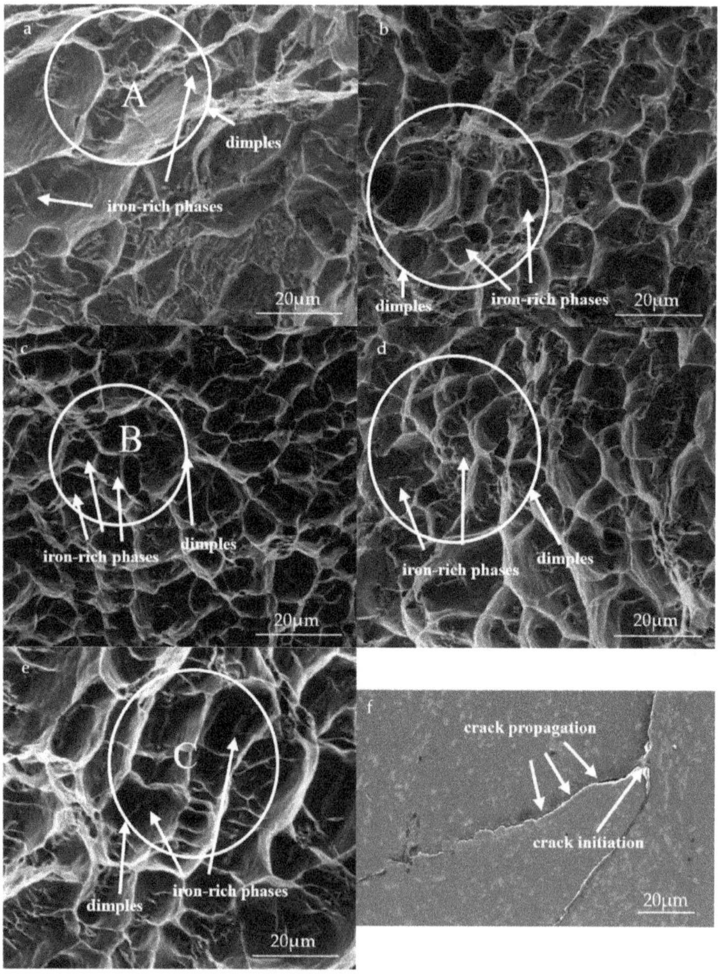

Figure 9. SEM image of tensile fracture morphology: (**a**) AA8176; (**b**) Alloy 03Y; (**c**) Alloy 05Y; (**d**) Alloy 07Y; (**e**) Alloy 09Y; (**f**) SEM image of tensile fracture side of Alloy 05Y.

https://doi.org/10.3390/cryst16010028

4. Discussion

4.1. Effect of Y Addition on α-Al Dendrite

As the total solid–liquid interface area between dendritic arms in the system is gradually decreased, neighboring dendritic arms are grown and merged, resulting in the formation of coarse dendritic structures during the solidification process. The atomic radii of the rare earth element Y and Al are 0.182 nm and 0.143 nm [35], respectively, resulting in a difference of 27%. According to the Hume-Rothery rule, the formation of solid solutions is inhibited by a radius difference exceeding 15% between solvent and solute [43]. Consequently, Y exhibits extremely low solubility in the Al matrix (less than 0.035 mol.% at 573 K [44]), and enrichment of Y atoms at the dendritic fronts tends to occur during solidification. The growth and merging of dendritic arms are restricted by this enrichment, thereby refining the SDAS of α-Al matrix [45]. The growth induced by element enrichment is described based on the growth restriction factor (*GRF*) as follows [20,46]:

$$GRF = C_Y \cdot m \cdot (k - 1)$$

where C_Y is the concentration of solute Y in the alloy melt, m is the liquidus slope, and k is the equilibrium partition coefficient [20]. For the Al–Y alloy system, $m = 4.98$ and $k = 0.5$ are given by the Al–Y binary phase diagram. For the AA8176 alloy with 0.5 wt.% Y, the calculated *GRF* is approximately 1.25 K, the growth of α-Al is effectively inhibited by the addition of Y.

4.2. Effect of Y Addition on Iron-Rich Phases

The $Al_{13}Fe_4$ phases prefer to grow along the [010] orientation, resulting in the formation of coarse needle-like and plate-like morphologies during solidification [24]. The influence of active solute atoms on crystal growth is theoretically explained through doping (substitutional incorporation) and adsorption (surface segregation). There exists a significant difference in atomic radius between Y atoms and Al/Fe atoms—specifically, the atomic radius of Y, Al, and Fe are 0.182 nm, 0.143 nm, and 0.128 nm, respectively [35,47]. Owing to this remarkable size mismatch, Y atoms cannot easily substitute Al or Fe atoms in the $Al_{13}Fe_4$ phases. According to result of EDS, rare Y element was detected in the eutectic $Al_{13}Fe_4$ phase, which directly confirms that the doping of Y atoms into $Al_{13}Fe_4$ phases is not the main mechanism for the refinement of $Al_{13}Fe_4$ phases. This experimental observation is consistent with the theoretical calculation results reported by Pang et al. [31], which show that Y doping into $Al_{13}Fe_4$ phases increases their formation enthalpy. In contrast, the adsorption of Y on the (010) plane is an exothermic reaction, Y atoms are easily adsorbed on the (010) plane during the solidification process. The adsorption of Y atoms modifies the surface structure of the (010) plane and impedes the preferential growth along the [010] direction. In the final solidification stage, Y atoms eventually form $Al_{10}Fe_2Y$ ternary phases [31,42]. When the Y content reaches 0.9 wt.%, excessive aggregation of Y atoms significantly increases the nucleation temperature of the $Al_{10}Fe_2Y$ phases. The nucleation and growth of coarse $Al_{10}Fe_2Y$ phases leads to a massive consumption of Y atoms. This reduces the amount of Y atoms for inhibiting the preferential growth of the $Al_{13}Fe_4$ phases, resulting in a weakened refinement effect [20].

4.3. Strengthening Mechanism

Due to the extremely low solid solubility of Y and Fe in the Al matrix and the fact that the mechanical test specimens used in this study underwent no plastic deformation processing, solid solution strengthening and dislocation strengthening are not considered to be the primary mechanisms responsible for the enhancement in mechanical properties.

https://doi.org/10.3390/cryst16010028

Thus, the enhancement of ultimate tensile strength can be attributed to two factors. First, $Al_{13}Fe_4$ intermetallic compounds are second phases with high hardness and brittleness, distributing on the α-Al matrix. During tensile process, deformation of the alloy is hindered by the $Al_{13}Fe_4$ phases distributed between the grains, resulting in the formation of dislocations [48].

However, in the AA8176 alloy without Y addition, the coarse $Al_{13}Fe_4$ phases can induce stress concentrations, which adversely affects the mechanical properties of the alloy. When the Y addition was increased to 0.5 wt.%, the morphology of the $Al_{13}Fe_4$ phases was transformed into short rod-like and particulate with an average length of 2.65 μm. Consequently, the stress concentration during plastic deformation is effectively mitigated by this refinement for $Al_{13}Fe_4$ phases. Meanwhile, the number of secondary phases in the alloy matrix is further increased by the formation of $Al_{10}Fe_2Y$ phases, which grow around the $Al_{13}Fe_4$ phases. The mechanical properties of the AA8176 alloy with 0.5 wt.% Y are enhanced by this dual effect. When the Y addition exceeded the optimal content of 0.5 wt.%, no further enhancement in the adsorption effect of Y on $Al_{13}Fe_4$ phases was observed. However, the nucleation and growth temperature of the $Al_{10}Fe_2Y$ phases was increased by the excessive addition of Y. The Y atoms that would otherwise inhibit the growth of $Al_{13}Fe_4$ are consumed by the growth of the $Al_{10}Fe_2Y$ phases, eventually leading to the formation of coarse $Al_{13}Fe_4$ and $Al_{10}Fe_2Y$ phases, which consequently results in the degradation of mechanical properties.

Secondly, the SDAS of the AA8176 alloys was significantly reduced by the addition of Y, leading to a more refined microstructure. Secondly, the addition of Y significantly reduces the SDAS of the AA8176 alloys, resulting in a more refined microstructure. Generally, finer grains lead to a higher density of grain boundaries. Dislocation movement is effectively hindered and dislocation pile-up is induced by this increased boundary density, resulting in a significant improvement in the strength of the AA8176 alloys.

Gerbe et al. [49] proposed that in hypoeutectic alloys, the dendritic regions are regarded as behaving similarly to grain boundaries in polycrystalline alloys, since the eutectic zones between dendrites are traversed by slip planes. The SDAS is considered analogous to grain size in the Hall–Petch relationship, whereby a reduction in the mean free path of dislocation motion during plastic deformation is produced by smaller dendritic regions, thus further enhancing the strength of the AA8176 alloys.

5. Conclusions

For the as-cast AA8176 alloys, an optimal microstructure was achieved by the addition of 0.5 wt.% Y. In the unmodified AA8176, the microstructure is predominantly composed of the α-Al phases and the eutectic structure. The SDAS of α-Al phases was measured to be approximately 31.33 μm, and a needle-like morphology with an average length of 10.01 μm was exhibited by the iron-rich phases. In contrast, after the addition of 0.5 wt.% Y, the SDAS of the alloy was decreased to 20.24 μm, and the iron-rich phases were transformed into short rod-like and particulate forms with an average length of 2.65 μm. When the Y content exceeded 0.5 wt.%, no significant change in the SDAS of α-Al phases was observed, while an increase in both the number and average size of the iron-rich phases occurred.

In the unmodified AA8176 alloy, the iron-rich phases consist of $Al_{13}Fe_4$ phases. The refinement of the $Al_{13}Fe_4$ phases is primarily attributed to the adsorption effect of Y atoms around these phases. Microstructural observations revealed that small-size $Al_{10}Fe_2Y$ ternary phases were attached to the surfaces of $Al_{13}Fe_4$ phases in the AA8176 alloy with 0.5 wt.% Y content. When the Y content exceeded 0.5 wt.%, the coarse $Al_{10}Fe_2Y$ phases were formed. And a substantial consumption of Y atoms is caused by the growth of these $Al_{10}Fe_2Y$ phases, resulting in a weakening of the refinement effect.

https://doi.org/10.3390/cryst16010028

With the Y content increased, a trend of initially increasing and then decreasing was exhibited by the mechanical properties of the cast AA8176 alloys. When the Y content was 0.5 wt.%, an optimal balance of strength and ductility was achieved, with ultimate tensile strength, yield strength, elongation, and Vickers hardness measured at 96.86 MPa, 57.21 MPa, 23.1%, and 30.55 HV, respectively. The primary strengthening mechanism of the AA8176 alloys is attributed to the refinement in the morphology of iron-rich phases and reduction in the SDAS of α-Al matrix. When the Y content exceeded 0.5 wt.%, coarse $Al_{10}Fe_2Y$ phases were formed, and the refining effect was reduced. The deterioration in the mechanical properties of the alloy is attributed to both coarse $Al_{13}Fe_4$ phases and $Al_{10}Fe_2Y$ phases lead to worse mechanical properties of the alloy.

Supplementary Materials: The following supporting information can be downloaded at: https://www.mdpi.com/article/10.3390/cryst16010028/s1, Figure S1. OM image of tensile fracture side of AA8176 alloy.

Author Contributions: Conceptualization, W.W. and C.L.; methodology, W.W.; formal analysis, W.W.; investigation, W.W., W.L. and Z.C.; resources, C.L. and M.Z.; data curation, W.W., W.L. and Z.C.; writing—original draft preparation, W.W.; writing—review and editing, W.W., C.L. and M.Z.; supervision, C.L. and M.Z.; funding acquisition, C.L. and M.Z. All authors have read and agreed to the published version of the manuscript.

Funding: This research was funded by National Natural Science Foundation of China under Grant Nos. 51672145, 52171229 and 51872034.

Data Availability Statement: The data presented in this study is available on request from the corresponding author. The data is not publicly available due to privacy restrictions.

Conflicts of Interest: The authors declare that they have no known competing financial interests or personal relationships that could have appeared to influence the work reported in this paper.

References

1. Medvedev, A.; Zhukova, O.; Enikeev, N.; Kazykhanov, V.; Timofeev, V.; Murashkin, M. The effect of casting technique and severe straining on the microstructure, electrical conductivity, mechanical properties and thermal stability of the Al–1.7 wt.% Fe alloy. *Materials* **2023**, *16*, 3067. [CrossRef]
2. Zhao, Q.; Qian, Z.; Cui, X.; Wu, Y.; Liu, X. Optimizing microstructures of dilute Al–Fe–Si alloys designed with enhanced electrical conductivity and tensile strength. *J. Alloys Compd.* **2015**, *650*, 768–776. [CrossRef]
3. Meignanamoorthy, M.; Ravichandran, M.; Mohanavel, V.; Afzal, A.; Sathish, T.; Alamri, S.; Khan, S.A.; Saleel, C.A. Microstructure, mechanical properties, and corrosion behavior of boron carbide reinforced Al alloy (Al-Fe-Si-Zn-Cu) matrix composites produced via powder metallurgy route. *Materials* **2021**, *14*, 4315. [CrossRef] [PubMed]
4. Trink, B.; Weißensteiner, I.; Uggowitzer, P.J.; Strobel, K.; Pogatscher, S. High Fe content in Al-Mg-Si wrought alloys facilitates excellent mechanical properties. *Scr. Mater.* **2022**, *215*, 114701. [CrossRef]
5. Wang, X.; Guan, R.G.; Wang, Y.; Misra, R.D.K.; Yang, B.W.; Li, Y.D.; Chen, T.J. Mechanistic understanding on the evolution of nanosized Al_3Fe phase in Al–Fe alloy during heat treatment and its effect on mechanical properties. *Mater. Sci. Eng. A* **2019**, *751*, 23–34. [CrossRef]
6. Que, Z.; Fang, C.; Mendis, C.L.; Wang, Y.; Fan, Z. Effects of Si solution in θ-Al13Fe4 on phase transformation between Fe-containing intermetallic compounds in Al alloys. *J. Alloys Compd.* **2023**, *932*, 167587. [CrossRef]
7. Chen, J.; Lengsdorf, R.; Henein, H.; Herlach, D.M.; Dahlborg, U.; Calvo-Dahlborg, M. Microstructure evolution in undercooled Al–8 wt% Fe melts: Comparison between terrestrial and parabolic flight conditions. *J. Alloys Compd.* **2013**, *556*, 243–251. [CrossRef]
8. Que, Z.; Wang, Y.; Mendis, C.L.; Fang, C.; Xia, J.; Zhou, X.; Fan, Z. Understanding Fe-containing intermetallic compounds in Al alloys: An overview of recent advances from the LiME research hub. *Metals* **2022**, *12*, 1677. [CrossRef]
9. Zhao, Y.; Zhang, W.; Song, D.; Lin, B.; Shen, F.; Zheng, D.; Xie, C.; Sun, Z.; Fu, Y.; Li, R. Nucleation and growth of Fe-rich phases in Al-5Ti-1B modified Al-Fe alloys investigated using synchrotron X-ray imaging and electron microscopy. *J. Mater. Sci. Technol.* **2021**, *80*, 84–99. [CrossRef]
10. Nayak, S.S.; Chang, H.J.; Kim, D.H.; Pabi, S.K.; Murty, B.S. Formation of metastable phases and nanocomposite structures in rapidly solidified Al–Fe alloys. *Mater. Sci. Eng. A* **2011**, *528*, 5967–5973. [CrossRef]

https://doi.org/10.3390/cryst16010028

11. Nayak, S.S.; Wollgarten, M.; Banhart, J.; Pabi, S.K.; Murty, B.S. Nanocomposites and an extremely hard nanocrystalline intermetallic of Al–Fe alloys prepared by mechanical alloying. *Mater. Sci. Eng. A* **2010**, *527*, 2370–2378. [CrossRef]
12. Lan, Q.; Le, Q.; Liao, Q.; Bao, L.; Zhang, J.; Cui, J. Effect of AC magnetic field on Al3Fe phase by measuring thermoelectric power. *Mater. Chem. Phys.* **2018**, *214*, 383–390. [CrossRef]
13. Duchaussoy, A.; Sauvage, X.; Edalati, K.; Horita, Z.; Renou, G.; Deschamps, A.; De Geuser, F. Structure and mechanical behavior of ultrafine-grained Al-iron alloy stabilized by nanoscaled intermetallic particles. *Acta Mater.* **2019**, *167*, 89–102. [CrossRef]
14. Medvedev, A.; Murashkin, M.; Enikeev, N.; Medvedev, E.; Sauvage, X. Influence of morphology of intermetallic particles on the microstructure and properties evolution in severely deformed Al-Fe alloys. *Metals* **2021**, *11*, 815. [CrossRef]
15. Song, D.; Zhao, Y.; Jia, Y.; Li, X.; Fu, Y.; Zhang, W. Synergistic effects of Mn and B on iron-rich intermetallic modification of recycled Al alloy. *J. Mater. Res. Technol.* **2023**, *24*, 527–541. [CrossRef]
16. Watanabe, Y.; Mihara-Narita, M.; Sato, H. Grain refinement of cast Al by heterogeneous nucleation site particles with high lattice matching. *Mater. Trans.* **2023**, *64*, 1083–1097. [CrossRef]
17. Qin, L.; Tang, P.; Meng, S. Effect of Ni addition on the microstructure, conductivities and mechanical properties of as-cast Al-Fe alloys. *J. Alloys Compd.* **2024**, *986*, 174160. [CrossRef]
18. Song, X.; Gao, M.; Yang, B.; Guan, R. Modification and refinement of Fe-containing phases, mechanical properties and strengthening mechanisms in Al–Fe alloys via Cr alloying and continuous rheo-extrusion. *Mater. Sci. Eng. A* **2022**, *850*, 143557. [CrossRef]
19. Xu, H.; Jia, H.; Ji, Z.; Li, M.L.; Yu, H.; Jiang, B.; Wang, Y.; Hu, M.L. Effect of La content on microstructure, tensile properties, and electrical conductivity of cast Al-Mg-Si-xLa alloys. *China Foundry* **2025**, *22*, 385–394. [CrossRef]
20. Liao, H.C.; Liu, Y.; Lü, C.L.; Wang, Q.G. Effect of Ce addition on castability, mechanical properties and electric conductivity of Al–0.3 Si–0.2 Mg alloy. *Int. J. Cast Met. Res.* **2015**, *28*, 213–220. [CrossRef]
21. Mahmoud, M.G.; Mosleh, A.O.; Mohamed, M.S.; El-Moayed, M.H.; Khalifa, W.; Pozdniakov, A.V.; Salem, S. The impact of Ce-containing precipitates on the solidification behavior, microstructure, and mechanical properties of Al-6063. *J. Alloys Compd.* **2023**, *948*, 169805. [CrossRef]
22. Liang, Y.H.; Shi, Z.M.; Li, G.W.; Zhang, R.Y.; Zhao, G. Effects of Er addition on the crystallization characteristic and microstructure of Al-2wt% Fe cast alloy. *J. Alloys Compd.* **2019**, *781*, 235–244. [CrossRef]
23. Li, Y.; Yu, Q.; Chen, F.; Yang, H.M.; Li, M.N. Effect of Sc on Al3Fe phase and mechanical properties of as-cast AA5052 Al alloy. *China Foundry* **2024**, *21*, 257–264. [CrossRef]
24. Dai, K.; Ye, J.; Wang, Z.; Gao, M.; Chen, J.; Guan, R. Effects of Sc and Zr addition on the solidification and mechanical properties of Al–Fe alloys. *J. Mater. Res. Technol.* **2022**, *18*, 112–121. [CrossRef]
25. Mo, L.; Jiang, M.; Zhou, X.; Zhao, Y.J.; Du, J. Modification mechanisms of hypereutectic Al-Fe alloys treated by Sm/Yb addition: Experiments and first-principles calculations. *J. Alloys Compd.* **2023**, *948*, 169786. [CrossRef]
26. Tariq, H.M.R.; Ishtiaq, M.; Kang, H.H.; Chaudry, U.M.; Jun, T.S. A Critical Review on the Comparative Assessment of Rare-Earth and Non-Rare-Earth Alloying in Magnesium Alloy. *Metals* **2025**, *15*, 128. [CrossRef]
27. Mansoor, A.; Zulfiqar, M.; Ishtiaq, M.; Hussain, M.A.; Younes, M.H.; Inam, A.; Afifi, M.A.; Khan, M.A.; Jabar, B. Tuning the microstructure and strength of Mg–RE alloys through the rare-earth elements. *J. Mater. Sci.* **2025**, *60*, 4808–4819. [CrossRef]
28. Sahin, H.; Dispinar, D. Effect of rare earth elements erbium and europium addition on microstructure and mechanical properties of A356 (Al–7Si–0.3 Mg) alloy. *Int. J. Met.* **2023**, *17*, 2612–2621. [CrossRef]
29. Luo, S.X.; Shi, Z.M.; Li, N.Y.; Lin, Y.M.; Liang, Y.H.; Zeng, Y.D. Crystallization inhibition and microstructure refinement of Al-5Fe alloys by addition of rare earth elements. *J. Alloys Compd.* **2019**, *789*, 90–99. [CrossRef]
30. Pang, N.; Shi, Z.; Wang, C.; Lian, H.; Liu, J. Modification mechanisms of the primary Al13Fe4 phase in hypereutectic Al–Fe alloys by La element: Experiments and first–principles study. *J. Cryst. Growth* **2025**, *659*, 128136. [CrossRef]
31. Pang, N.; Shi, Z.; Wang, C.; Wu, Y.; Zhang, J.; Sun, J. Doping and adsorption mechanism of the element Y modifying the primary $Al_{13}Fe_4$ phase in hypereutectic Al–Fe alloys from first-principles. *CrystEngComm* **2021**, *23*, 3802–3811. [CrossRef]
32. Mao, G.; Liu, S.; Wu, Z.; Zhu, C.; Gao, W. The effects of Y on primary α-Al and precipitation of hypoeutectic Al-Si alloy. *Mater. Lett.* **2020**, *271*, 127795. [CrossRef]
33. Mao, G.; Yan, H.; Zhu, C.; Wu, Z.; Gao, W. The varied mechanisms of yttrium (Y) modifying a hypoeutectic Al–Si alloy under conditions of different cooling rates. *J. Alloys Compd.* **2019**, *806*, 909–916. [CrossRef]
34. Ding, W.; Zhao, X.; Chen, T.; Zhang, H.; Liu, X.; Cheng, Y.; Lei, D. Effect of rare earth Y and Al–Ti–B master alloy on the microstructure and mechanical properties of 6063 Al alloy. *J. Alloys Compd.* **2020**, *830*, 154685. [CrossRef]
35. Liu, J.; Wu, Q.; Yan, H.; Zhong, S.; Huang, Z. Effect of trace yttrium addition and heat treatment on the microstructure and mechanical properties of as-cast ADC12 Al alloy. *Appl. Sci.* **2018**, *9*, 53. [CrossRef]
36. Wan, B.; Chen, W.; Liu, L.; Cao, X.; Zhou, L.; Fu, Z. Effect of trace yttrium addition on the microstructure and tensile properties of recycled Al–7Si–0.3 Mg–1.0 Fe casting alloys. *Mater. Sci. Eng. A* **2016**, *666*, 165–175. [CrossRef]

https://doi.org/10.3390/cryst16010028

37. Jiang, H.; Li, S.; Zhang, L.; He, J.; Zheng, Q.; Song, Y.; Li, Y.; Zhao, J. The influence of rare earth element lanthanum on the microstructures and properties of as-cast 8176 (Al-0.5 Fe) Al alloy. *J. Alloys Compd.* **2021**, *859*, 157804. [CrossRef]
38. Nemati, N.; Emamy, M. Evaluating Microstructure and High-Temperature Shear Behavior of Hot Extruded Al-Al13Fe4 Nanocomposite. *Mater. Trans.* **2016**, *57*, 1236–1245. [CrossRef]
39. Waerenborgh, J.C.; Salamakha, P.; Sologub, O.; Sério, S.; Godinho, M.; Gonçalves, A.P.; Almeida, M. Y–Fe–Al ternary system: Partial isothermal section at 1070 K: Powder X-ray diffraction and Mössbauer spectroscopy study. *J. Alloys Compd.* **2001**, *323*, 78–82. [CrossRef]
40. da Silva Leal, J.R.; Saldanha, F.E.; Ganju, E.; Spinelli, J.E.; de Gouveia, G.L. Effect of cooling rate on AlFe primary and eutectic phase growth evolution in an Al-2Fe-1Mn alloy. *J. Alloys Compd.* **2025**, *1010*, 177870. [CrossRef]
41. Mao, F.; Li, J.; Yan, G.; Zou, L.; Cao, Z.; Wang, T. Effect of Eu on the silicon phase in Al-40Zn-5Si alloys. *J. Alloys Compd.* **2017**, *722*, 116–130. [CrossRef]
42. Wang, M.; Knezevic, M.; Chen, M. Effects of Fe content on microstructure and mechanical properties of Al-Y eutectic alloys. *Mater. Sci. Eng. A* **2023**, *863*, 144558. [CrossRef]
43. Hume-Rothery, W.; Raynor, G.V. The Structure of Metals and Alloys. *Nature* **1936**, *138*, 7–8. [CrossRef]
44. Li, B.; Wang, H.; Jie, J.; Wei, Z. Effects of yttrium and heat treatment on the microstructure and tensile properties of Al–7.5 Si–0.5 Mg alloy. *Mater. Des.* **2011**, *32*, 1617–1622. [CrossRef]
45. Mao, F.; Yan, G.; Xuan, Z.; Cao, Z.; Wang, T. Effect of Eu addition on the microstructures and mechanical properties of A356 Al alloys. *J. Alloys Compd.* **2015**, *650*, 896–906. [CrossRef]
46. Zheng, Q.; Zhang, L.; Jiang, H.; Zhao, J.; He, J. Effect mechanisms of micro-alloying element La on microstructure and mechanical properties of hypoeutectic Al-Si alloys. *J. Mater. Sci. Technol.* **2020**, *47*, 142–151. [CrossRef]
47. He, J.; Liao, X.; Lan, X.; Qiu, W.; Yu, H.; Zhang, J.; Fan, W.; Zhong, X.; Liu, Z. Annealed Al-Cr coating: A hard anti-corrosion coating with grain boundary modification effect for Nd-Fe-B magnets. *J. Alloys Compd.* **2021**, *870*, 159229. [CrossRef]
48. Fuller, C.B.; Seidman, D.N.; Dunand, D.C. Mechanical properties of Al (Sc, Zr) alloys at ambient and elevated temperatures. *Acta Mater.* **2003**, *51*, 4803–4814. [CrossRef]
49. Gerbe, S.; Krupp, U.; Michels, W. Influence of secondary dendrite arm spacing (SDAS) on the fatigue properties of different conventional automotive Al cast alloys. *Fract. Struct. Integr.* **2019**, *13*, 105–115. [CrossRef]

https://doi.org/10.3390/cryst16010028

crystals

MDPI

Article

Effect of Fe Content on the Microstructure and Properties of 5083 Aluminum Alloy

Jun Cao [1], Wenjia Zhao [1], Jiaxing Li [1], Hongqun Tang [1,2,*], Xu Zheng [3], Kezhun He [3], Qizhong Zhao [3,*], Hongchi Yang [3], Xianye Lu [3], Shengyuan Lei [1,2] and Chunhua Wei [1,2]

1 School of Resources, Environment and Materials, Guangxi University, Nanning 530004, China; cj15172048283@163.com (J.C.); 13593391579@163.com (W.Z.); 15655513985@163.com (J.L.); lsy@gxu.edu.cn (S.L.); weichunhua@gxu.edu.cn (C.W.)

2 Key Laboratory of High-Performance Structural Materials and Thermal-Surface Processing of Education Department of Guangxi Zhuang Autonomous Region, Guangxi University, Nanning 530004, China

3 Guangxi Key Laboratory of Aluminum Alloy Materials and Processing, Guangxi Nannan Aluminum Processing Co., Ltd., Nanning 530031, China; zhengxu@algig.cn (X.Z.); hekezhun@algig.cn (K.H.); yanghongchi@algig.cn (H.Y.); luxianye@algig.cn (X.L.)

* Correspondence: hqtang@gxu.edu.cn (H.T.); zhaoqizhong@algig.cn (Q.Z.)

Abstract

To address the challenge of controlling Fe impurity content during the recycling of aluminum alloys, this study utilized commercial 5083 aluminum alloy as a matrix to prepare alloy samples with four different Fe contents via smelting. The effects of Fe content on the microstructure, mechanical properties, and corrosion resistance of the as-cast 5083 aluminum alloy were systematically investigated. The results indicate that increasing the Fe content induces a significant morphological evolution of the Fe-rich phases, transitioning from compact Chinese-script α-Al(Fe,Mn)Si phases at low Fe levels to coarse needle-like β-AlFeSi phases. Concurrently, both the quantity and size of the second phases increase significantly. Mechanical testing reveals that the hardness of the alloy gradually rises from 67 HV to 72 HV due to second-phase strengthening. The tensile strength shows a trend of initially increasing and then decreasing, peaking at 0.45 wt.% Fe; however, excessive Fe leads to the formation of needle-like phases that cause stress concentration, resulting in a decline in tensile strength. The elongation decreases gradually with increasing Fe content, with a maximum reduction of 19.7%. Electrochemical tests show that higher Fe content increases the self-corrosion current density and decreases the capacitive loop radius, indicating a significant degradation in the alloy's corrosion resistance. This work provides an experimental basis for the tolerance control of Fe impurities and the performance optimization of recycled 5083 aluminum alloys.

Keywords: 5083 aluminum alloy; iron-rich phase; microstructure; mechanical properties; corrosion resistance

1. Introduction

The global aluminum industry stands at a pivotal juncture in the mid-2020s, driven by an urgent necessity to decarbonize metallurgical supply chains and transition towards a truly circular economy. As of 2025, the demand for aluminum products, particularly in the transportation and marine sectors, has surged, with global consumption projected to increase by over 80% by 2050. This growth is inextricably linked to the "lightweighting" revolution in automotive and aerospace engineering, where high-strength-to-weight ratio materials are essential for improving fuel efficiency and extending the range of electric vehicles. However, primary aluminum production remains energy-intensive, consuming approximately 14,000–17,000 kWh per tonne and generating significant perfluorocarbon and CO_2 emissions [1,2]. However, secondary aluminum production—recycling—consumes only about 5% of the energy required for primary extraction from bauxite ore, releasing approximately 95% fewer greenhouse gases. The integration of recycled material, or "scrap," into high-value wrought alloy production is therefore not merely an economic opportunity but an environmental imperative [3].

5083 aluminum alloy, a typical Al-Mg series corrosion-resistant alloy, has been widely used in aerospace and transportation industries owing to its low density, excellent thermal and electrical conductivity, vibration damping, and electromagnetic shielding properties [4]. However, with the surging demand for aluminum recycling, the development and utilization of recycled aluminum have become a critical pathway for achieving material sustainability. During the recycling process, iron (Fe) is the most difficult impurity element to remove and tends to accumulate. Although the equilibrium solid solubility of Fe in pure aluminum is extremely low (0.05 wt.%) [5], the precipitation behavior in 5083 alloys is far more complex due to the presence of Mn and Si. In this multicomponent system, the phase selection is governed by the competitive formation between the equilibrium α-phase and the metastable β-phase. Importantly, the interaction between Mn, Si, and Fe plays a decisive role. Mn is known to substitute for Fe atoms in the intermetallic lattice, promoting the formation of the compact, body-centered cubic α-Al(Fe,Mn)Si phase. Si is also essential for stabilizing this phase. However, when the Fe content exceeds the capacity of available Mn, the thermodynamic stability shifts. The insufficient Mn/Si interaction leads to the precipitation of the monoclinic β-AlFeSi phase (or Al_3Fe), which shows a detrimental needle-like morphology. Therefore, the microstructural evolution is not merely a function of Fe supersaturation but is dictated by the stoichiometric balance between Fe, Mn, and Si.

Research indicates that the morphology, size, and distribution of Fe-rich phases are critical factors determining alloy performance. Generally, Fe forms polygonal or skeleton-like α-Fe phases (commonly referred to as "Chinese-script" morphology in the literature due to their intricate branching shape), as well as needle-like or platelet-like β-Fe phases in aluminum alloys [6]. Among them, the Chinese-script α-phase shows good bonding with the matrix and is less detrimental to properties, whereas the coarse needle-like β-phase is considered a harmful phase. In terms of mechanical properties, the hard and brittle needle-like β-phase severely severs the continuity of the aluminum matrix. Under stress, it easily causes stress concentration at its tips, acting as a source for crack initiation and accelerating crack propagation, thereby significantly reducing the strength, ductility, and fatigue life of the alloy [7–12]. From a casting perspective, needle-like phases can impede the flow of the melt and induce micro-porosity between dendrites, increasing the porosity rate [13]. In corrosive environments, Fe-rich phases act as cathodes to form micro-galvanic corrosion couples with the aluminum matrix, accelerating matrix dissolution and deteriorating corrosion resistance [14]. Therefore, controlling the morphology and evolution of Fe-rich phases is central to enhancing the performance of recycled aluminum alloys.

While the behaviors of Fe impurities have been extensively studied in Al-Si casting alloys [10] and high-pressure die-cast Al-Mg systems [15], significant knowledge gaps remain regarding the specific evolution of Fe-rich phases in recycled 5083 wrought aluminum alloys. Unlike Al-Si alloys dominated by eutectic Si, or die-cast Al-Mg alloys where Fe is often tolerated to prevent die soldering under rapid cooling, the 5083 system relies heavily on Mn alloying for strengthening under gravity casting conditions. Consequently, the specific evolution mechanism of Fe-rich phases and their correlation with microstructural and

https://doi.org/10.3390/cryst16030192

property degradation in this system have not been comprehensively understood. Distinct from the grain size control in previous works [16], this study systematically investigates the effects of Fe elements on the microstructural evolution and comprehensive properties of 5083 aluminum alloy by synthesizing alloys with varying Fe contents. The aim is to provide a theoretical basis for the tolerance control of composition and impurity phase for high-quality recycled 5083 aluminum alloys.

2. Materials and Methods

Commercial 5083 aluminum alloy and Al-10Fe master alloy were selected as raw materials. The melting process was conducted in a well-type resistance furnace (SG2-7.5-12, Shanghai Shiyan Electric Furnace Co., Ltd., Shanghai, China). The melt temperature was controlled within the range of 750–760 °C. After refining and degassing with C_2Cl_6, followed by a holding period and slag skimming, the melt was cast into molds. The chemical compositions of the commercial 5083 aluminum alloy and the Al-10Fe master alloy are listed in Tables 1 and 2, respectively. To ensure the accuracy of the alloying process, the actual chemical compositions of the as-cast ingots were quantitatively analyzed using XRF spectrometry (S8 TIGER, Bruker, Karlsruhe, Germany). The measured compositions are presented in Table 3. The analysis confirms that the actual Fe contents (0.262, 0.463, 0.664, and 0.871 wt.%) align closely with the nominal design values (0.25, 0.45, 0.65, and 0.85 wt.%).

Table 1. 5083 aluminum alloy chemical composition (wt.%).

Si	Fe	Cu	Mn	Mg	Cr	Ni	Zn	Ti	Na	Al
0.38	0.25	0.03	0.51	4.32	0.11	-	0.04	0.02	-	Balance

Table 2. Chemical composition of AlFe10 intermediate alloy (wt.%).

Si	Fe	Ni	Cu	Zn	Cr	Mn	Ti	Pb	Al
0.12	10.11	-	0.01	0.01	0.01	0.001	-	-	Balance

Table 3. Chemical compositions of the experimental alloys measured by XRF (wt.%).

Alloy Group	Fe (Nominal)	Fe (Actual)
Low-Fe	0.25	0.242
Med-Fe	0.45	0.463
High-Fe A	0.65	0.664
High-Fe B	0.85	0.871

Four distinct experimental alloys were designed to simulate varying degrees of contamination in recycled aluminum streams, ranging from high-purity primary metal equivalents to heavily contaminated scrap:

Alloy A (Baseline): 0.25 wt.% Fe—Represents high-quality recycled or primary 5083.

Alloy B: 0.45 wt.% Fe—Represents the upper limit of many current industrial specifications.

Alloy C: 0.65 wt.% Fe—Represents a "mixed scrap" scenario common in unrefined recycling streams.

Alloy D: 0.85 wt.% Fe—Represents heavily contaminated material, often categorized as "downcycling" grade.

The alloy preparation was conducted in a laboratory-scale well-type resistance furnace using high-density graphite crucibles (SG2-7.5-12, Shanghai Shiyan Electric Furnace Co.,

https://doi.org/10.3390/cryst16030192

Ltd., Shanghai, China), which were coated with a boron nitride wash and pre-fired at 300 °C for 2 h to prevent iron contamination and eliminate volatiles. Commercial 5083 ingots were subsequently melted at a controlled temperature of 750–760 °C to optimize fluidity while minimizing magnesium oxidation; once the melt stabilized, calculated quantities of Al-10Fe master alloy were added and manually stirred with a preheated high-purity graphite rod for 5 min to ensure homogeneity. To remove dissolved hydrogen and non-metallic inclusions, the melt was treated with 0.5 wt.% hexachloroethane (C_2Cl_6) flux and held for 10 min, a method selected for its high laboratory efficiency despite shifting industrial preferences toward inert gas degassing. Finally, after skimming the surface dross, the melt was poured into a permanent steel mold preheated to 250 °C to control cooling rates and prevent defects such as cold shuts, after which the castings were air-cooled to room temperature.

Specimens (15 mm × 15 mm × 5 mm) were sectioned from the geometric center of the solidified ingots to minimize artifacts associated with the chill zone and macro-segregation near the riser, and subsequently mounted in a conductive thermosetting phenolic resin. The samples underwent a standard metallographic preparation sequence, beginning with planar grinding using SiC papers of increasing fineness (240 to 2000 grit) under water lubrication, followed by rough polishing with a 3 mm diamond suspension on a napless cloth, and concluding with a 0.05 mm colloidal silica suspension to achieve a deformation-free mirror finish. To reveal grain boundaries and second-phase intermetallics, the polished surfaces were chemically etched for 10–20 s using Keller's Reagent, immediately followed by warm water rinsing and ethanol drying. Microstructural characterization was performed using optical microscopy (PA53MET, Motic, Xiamen, China) and scanning electron microscopy (TM4000, Hitachi, Tokyo, Japan) at an accelerating voltage of 20 kV, while energy-dispersive X-ray spectroscopy (Rigaku D/MAX 2500V, Rigaku, Tokyo, Japan) was employed for semi-quantitative elemental analysis to discriminate between Al-Fe-Mn-Si and Al-Fe-Si phases.

Tensile tests were conducted in accordance with GB/T 16865-2023 [17]. Standard tensile specimens with a gauge length of 20 mm were extracted from the exact center of the ingot using wire electrical discharge machining, with specific dimensions illustrated in Figure 1. After sectioning, the specimens were cleaned with kerosene and anhydrous ethanol to remove surface contaminants, followed by light grinding with sandpaper to eliminate machining marks. Tensile properties of alloys with different Fe contents were evaluated at room temperature using an Instron 8801 universal testing machine (Instron, Norwood, MA, USA), with a crosshead speed set to 1 mm/min. For each alloy composition, three parallel specimens (n = 3) were tested to ensure statistical reliability. The reported tensile strength and elongation values represent the average of these three measurements, with error bars indicating the standard deviation. Microhardness testing was performed using an NVM-1000A nano/micro indentation hardness tester (Shanghai Lunjie Electromechanical Instrument Co., Ltd., Shanghai, China). The samples were ground and polished to a mirror finish. To ensure statistical reliability and minimize the influence of microstructural heterogeneity, seven uniformly distributed indentations (n = 7) were performed on each sample. The reported hardness values represent the mean ± standard deviation of these seven measurements, providing a robust assessment of the alloy's mechanical properties. The test parameters were set as follows: a load duration of 15 s and an applied load of 0.5 kgf. Experimental data were analyzed using Origin software (Version 2024, OriginLab Corporation, Northampton, MA, USA) to generate a plot illustrating the variation in hardness of the 5083 aluminum alloys with different iron contents.

Electrochemical performance testing was conducted using a CHI750E electrochemical workstation (Shanghai Chenhua Instrument Co., Ltd., Shanghai, China) configured with a standard three-electrode cell. A 3.5 wt.% NaCl solution served as the corrosive electrolyte. The aluminum alloy sample acted as the working electrode, a platinum plate as the counter

https://doi.org/10.3390/cryst16030192

electrode, and a saturated calomel electrode (SCE) as the reference electrode. The exposed testing area was 1 cm^2. Electrochemical impedance spectroscopy (EIS) measurements were first performed at the open circuit potential, with an applied sinusoidal perturbation signal frequency sweeping from 10^{-2} Hz to 10^5 Hz and an amplitude of 5 mV. This was followed by potentiodynamic polarization curve testing, conducted at a constant scan rate of 1 mV/s. The entire testing procedure was carried out at room temperature. The acquired EIS data and polarization curves were subsequently fitted and analyzed.

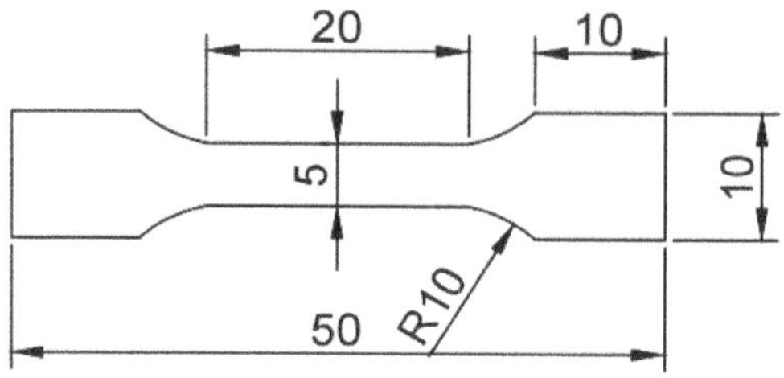

Figure 1. Size specification of tensile specimen.

3. Discussion

3.1. Microstructure of the As-Cast Alloy

The as-cast phases of the 5083 aluminum alloy consist primarily of the Al matrix, (Fe,Mn)Al_6, Al_3Mg_2, and a small amount of Mg_2Si [16]. In this study, four groups of alloys with varying Fe contents (0.25 wt.%, 0.45 wt.%, 0.65 wt.%, and 0.85 wt.%) were prepared. Their corresponding as-cast metallographic microstructures are presented in Figure 2a–d, all obtained at a magnification of 200×.

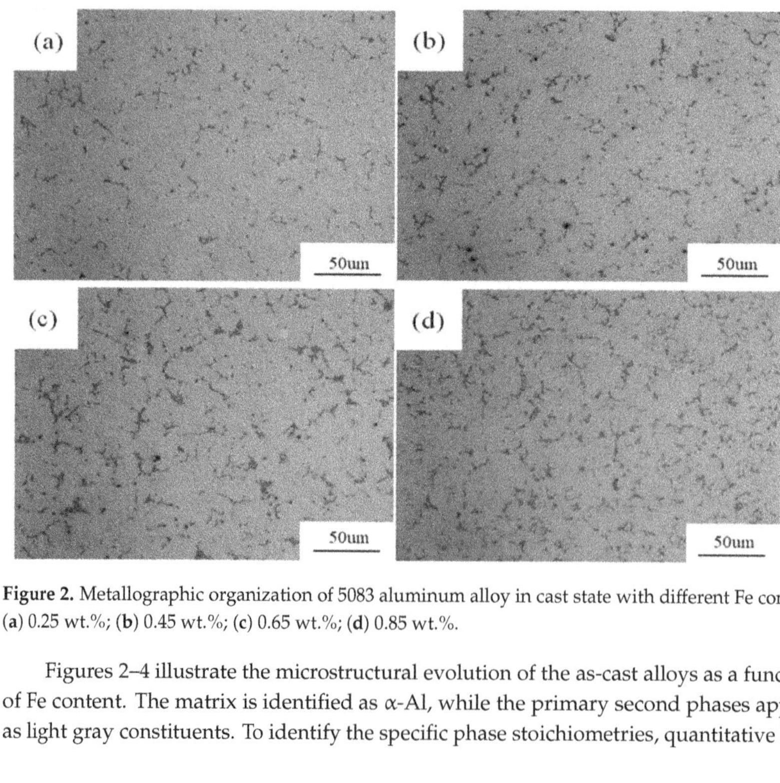

Figure 2. Metallographic organization of 5083 aluminum alloy in cast state with different Fe content: (**a**) 0.25 wt.%; (**b**) 0.45 wt.%; (**c**) 0.65 wt.%; (**d**) 0.85 wt.%.

Figures 2–4 illustrate the microstructural evolution of the as-cast alloys as a function of Fe content. The matrix is identified as α-Al, while the primary second phases appear as light gray constituents. To identify the specific phase stoichiometries, quantitative EDS

https://doi.org/10.3390/cryst16030192

point analysis was performed on the marked regions in Figure 3, and the detailed chemical compositions are listed in Table 4.

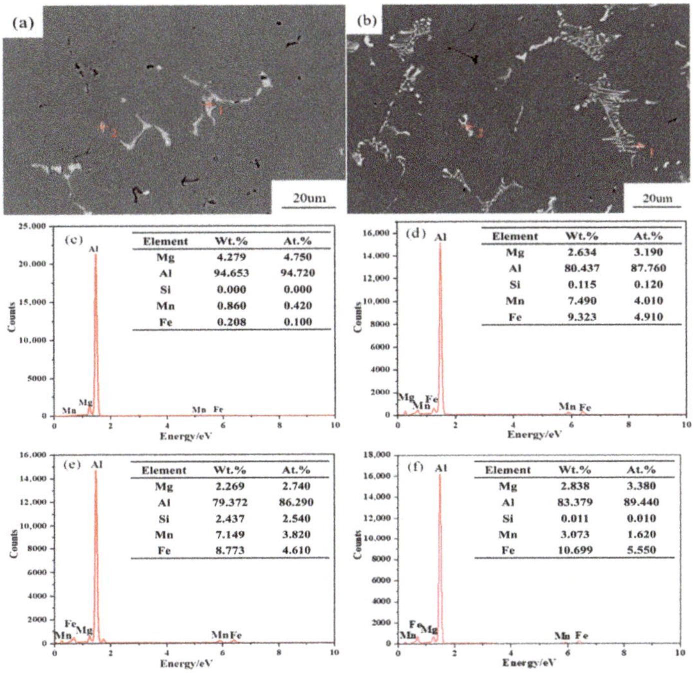

Figure 3. SEM morphologies and specific EDS phase analyses of the as-cast 5083 aluminum alloys: (**a**) SEM image of the alloy with 0.25 wt.% Fe, showing predominantly Chinese-script phases; (**b**) SEM image of the alloy with 0.85 wt.% Fe, showing a mix of coarse needle-like and skeleton-like phases; (**c**,**d**) EDS spectra corresponding to Point 1 and Point 2 marked in (**a**), confirming the composition of the Mn-rich α-Al(Fe,Mn)Si phases; (**e**,**f**) EDS spectra corresponding to Point 1 and Point 2 marked in (**b**). Note that Point 1 retains a higher Mn/Fe ratio, whereas Point 2 shows a high Fe concentration characteristic of the needle-like β-AlFeSi phase, evidencing the phase transformation mechanism.

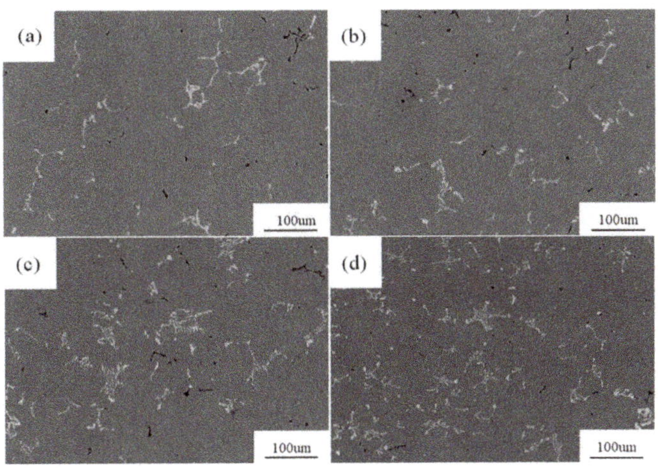

Figure 4. Scanning images of 5083 aluminum alloy with different Fe content in cast state: (**a**) 0.25 wt.%; (**b**) 0.45 wt.%; (**c**) 0.65 wt.%; (**d**) 0.85 wt.%.

https://doi.org/10.3390/cryst16030192

Table 4. Chemical composition of the typical phases in as-cast 5083 alloys analyzed by EDS (corresponding to points in Figure 3).

Alloy (wt.% Fe)	Point Location	Mg	Al	Si	Fe	Mn
0.25 wt.%	1	4.75	94.72	0.00	0.10	0.42
0.25 wt.%	2	3.19	87.76	0.12	4.91	4.01
0.85 wt.%	1	2.74	86.29	2.54	4.61	3.82
0.85 wt.%	2	3.38	89.44	0.01	5.55	1.62

At a low Fe content (0.25 wt.%), the Fe-rich phases show irregular Chinese-script or skeleton-like morphologies. As shown in Table 4 (Point 2), this phase contains 4.91 wt.% Fe and 4.01 wt.% Mn, yielding an Fe/Mn mass ratio of approximately 1.2. This ratio is consistent with the stoichiometry of the α-Al(Fe,Mn)Si phase, which is relatively compact and causes minimal disruption to the continuity of the matrix.

As the Fe content increases to 0.85 wt.%, a large amount of coarse needle-like phases precipitate within the microstructure. According to Table 4 (Point 2), these needle-like phases show a significantly higher Fe content (5.55 wt.%) and lower Mn content (1.62 wt.%), resulting in an increased Fe/Mn ratio of 3.4. This distinct chemical composition identifies them as the β-phase. This morphological and chemical evolution suggests that excessive Fe induces a transition of the Fe-rich phases from the favorable α-phase to the detrimental needle-like β-phase. It is worth noting that the base alloy contains 0.38 wt.% Si (Table 1), which plays a fundamental role in the phase selection of the 5083 system. In Al-Mg alloys, the presence of Si is the prerequisite for the formation of ternary or quaternary Al-Fe-Si-(Mn) intermetallics rather than binary Al-Fe phases. The Si content was constant in all samples. However, it provided the thermodynamic environment for the precipitation of both α-Al(Fe,Mn)Si and β-AlFeSi phases. At low Fe levels (0.25 wt.%), the Si interacts with the abundant Mn to stabilize the compact, body-centered cubic α-phase. However, as the Fe content increases to 0.85 wt.%, the relative surplus of Fe (and the local depletion of Mn) drives the reaction with Si to form the monoclinic β-AlFeSi phase. Therefore, the phase evolution observed in this study is a result of the synergistic interaction between the increasing Fe content and the available Si, regulated by the Mn/Fe ratio.

The quantity of the gray Fe-containing phases in the metallographic structure increases visibly with increasing Fe content. Although quantitative phase fraction analysis was not performed, the significant morphological transition from compact α-phase to coarse, acicular β-phase (as confirmed by the EDS ratios in Table 4) is considered the dominant microstructural factor responsible for the deterioration of mechanical properties, outweighing the influence of phase volume fraction alone.

3.2. *Effect of Iron Content on the Properties of Aluminum Alloys*

3.2.1. Microhardness

As illustrated in Figure 5, the hardness of the aluminum alloy increased with Fe content. The hardness value increases from 67 HV at 0.25 wt.% Fe to 72 HV at the maximum Fe content of 0.85 wt.%, representing an increment of 7.5%. These results suggest that Fe content exerts a significant influence on the hardness of the alloy. The observed improvement is mainly due to the second-phase strengthening mechanism.

Since the maximum solid solubility of Fe in pure aluminum is limited to 0.05% [18], Fe predominantly exists in the form of alloy phases. In the presence of alloying elements or additional impurities at room temperature, the solubility of Fe decreases further to approximately 0.01% [18]. Consequently, during solidification, a significant amount of Fe precipitates as intermetallic compounds, serving as second-phase particles, which are

https://doi.org/10.3390/cryst16030192

typically Al-Fe and Al-Mg-Fe phases. These Fe-rich phases inherently possess higher hardness than the aluminum matrix, acting as hard obstacles to dislocation motion [15]. As the Fe content increases, the volume fraction and density of these hard second-phase particles within the matrix rise significantly. These hard phases, distributed dispersedly or in a reticular pattern, effectively impede dislocation slip and motion, thereby improving the overall hardness of the alloy. Although a minor amount of solid solution Fe atoms may induce lattice distortion, second-phase strengthening plays the dominant role under the high-Fe conditions examined in this study.

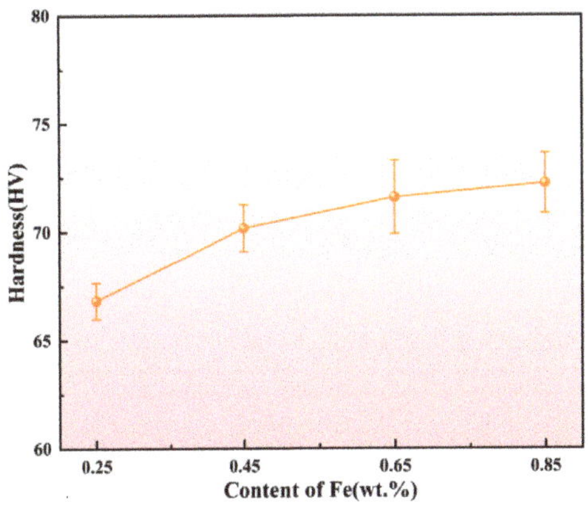

Figure 5. Hardness distribution.

3.2.2. Tensile Properties

Figure 6a shows that the tensile strength increased at first and then decreased with adding Fe. At 0.25 wt.% Fe, the tensile strength was 262.3 MPa and reaches a peak value of 267.5 MPa as the Fe content increases to 0.45 wt.%. However, with a further increase in Fe content to 0.65 wt.% and 0.85 wt.%, the tensile strength begins to decline, dropping from 267.5 MPa to 250.5 MPa, and finally to 239.3 MPa. At lower Fe contents (0.25–0.45 wt.%), the improvement in tensile strength is primarily attributed to the second-phase strengthening mechanism. During this stage, the precipitated fine Fe-rich phases are dispersedly distributed within the aluminum matrix, effectively hindering dislocation motion. However, as the Fe content increases further, the Fe-rich phases gradually coarsen and evolve into the β-phase with a long needle-like structure. These coarse, brittle phases severely sever the continuity of the matrix and act as crack initiation sites during tensile deformation, resulting in a sharp deterioration in both strength and ductility [15]. As the tensile strength gradually decreases from 267.5 MPa to 239.3 MPa with continuing Fe addition, a critical threshold is identified at 0.45 wt.% Fe. Beyond this limit, the strengthening effect of the second phases is overridden by the stress concentration induced by coarse needle-like β-phases, resulting in a significant deterioration of mechanical properties.

As illustrated in Figure 6b, the elongation of the aluminum alloy shows a continuous downward trend with increasing Fe content. The elongation reaches a maximum of 18.3% at 0.25 wt.% Fe. Subsequently, it decreases to 16.9% at 0.45 wt.% Fe (a reduction of 7.6%), 15.8% at 0.65 wt.% Fe (a reduction of 13.8%), and finally drops to a minimum of 14.7% at 0.85 wt.% Fe, representing a total reduction of 19.7%.

At elevated Fe levels, Fe-rich phases precipitate within the as-cast 5083 aluminum alloy. These phases, characterized as hard and brittle constituents, possess inherently poor plasticity. This observation aligns with the 'pore nucleation theory' [19], which emphasizes that the β-AlFeSi phase acts as a heterogeneous nucleation site during solidification. These

https://doi.org/10.3390/cryst16030192

phases impede melt feeding, thereby promoting the formation of solidification defects such as shrinkage cavities and porosity. During subsequent tensile deformation, these brittle phases act as stress concentration sites, facilitating crack initiation and thereby reducing the elongation of the alloy [20]. At higher Fe contents, Fe-rich intermetallic compounds—particularly the needle-like phases—induce significant stress concentration in their vicinity under applied stress, which triggers crack initiation and accelerates crack propagation. The porosity and gas bubbles associated with these Fe-rich phases within the matrix reduce the effective load-bearing area of the recycled aluminum alloy. This exerts a detrimental effect on the ultimate tensile strength, yield strength, and elongation. The impact on elongation is significantly more pronounced than on tensile or yield strength [21,22].

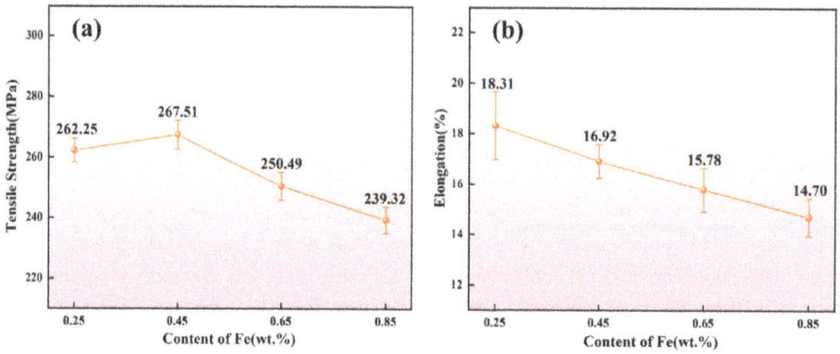

Figure 6. Tensile strength and Elongation: (**a**) Tensile strength; (**b**) Elongation.

To deeply elucidate the mechanism behind the peak performance at 0.45 wt.% Fe, the role of the Mn/Fe ratio is critical. The experimental alloy contains 0.51 wt.% Mn. At 0.25 wt.% Fe, the Mn/Fe mass ratio is approximately 2.0, which effectively promotes the formation of the compact α-Al(Fe,Mn)Si phase. Interestingly, the peak tensile strength observed at 0.45 wt.% Fe corresponds to a Mn/Fe ratio of 1.1. This suggests that a Mn/Fe ratio near 1.1 serves as a critical thermodynamic threshold for the 5083 system. Above this Fe level, the available Mn is insufficient to modify the Fe-rich phases, triggering the dominant precipitation of coarse needle-like β-AlFeSi phases. In addition, compared to the standard industrial limit for Fe in 5083 alloys, our findings indicate that recycled 5083 alloys can tolerate a slightly higher Fe content (up to 0.45 wt.%) without compromising mechanical integrity, provided the Mn content is sufficient to maintain the Mn/Fe ratio above unity. This finding offers a quantifiable tolerance window for the utilization of high-Fe scrap in secondary aluminum production.

3.2.3. Corrosion Resistance

Fe content exerts a significant influence on the corrosion resistance of the alloy. During the solidification of the aluminum alloy, the precipitation of Fe leads to the formation of various intermetallic compounds. These second phases show a distinct potential difference relative to the aluminum matrix, which can induce micro-galvanic corrosion [14], thereby exacerbating the tendency for localized corrosion. Fe-rich phases typically grow with needle-like or platelet-like morphologies, disrupting the continuity of both the aluminum matrix and the surface oxide film, which further impairs the corrosion resistance of the material [23]. Figure 7 presents the potentiodynamic polarization curves of the four groups of 5083 aluminum alloy specimens with varying Fe contents in a 3.5 wt.% NaCl solution. As observed from the Tafel curves, the cathodic branches of the different samples are remarkably similar, indicating consistency in their cathodic reduction behavior. However, significant differences exist in the anodic branches, suggesting that the samples show distinct anodic corrosion behaviors. Additionally, it is evident that all four groups

https://doi.org/10.3390/cryst16030192

display a passivation region followed by a breakdown region. Table 5 lists the electrochemical parameters from the polarization curves and EIS. As the Fe content increases from 0.25 wt.% to 0.85 wt.%, the corrosion current density, which serves as a kinetic indicator of the corrosion rate, shows a distinct upward trend, rising from 2.753×10^{-2} μA·cm^{-2} to 4.699×10^{-2} μA·cm^{-2}. This indicates a progressive degradation in the corrosion resistance of the alloy matrix. Importantly, this trend is corroborated by the electrochemical impedance spectroscopy (EIS) analysis. The charge transfer resistance (R_t), extracted from the equivalent circuit modeling, represents the energy barrier for the corrosion reaction at the electrolyte/electrode interface. As listed in Table 5 the Rt values decrease monotonically from 1.837×10^4 Ω·cm^2 to 1.271×10^4 Ω·cm^2 with increasing Fe content. According to the Stern-Geary relationship, the reduction in Rt facilitates the electron transfer process, directly accounting for the observed increase in I_{corr}. This consistency between the thermodynamic (E_{corr}) and kinetic (R_t) parameters confirms that the potential difference between the noble Fe-rich phases and the α-Al matrix promotes micro-galvanic coupling, thereby accelerating the dissolution of the aluminum matrix.

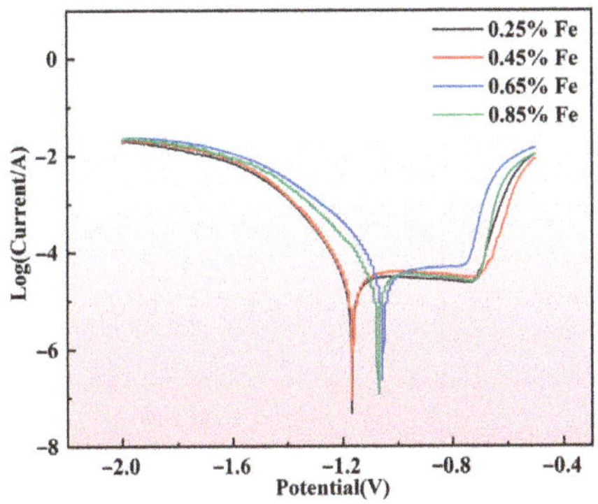

Figure 7. Tafel curves of different samples.

Table 5. Electrochemical corrosion parameters of 5083 alloys derived from potentiodynamic polarization curves and EIS spectra in 3.5 wt.% NaCl solution.

Samples	I_{corr} (μA·cm^{-2})	E_{corr} (V)	Cat.Slp (V^{-1})	Ano.Slp(V^{-1})	R_t (Ω·cm^2)
0.25 wt.%Fe	2.753×10^{-2}	−1.169	8.961	1.541	1.837×10^4
0.45 wt.%Fe	3.224×10^{-2}	−1.17	8.729	1.881	1.692×10^4
0.65 wt.%Fe	4.002×10^{-2}	−1.056	8.164	1.808	1.566×10^4
0.85 wt.%Fe	4.699×10^{-2}	−1.07	6.977	1.046	1.271×10^4

To accurately analyze the electrochemical behavior of the samples with varying Fe contents in a 3.5 wt.% NaCl solution, the equivalent circuit model illustrated in Figure 8 was employed for fitting the Electrochemical Impedance Spectroscopy (EIS) data. In this model, R_s represents the solution resistance between the working electrode and the reference electrode, R_c denotes the resistance of the metal oxide film, and R_t corresponds to the charge transfer resistance. The R_t value reflects the resistance encountered by electron transfer during the electrochemical reaction; generally, a higher charge transfer resistance indicates a greater tendency for the metal to undergo polarization (implying higher corrosion resistance). Due to the porous structure of the metal oxide film, a Constant Phase Element (CPE), denoted as Q_c, is utilized to characterize the electrochemical behavior at the interface between the electrolyte and the metal oxide film. Similarly, Q_t is employed to characterize the electrochemical response at the interface between the metal oxide film and

https://doi.org/10.3390/cryst16030192

the substrate matrix. Based on this equivalent circuit, Figure 9 presents the experimental and fitted EIS spectra (Nyquist and Bode plots). The experimental data were fitted using ZSimpWin software (Version 3.60, EChem Software, Ann Arbor, MI, USA) to ensure the reliability of the parameters.

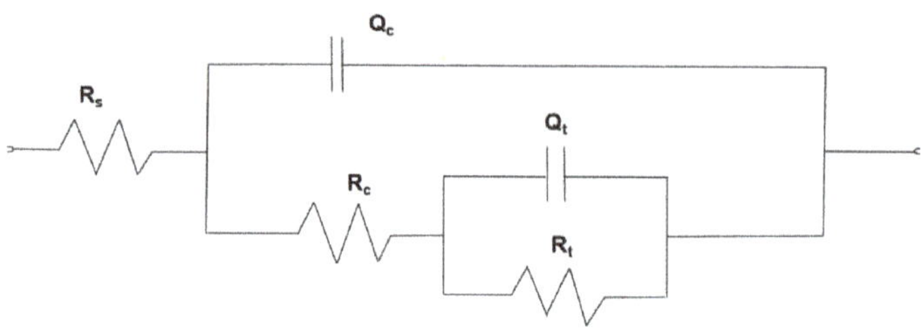

Figure 8. Electrochemical test impedance equivalent circuit mode.

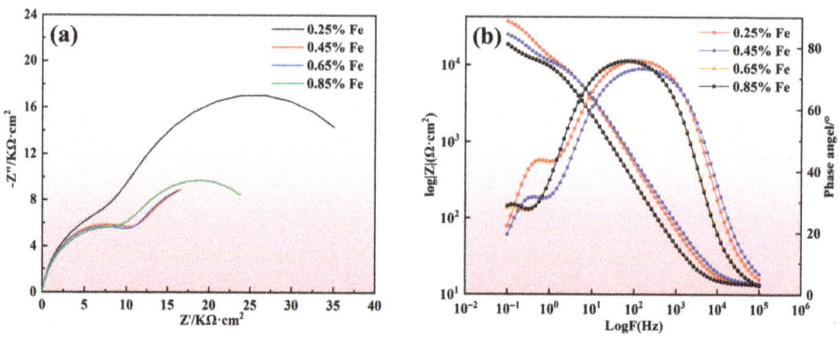

Figure 9. Impedance spectra of different samples: (**a**) Nyquist; (**b**) Bode.

Observation of the Nyquist data reveals that all specimens show capacitive behavior during the immersion corrosion process. In the Nyquist plots, a larger diameter of the capacitive loop corresponds to superior corrosion resistance. Concurrently, it is observed that as the Fe content increases from 0.25 wt.% to 0.85 wt.%, the diameter of the capacitive loop decreased continuously. This indicates that adding Fe reduced the corrosion resistance in the corrosion resistance of the alloy. Correspondingly, in the Bode plots, a higher impedance modulus value (|Z|) at low frequencies indicates better corrosion resistance. As shown in Figure 9b, the specimen with 0.25 wt.% Fe shows the highest |Z| value at low frequencies, and |Z| follows a decreasing trend with increasing Fe content. This electrochemical degradation is visually confirmed by the corrosion morphologies shown in Figure 10. The corrosion morphology presented in Figure 10 clearly reveals that corrosion pits preferentially nucleate along the periphery of Fe-rich phases and propagate deep into the matrix, ultimately resulting in severe localized corrosion. When the Fe content reaches 0.65 wt.%, distinct changes in the morphology and quantity of Fe-rich phases are visible. The Fe-rich phases transition from initially sparse blocky or Chinese-script shapes into coarse needle-like and fish-bone morphologies. Concurrently, these fish-bone structures tend to fragment and evolve into numerous independent needle-like phases. The initiation of intergranular corrosion cracks is clearly visible, with the majority of these cracks clustered in the vicinity of the intermetallic phases. When the Fe content reaches 0.85 wt.%, deep pits and cavities are clearly evident beneath the Fe-rich phases. Therefore, as the Fe content increases, these phenomena become increasingly pronounced, accompanied by the emergence of distinct intergranular corrosion cracks, which serves as a manifestation of the deteriorated corrosion resistance.

https://doi.org/10.3390/cryst16030192

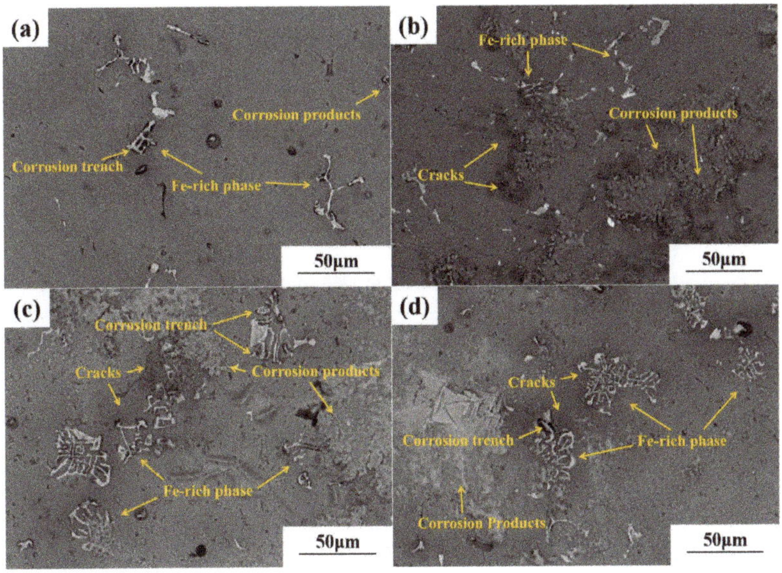

Figure 10. Electrochemical corrosion morphology of 5083 aluminum alloy with different Fe content in cast state: (**a**) 0.25 wt.%; (**b**) 0.45 wt.%; (**c**) 0.65 wt.%; (**d**) 0.85 wt.%.

3.3. Implications for Downstream Processing

Although 5083 aluminum alloy is primarily utilized in wrought tempers rather than the as-cast state, the microstructural quality of the initial ingot is decisive for its downstream processability. The coarse, needle-like β-AlFeSi phases identified in the high-Fe alloys (≥0.65 wt.%) possess inherent brittleness and distinct hardness mismatch with the α-Al matrix. During subsequent hot or cold rolling processes, these phases cannot deform coordinately with the matrix; instead, they tend to fracture and decohere from the matrix, serving as nucleation sites for voids and cracks. This 'hereditary effect' significantly increases the risk of edge cracking and surface defects during rolling, severely limiting the hot workability of recycled ingots. Therefore, strictly controlling the Fe content below the critical threshold of 0.45 wt.% (or maintaining a Mn/Fe ratio > 1.1) in the as-cast stage is not only essential for mechanical integrity but also a prerequisite for ensuring the smooth production of high-quality wrought sheets. This study provides a foundational guideline for optimizing the impurity tolerance of recycled 5083 ingots prior to deformation processing.

4. Conclusions

(1) In the 5083 aluminum alloy, the quantity of Fe-rich phases increases continuously with rising Fe content. Since these Fe-rich phases impede lattice slip and dislocation motion, thereby enhancing the crystal hardness, the hardness value increases from 67 HV to 72 HV as the Fe content rises from 0.25 wt.% to 0.85 wt.%.

(2) The increase in Fe content significantly alters the morphology and composition of the second phases in the 5083 aluminum alloy. At low Fe contents, the Fe-rich phases predominantly appear as the Chinese-script α-Al(Fe,Mn)Si phase; conversely, at high Fe contents, these phases gradually transform into the coarse, needle-like β-AlFeSi phase.

(3) The Fe content exerts a dual influence on the mechanical properties of the 5083 aluminum alloy. At low Fe levels (≤0.45 wt.%), second-phase strengthening plays a dominant role, enhancing the alloy's strength with increasing Fe content. Con-

versely, when the Fe content is excessive (>0.45 wt.%), the detrimental effects of coarse, needle-like Fe-rich phases become pronounced, resulting in a significant decline in both tensile strength and elongation. Elongation is the most sensitive parameter to Fe content, with a maximum reduction reaching 19.7%.

(4) As the Fe content increases from 0.25 wt.% to 0.85 wt.%, the corrosion resistance of the 5083 aluminum alloy gradually decreases. The number of corrosion-induced cracks and pits increases progressively, and the cracks predominantly propagate along the periphery of the Fe-rich phases.

Author Contributions: Conceptualization, J.C. and H.T.; methodology, J.C.; validation, J.C., W.Z. and J.L.; investigation, J.L. and X.Z.; resources, J.C.; data curation, H.T.; writing—original draft preparation, J.C.; writing—review and editing, J.C.; visualization, X.L. and Q.Z.; supervision, H.Y., K.H., S.L. and C.W.; project administration, H.T.; funding acquisition, H.T. All authors have read and agreed to the published version of the manuscript.

Funding: This research was funded by Guangxi Science and Technology Major Program (AA23062027) (AA24206035), National Undergraduate Innovation and Entrepreneurship Training Program (S202510593003) (S202510593199).

Data Availability Statement: The data presented in this study are available on request from the corresponding author.

Acknowledgments: The authors gratefully acknowledge the Guangxi Key Laboratory of Processing for Non-ferrous Metals and Featured Materials, Guangxi University, for their support and for providing the experimental facilities used in this research.

Conflicts of Interest: Xu Zheng, Kezhun He, Qizhong Zhao, Hongchi Yang, and Xianye Lu were employed by the Guangxi Nannan Aluminum Processing Co., Ltd. The remaining authors declare that the research was conducted in the absence of any commercial or financial relationships that could be construed as a potential conflict of interest.

References

1. International Aluminium Institute. Global Aluminium Cycle. 2016. Available online: https://international-aluminium.org/ (accessed on 10 February 2026).
2. Ritchie, H.; Rosado, P.; Roser, M. CO_2 and Greenhouse Gas Emissions. *Our World in Data*. 2023. Available online: https://ourworldindata.org/co2-and-greenhouse-gas-emissions/ (accessed on 10 February 2026).
3. Raabe, D.; Ponge, D.; Uggowitzer, P.J.; Roscher, M.; Paolantonio, M.; Liu, C.; Antrekowitsch, H.; Kozeschnik, E.; Seidmann, D.; Gault, B.; et al. Making sustainable aluminum by recycling scrap: The science of "dirty" alloys. *Prog. Mater. Sci.* **2022**, *128*, 100947. [CrossRef]
4. Zhou, K.; Jia, X.-L.; Zhang, H.; Zhang, X.; Sun, H. Experimentation and Optimization of Integrated CNC Milling for Thin-Walled 5083 Aluminum-Magnesium Alloy Components. *Equip. Manuf. Technol.* **2018**, *9*, 116–120+127.
5. Jiang, L.; Marceau, R.K.W.; Dorin, T. Improved Iron-Tolerance in Recycled Aluminum Alloys via Direct Strip Casting Process. *J. Sustain. Metall.* **2024**, *10*, 435–441. [CrossRef]
6. Hatch, J.E. *Aluminum: Properties and Physical Metallurgy*; ASM International: Metal Park, OH, USA, 1984.
7. Liang, K.Y.; Huang, H.C.; Tseng, C.Y.; Chen, M.C.; Lee, S.L.; Lin, C.C.; Su, T.C. Effects of Fe, Si, Cu, and TiB_2 Grain Refiner Amounts on the Hot Tearing Susceptibility of 5083, 6061, and 7075 Aluminum Ingots. *Metals* **2024**, *14*, 15. [CrossRef]
8. Qian, W.-T.; Song, F.-C.; Ge, S.-J.; Wang, J.; Sun, F.-F.; Han, G.-P. Microstructural Characteristics and Chemical Composition of Typical Inclusions in Cast Aluminum Alloys. *China Foundry* **2023**, *72*, 153–159.
9. Zhou, Y.-L.; Zhang, J.-C.; Lin, S.-P.; Zhong, G. Effect of Iron Content on Thermal Conductivity and Mechanical Properties of Cast Aluminum Alloys. *Spec. Cast. Nonferrous Alloys* **2021**, *41*, 416–419.
10. Sun, Y.-Z.; Yu, C.; Li, S.-C.; Qiu, H.-X.; Zhang, G.-W. Analysis of the Morphology and Effects of Iron in Aluminum-Silicon Alloys. *China Foundry* **1998**, *7*, 42–46.
11. Chen, X.-W.; Kuang, Z.-Q.; Wang, S.-C.; Fu, Y.-C.; Fan, W.-Z. Research on Melt Purification, Refinement, and Modification Technologies for Recycled Aluminum Alloys. *Eng. Technol. Res.* **2023**, *8*, 4–7.

https://doi.org/10.3390/cryst16030192

12. He, X.-F.; Liu, B.; Zhang, S.-G. Current Status of Control Technologies for Fe Impurities in Recycled Aluminum Alloys. *Chem. Ind. Eng. Prog.* **2021**, *40*, 5251–5269.
13. Yu, M.-X.; Fan, C.; Zhang, J.; Wang, T.-Z.; Wei, W.; Tang, Q.-C. The Effect of Fe Content on the Microstructure and Properties of Recycled Aluminum-Silicon Alloys. *Hot Work. Technol.* **2024**, *53*, 108–113+117.
14. Ambat, R.; Davenport, A.J.; Scamans, G.M.; Afseth, A. Effect of iron-containing intermetallic particles on the corrosion behaviour of aluminium. *Corros. Sci.* **2006**, *48*, 3455–3471. [CrossRef]
15. Zhu, X.; Blake, P.; Dou, K.; Ji, S. Strengthening die-cast Al-Mg and Al-Mg-Mn alloys with Fe as a beneficial element. *Mater. Sci. Eng. A-Struct. Mater. Prop. Microstruct. Process.* **2018**, *732*, 240–250. [CrossRef]
16. Huang, L.; Zhao, Q.-Z.; Yang, H.-C.; He, K.-Z. Comparative Study of Different Grain Refiners on Grain Refinement of 5083 Aluminum Alloy Ingots. *Equip. Manuf. Technol.* **2023**, *8*, 17–20.
17. *GB/T 16865-2023*; Test Pieces and Method for Tensile Testing for Wrought Aluminum and Magnesium Alloys Products. Standardization Administration of the People's Republic of China: Beijing, China, 2023.
18. Chanyathunyaroj, K.; Moonrin, N.; Laungsopapun, G.; Phetchcrai, S. Corrosion Fatigue Study of 6061 Aluminum Alloy: The Effect of Coatings on the Fatigue Characteristics. *Metall. Mater. Trans. A* **2022**, *53*, 2874–2889. [CrossRef]
19. Zhang, H.-T.; Tang, W.; Wang, D.-T.; Liu, X.-T.; Li, X.-Z.; Changhai, B.-W. Research Progress on the Detrimental Effects and Removal Methods of Fe in Aluminum Alloys. *Light Alloy Fabr. Technol.* **2021**, *49*, 7–14.
20. Liu, Y.; Luo, L.; Han, C.; Ou, L.; Wang, J.; Liu, C. Effect of Fe, Si and Cooling Rate on the Formation of Fe- and Mn-rich Intermetallics in Al–5Mg–0.8Mn Alloy. *J. Mater. Sci. Technol.* **2016**, *4*, 305–312. [CrossRef]
21. Zhang, Y.; Wang, T.; Li, T. Crystallization Behavior of Iron-Containing Intermetallic Compounds in Al-Si Alloy Under Ultrasonic Treatment. *Light Met.* **2019**, *6*, 1631–1635.
22. Musza, A.; Ugi, D.; Vida, Á.; Chinh, N.Q. Study of Anodic Film's Surface and Hardness on A356 Aluminum Alloys. Using Scanning Electron Microscope and In-Situ Nanoindentation. *Coatings* **2022**, *12*, 1528. [CrossRef]
23. Meng, D.; Xia, D.-Z.; Fan, Y.-H.; Xue, S.; Yan, L.-L. Study on the Influence Mechanism of Fe on Microstructure and Properties of Al-Si Recycled Aluminum Alloys and Their Control Technologies. *Hot Work. Technol.* **2024**, *53*, 7–13.

https://doi.org/10.3390/cryst16030192

crystals

MDPI

Article

Micro-Alloying with Samarium to Enhance the Thermal Conductivity and Mechanical Properties of Al-5Ni Cast Alloys

Yi Huang [1], Kang Wang [2], Zhisheng Zhu [1], Wenxue Cheng [1] and Wenfang Li [2,*]

[1] Analytical and Testing Center, Dongguan University of Technology, Dongguan 523000, China; huangyi@dgut.edu.cn (Y.H.); zhuzs@dgut.edu.cn (Z.Z.); chengwx@dgut.edu.cn (W.C.)

[2] School of Materials Science and Engineering, Dongguan University of Technology, Dongguan 523000, China; wangk@dgut.edu.cn

* Correspondence: mewfli@163.com; Tel./Fax: +86-18802566988

Abstract

The research focuses on how the samarium (Sm) addition amount affects the microstructural development, mechanical characteristics, and thermal conductivity of Al-5wt%Ni alloys. Microstructural characterization revealed that minor Sm additions concurrently refined α-Al grains and eutectic structures (α-Al + Al_3Ni), attributed to the reduced initial nucleation temperature of α-Al and an adsorption inhibition effect on Al_3Ni phase formation. The experimental findings revealed that the integration of Sm improved the yield strength, maximum tensile strength, and heat conduction. The enhancements were chiefly driven by three processes: the refinement of grains, the improvement of the eutectic structure's morphology, and diminishing the dissolvability of Fe/Si impurities within the α-Al matrix. The combined improvement of mechanical and thermal characteristics via regulated Sm addition offers a hopeful approach for the microstructural design of Al-Ni alloys. This study provides a crucial basis for the production of Al-Ni alloys that possess high thermal conductivity and satisfactory mechanical properties.

Keywords: Al-5Ni alloy; micro-alloying element Sm; microstructure; mechanical properties; thermal conductivity

1. Introduction

The rapid progress in power semiconductor technology and electric vehicle battery development has led to a growing demand for lightweight structural materials with superior thermal conductivity and strength [1,2]. Aluminum alloys are widely used in the field of thermal management due to their excellent thermal conductivity, lightweight, and good machinability [3,4]. However, traditional cast aluminum alloys such as Al-Si, Al-Cu, and Al-Mg systems exhibit relatively low thermal conductivity (well below 200 W/(m·K)), which makes it increasingly difficult to meet the thermal management requirements of high-power electronic devices [5]. Traditional cast aluminum alloys, owing to their high fluidity, are well-suited for manufacturing complex structural thermal management components. However, the high solid solubility of elements such as Si, Mg, and Cu in the aluminum lattice typically limits their thermal conductivity [6]. Therefore, it is of great significance to develop a novel casting aluminum alloy that possesses both high strength and good thermal conductivity.

The solid solubilities of Ni, Co, and Fe in aluminum alloys are extremely low [7–9], making them promising to be major alloying elements. Luo et al. [9] found that the excellent thermal conductivity of the Al-2Fe-0.3Co hypereutectic alloy reaches 235 W/(m·K), which is attributed to the low solid solubility of iron and cobalt in the aluminum lattice. However, the Al-2Fe alloy exhibits insufficient fluidity during casting. Al-Ni-based alloys possess a relatively high eutectic composition and low solid solubility of Ni, endowing them with excellent formability and thermal conductivity. These alloys are well-suited for manufacturing structural components requiring specific thermophysical properties [7,10,11]. Despite this, the Al-Ni eutectic alloy's microstructure consists of a rough eutectic form (α-Al + Al_3Ni), leading to a decline in its mechanical characteristics [12]. Modification represents an economical and highly efficient approach for regulating the morphologies and distributions of precipitates that form during the solidification process [13–16]. Therefore, through appropriate modification, it is feasible to enhance both the thermal conductivity (TC) and mechanical properties of Al-Ni-based alloys concurrently.

Aluminum alloys are usually altered by adding a minor amount of rare earth elements like Er, Ce, Sm, among others. Pozdniakov et al. [17] found that Ni and rare earth elements (Y, Er, Yb, Gd) can reduce the volume fraction of the Al_2Cu phase in Al-Cu alloys at 180–250 °C, achieving a good combination of tensile strength and electrical conductivity. For Al-Ni alloys, Suwanpreecha et al. [18] introduced scandium (Sc) into near-eutectic Al-Ni alloys. The study found that the precipitated Al_3Sc phase contributes to enhancing the material's strength. However, the research also pointed out that the solubility of Sc in the aluminum matrix may have an adverse effect on electrical and thermal conductivity (ETC) performance. There were some reports indicating that adding Sm element would not make the TC of aluminum alloys worse as its solubility in α-Al matrix was low [19]. Rao et al. [16] mentioned that introducing Sm might alter both the rough acicular eutectic Si and the needle-shaped β-Al_5FeSi phase in ADC12 alloy by lowering the eutectic melting point. Qiu et al. [20] investigated that Sm had a remarkable effect on the microstructures of eutectic Si, which was transformed from coarse plates to fine and fibrous. The alloy of Al-7Si-0.7Mg, altered with Sm, exhibited an excellent mix of maximum tensile strength (UTS) and stretching.

Present studies are motivated by the scarcity of data in the existing literature on how Sm addition affects the microstructures, mechanical characteristics, and thermal conductivity of almost eutectic Al-5Ni alloys. In this research, the intention was to figure out the inner mechanism of Sm modification and explore the relationship between thermal conductivity and Sm content. Through comprehending how Sm modification influences the microstructure and properties and their mutual relationship, alloy design and processing procedures can be refined for better performance. The study assists in making educated decisions regarding the alteration of Al-Ni-based alloys to produce heat-dissipating materials with enhanced mechanical characteristics and heat transfer efficiency.

2. Materials and Methods

2.1. Preparation of Samples

The permanent mold casting was conducted with the Al-5Ni and the Al-5Ni-xSm (x = 0.1, 0.2, 0.3, 0.5) alloys. Initial melting was performed in a resistance furnace by heating 99.7% purity aluminum ingots (containing 0.08 wt.% Si and 0.05 wt.% Fe impurities) to 1023 K. The Al-10Ni master alloy was then added to reach the desired 5 wt.% Ni composition, followed by 5 min of mechanical stirring to ensure melt homogenization. Controlled additions of Al-20Sm master alloy were then incorporated to produce the designed Sm-containing samples. The ultimate molten material was moved into a steel mold that had been preheated to 473 K to solidify. The casting is processed into a tensile specimen.

2.2. Measurements

The metallographic specimens were first ground in stages with abrasive papers and then polished mechanically to obtain smooth, mirror-like surfaces. Etching was carried out using a 0.5 vol.% solution of hydrofluoric acid. Microstructural analysis involved three different techniques: optical microscopy (OM, Leica DMI3000, Wetzlar, Germany), field-emission scanning electron microscopy (FE-SEM, Thermo Fisher Verios G4, Waltham, MA, USA), and energy-dispersive X-ray spectroscopy (EDS, Oxford x-Max, Oxfordshire, UK). Phase composition was analyzed using X-ray diffraction (XRD, Bruker D8 Advance, Karlsruhe, Germany) with Cu-K radiation, at a scan rate of 5°/min. For tensile tests, the specimens were tested at room temperature with an Instron 5982 testing machine, maintaining a constant strain speed of 0.5 mm/min. The as-cast Al-5Ni-xSm alloys were machined into disk-shaped specimens (φ12.7 × 3 mm^2) for thermal property characterization. The Netzsch LFA457 laser flash analyzer was used to measure thermal diffusivity at room temperature. The densities of the samples were ascertained using the Archimedes method, employing deionized water as the immersion medium. The Neumann–Kopp rule was employed to determine specific heat capacities [21,22]. Consequently, we calculated the thermal conductivity (λ) by employing this specific formula:

$$\lambda = \alpha \cdot \rho \cdot C_p \quad (1)$$

Here, α represents the thermal diffusivity (mm^2/s), ρ denotes the density (g/cm^3), and C_p signifies the specific heat capacity (J/(g·K)). The total measurement uncertainty for thermal conductivity was maintained below 5%. To ensure experimental reproducibility, triplicate measurements were conducted for each specimen, with results reported as the statistical mean. Furthermore, the samples' electrical conductivity was gauged through the vortex technique utilizing the PZ-60A apparatus.

Compared with DSC, the cooling curve thermal analysis (CCTA) technique [23,24], with dynamic temperature change as its core, is more suitable for industrial process simulation and macroscopic characterization of non-equilibrium solidification behavior. Therefore, in this study, a CCTA device was used to collect the cooling curves of molten Al-5Ni-xSm alloy samples. Prior to the experiment, K-type thermocouples were positioned at the center and edge of the melt for temperature collection. During solidification, the temperature evolution was recorded using K-type thermocouples (measurement accuracy: ±0.01 °C) and an NI 9212 high-speed data acquisition system.

3. Results and Discussion

3.1. Microstructure

Figure 1 illustrates the evolution of the microstructure in Al-5Ni alloys with different levels of samarium addition. The base alloy (Figure 1a) exhibits characteristic dendritic α-Al grains, while incremental Sm incorporation induces progressive microstructural refinement. As can be seen from Figure 1, a distinct trend of secondary dendrite arm spacing (SDAS) reduction can be observed. The secondary dendrite arm spacing (SDAS) of Al-5Ni-xSm alloys was measured by using a digital image analysis software (Nano measurement) [13], is presented in Figure 2. As depicted in this figure, the secondary dendrite arm spacing (SDAS) measures 45.7 ± 8.2 μm in the absence of Sm. When the additional amount of samarium is 0.1 wt.%, the SDAS decreases markedly to 23.4 ± 4.4 μm. Upon increasing the Sm content to 0.2 wt.%, the SDAS shows a slight increment to 25.4 ± 3.8 μm, followed by no discernible change at 0.3 wt.% Sm. An increase in SDAS to 27.5 ± 3.9 μm is observed when the Sm content reaches 0.5 wt.%. Results indicate that the SDAS reaches the minimum value when the addition amount of Sm reaches 0.1 wt.%. However, this effect gradually diminishes as the Sm content continues to increase. This microstructural refinement, accompanied by homogenization of α-Al and eutectic structure (α-Al + Al_3Ni) morphology, provides favorable conditions for mechanical property enhancement. The observed microstructural modification indicates that Sm plays a distinct role as a phase morphology modifier in the Al-Ni alloy system.

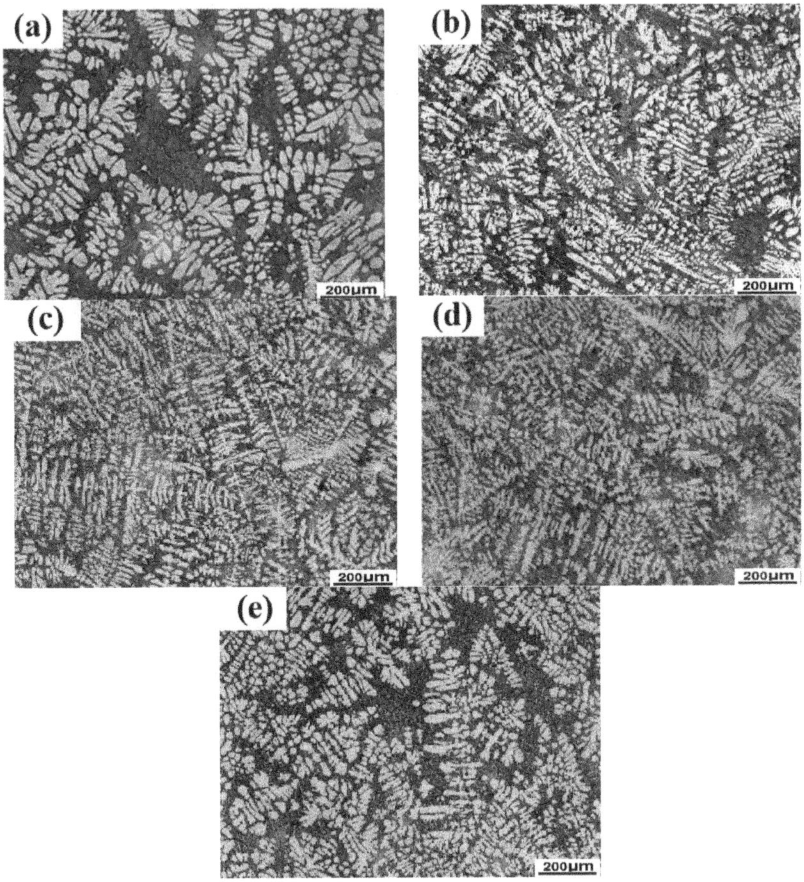

Figure 1. Microstructures of Al-5Ni alloys modified with different Sm additions: (**a**) 0 wt.%; (**b**) 0.1 wt.%; (**c**) 0.2 wt.%; (**d**) 0.3 wt.%; (**e**) 0.5 wt.%.

X-ray diffraction analysis of Al-5Ni-xSm alloys (Figure 3) revealed consistent α-Al and Al_3Ni phase formation across all compositions, with Sm-containing variants exhibiting additional Al_3Sm intermetallic phases (Figure 3a). High-resolution examination of Al_3Ni's strongest diffraction peaks (Figure 3b) showed invariant peak positions regardless of Sm content, indicating preserved lattice parameters. This crystallographic stability suggests Sm primarily interacts through surface segregation at Al_3Ni interfaces rather than bulk solid solution formation. The combined XRD evidence supports a microstructure evolution mechanism involving independent Al_3Sm precipitation alongside unmodified Al_3Ni phase formation.

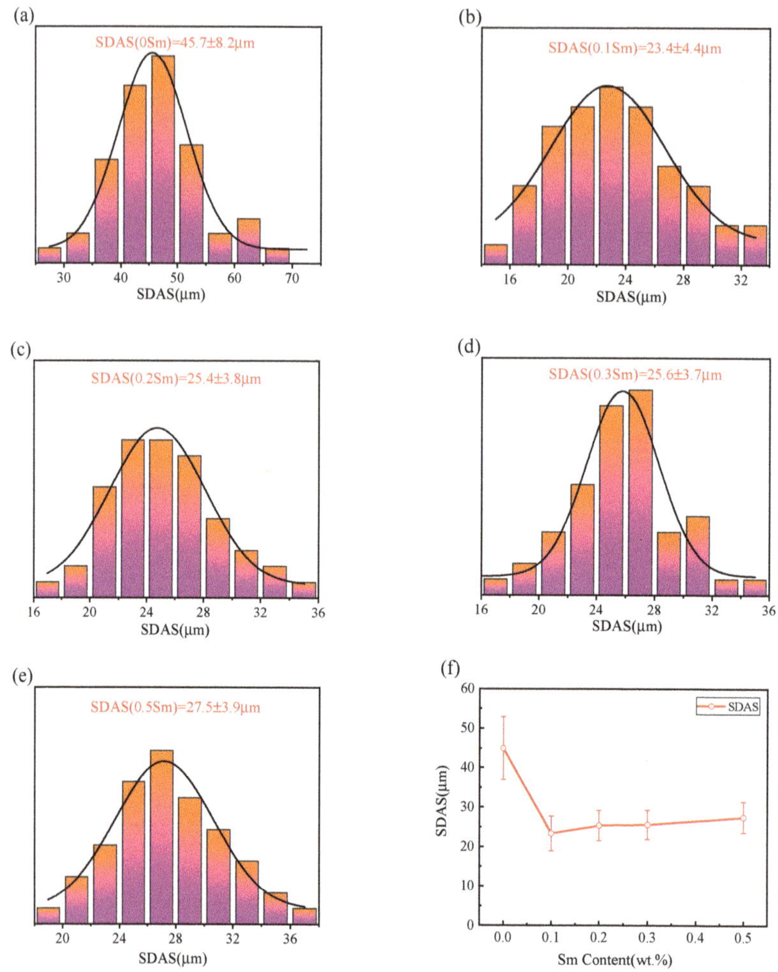

Figure 2. The secondary dendrite arm spacing (SDAS) of Al-5Ni-xSm alloys: (**a**) 0 wt.%; (**b**) 0.1 wt.%; (**c**) 0.2 wt.%; (**d**) 0.3 wt.%; (**e**) 0.5 wt.%. (**f**) The variation trend of SDAS with different Sm contents.

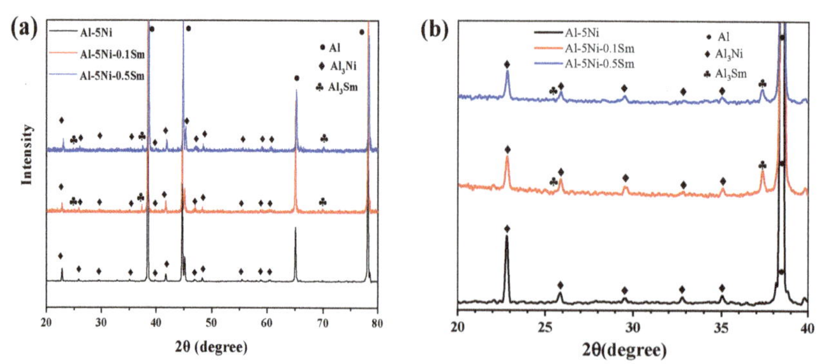

Figure 3. (**a**) X-ray diffraction patterns of as-cast Al-5Ni-*x*Sm alloys and (**b**) the zoomed-in XRD spectra in the angular range 20~40°.

3.2. Crystallization

Figure 4 presents the cooling curves of Al-5Ni alloys with varying Sm concentrations, accompanied by their corresponding first and second derivative curves. The solidification parameters are tabulated in Table 1. The base Al-5Ni alloy exhibited α-Al nucleation temperature (**T_N**) at 656.8 °C, eutectic growth temperature (**T_{EG}**) at 640.4 °C, and solidus temperature at 621.6 °C. Sm-modified alloys displayed significant solidification characteristic parameter modifications, including a secondary exothermic peak following the (α-Al + Al_3Ni) eutectic reaction (Figure 4b,c). This appearing peak, attributed to the formation of the Al_3Sm phase as confirmed by XRD analysis and the Al-Sm phase diagram [19], shows that the nucleation temperatures of the Al_3Sm phase are 615.2 °C and 606.3 °C for the 0.1 wt.% and 0.5 wt.% Sm-containing alloys, respectively. The progressive intensity enhancement of this peak with increasing Sm content reflects elevated Al_3Sm phase fraction.

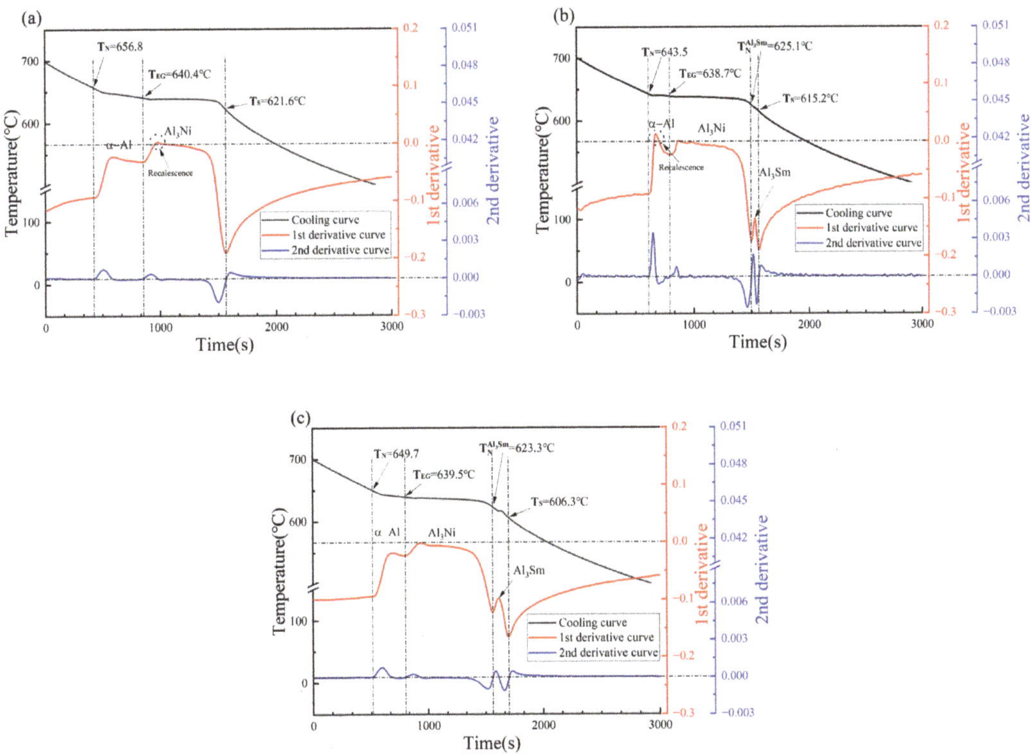

Figure 4. The cooling curves with corresponding first and second derivative curves of (**a**) Al-5Ni alloy, (**b**) Al-5Ni-0.1Sm alloy, (**c**) Al-5Ni-0.5Sm alloy.

Table 1. The solidification characteristic parameters of Al-5Ni alloys containing different contents of Sm.

Sm Content (wt.%)	T_N (°C)	T_{EG} (°C)	$T_N^{Al_3Sm}$ (°C)	T_S (°C)	ΔT (°C)
0	656.8	640.4	-	621.6	35.2
0.1	643.5	638.7	625.1	615.2	28.3
0.5	649.7	639.5	623.3	606.3	43.4

Analysis of solidification parameters (Table 1) reveals a non-monotonic dependence of α-Al nucleation temperature (**T_N**) on Sm concentration in Al-5Ni-xSm alloys. **T_N** decreases from 656.8 °C (0 wt.% Sm) to 643.5 °C (0.1 wt.% Sm), followed by a marginal increase to 649.7 °C at higher Sm levels. And the nucleation temperature of the eutectic structure (**T_{EG}**) exhibits the same trend. The decrease in the initial nucleation temperature implies a greater degree of undercooling to promote the nucleation, thereby refining the grain size.

The most interesting results obtained from the first-derivative curves show that recalescence occurs in both the Al-5Ni alloy and Al-5Ni-0.1Sm alloy (see Figure 4a,b for details). Recalescence is a specific thermal effect observed during the solidification of metals or alloys, manifesting as a temporary temperature rise on the cooling curve [24]. Fundamentally, it results from the latent heat released during solidification, which causes a transient local temperature increase. The presence of recalescence is indicated by the region where the first-order derivative of the cooling curve is greater than zero. In the Al-5Ni alloy, recalescence occurs at the onset of the (α-Al + Al_3Ni) eutectic reaction. This indicates that during the initial stage of the eutectic reaction, nuclei of the eutectic structure form and grow rapidly, releasing substantial latent heat of crystallization. When the latent heat release rate exceeds the system's heat dissipation rate, a localized temporary temperature rise creates a "recalescence peak". During recalescence, the undercooling decreases as temperature rebounds, reducing the nucleation driving force and slowing the crystallization rate, thereby forming a large amount of coarse and brittle eutectic structures.

The addition of Sm elements to the Al-5Ni melt can eliminate the recalescence phenomenon during the eutectic reaction. This elimination helps hinder the development of (α-Al + Al_3Ni) eutectic structures, thereby leading to the refinement of eutectic morphology. During solidification, Sm undergoes solute redistribution, accumulating preferentially at solid–liquid interfaces. This interfacial segregation enhances liquid-phase compositional gradients, increasing composition undercooling that promotes eutectic nucleation while simultaneously restricting eutectic growth through solute redistribute effects. The coupled mechanisms of nucleation site multiplication and diffusion-limited growth kinetics facilitate the development of refined, uniformly distributed eutectic microstructures.

Meanwhile, a pronounced recalescence phenomenon of α-Al was also observed in the Al-5Ni-0.1Sm alloy. Due to the addition of 0.1% Sm, the initial nucleation temperature (**T_N**) decreases from 656.8 °C to 643.5 °C, generating a higher degree of undercooling. When the undercooling reaches the critical value, massive nucleation and growth of α-Al occur, releasing substantial latent heat of crystallization. However, compared with the Al-5Ni alloy, the growth duration of α-Al is significantly shortened, leading to finer and more uniform α-Al in the Al-5Ni-0.1Sm alloy. As shown in Figure 4c, the recalescence phenomenon of both α-Al and eutectic structures disappears in the Al-5Ni-0.5Sm alloy. Compared with the Al-5Ni-0.1Sm alloy, the growth duration of α-Al increases and its nucleation temperature rises, which leads to the coarsening of α-Al.

The difference (**ΔT**) between the liquidus temperature (**T_N**) and solidus temperature (**T_S**) is the non-equilibrium solidification range of the alloy melt. As shown in Table 1, the addition of 0.1wt% Sm to the Al-5Ni alloy reduces the non-equilibrium solidification range of the melt from 35.2 °C to 28.3 °C, indicating improved fluidity of the alloy. However, when the Sm content increases to 0.5 wt.%, the non-equilibrium solidification range expands to 43.4 °C, leading to deteriorated fluidity of the melt.

3.3. Effect of Sm Addition on Mechanical Properties

Figure 5 displays the tensile stress–strain profiles and mechanical property statistics for Al-5Ni alloys with varying Sm concentrations. The base alloy (0 wt.% Sm) demonstrates yield strength (YS) of 93.2 ± 1.5 MPa, ultimate tensile strength (UTS) of 130.5 ± 2.2 MPa, and elongation (EL) of 19.5 ± 0.5%. Sm-modified alloys exhibit significant mechanical enhancement, with optimal performance achieved at 0.1 wt.% Sm: YS = 125.5 ± 1.4 MPa (+34.7%), UTS = 172.4 ± 2.5 (+32.1%), and EL = 20.0 ± 0.5% (+2.5%). Further Sm addition to 0.5 wt.% induces progressive property reduction (YS = 112.5 ± 1.3 MPa, UTS = 158.4 ± 1.9 MPa, EL = 18.5 ± 0.5%), demonstrating a concentration-dependent strengthening mechanism. This non-monotonic response suggests an optimal Sm threshold for mechanical performance enhancement in the Al-Ni system. Combining the microstructural observations in Figure 1 and the cooling curve thermal analysis, the addition of Sm reduces the nucleation temperatures of α-Al and eutectic structures, increases the degree of undercooling, and promotes nucleation. Meanwhile, the solute redistribution effect hinders the growth of eutectic structures, which is corroborated by the variation trend of SDAS in Figure 2. Sm addition facilitates the refinement of dendritic α-Al grains and optimization of the size, morphology, and distribution of eutectic structures, thereby significantly enhancing the material strength. Excessive incorporation of Sm induces a poisoning effect, suppressing the effect of lowering the nucleation temperature. This leads to the coarsening of eutectic structures (as illustrated in Figure 1e), thereby causing a gradual decline in mechanical properties.

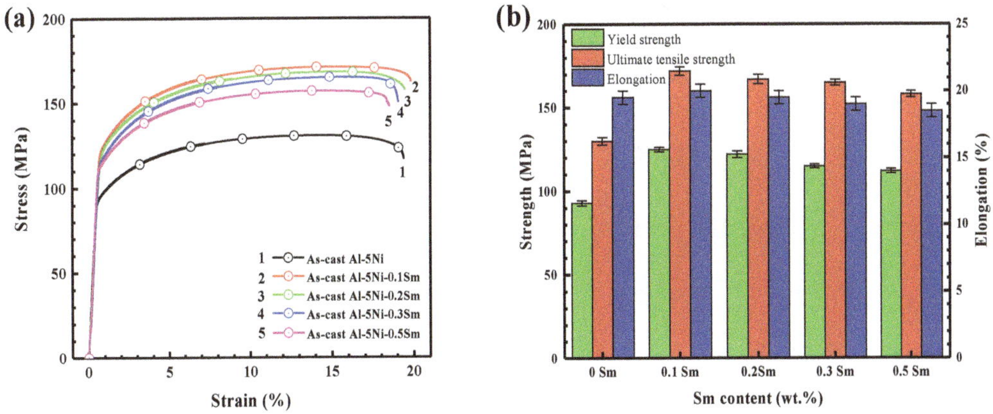

Figure 5. (**a**) Engineering stress–strain curves and (**b**) mechanical performance statistics of the Al-5Ni alloys with different Sm contents.

This research determines the Q value, a holistic quality metric, to evaluate the strength and elongation of Al-5Ni-xSm alloys. The Q value is denoted in the form of flow:

$$Q = UTS + a \times \log(\text{Elongation}\%) \quad (2)$$

For aluminum alloy, the value of *a* is 150 MPa [25,26]. For Al-5Ni alloys containing Sm at concentrations of 0%, 0.1%, 0.2%, 0.3%, and 0.5%, the respective Q values are 323.5 MPa, 367.2 MPa, 360.5 MPa, 356.8 MPa, and 348.1 MPa. Clearly, the Q value for the Al-5Ni-0.1Sm alloy stands at 367.2 MPa, marking it as the highest in the sample set. The figure is approximately 13.5% higher compared to the Sm-free alloy. Therefore, incorporating 0.1% Sm may lead to an impressive combination of strength and elongation performance.

Figure 6 illustrates the tensile fracture morphologies of Al-5Ni-xSm alloys. Examining the structure of tensile fractures allows for a clearer understanding of how these alloys break. In the sample without Sm (as clearly shown in Figure 6a), distinct quasi-cleavages and tear ridges are observable. These features are characteristic of brittle fracture. Moreover, the rod-like eutectic Al_3Ni phases are readily identifiable on the fracture surface. The experimental findings align well with the relatively inferior tensile performance of the Sm-free alloy. For the Al-5Ni-0.1Sm alloy, Figure 6b vividly illustrates how its fracture surface displays multiple small, uniformly distributed dimples. This morphological feature indicates a transition in the fracture mechanism from brittle to ductile. As the Sm concentration escalates, the dimples grow rougher. Additionally, the fracture surfaces display visible tear ridges, as illustrated in Figure 6c,d, which suggests a mixed fracture morphology characterized by quasi-cleavage and micropore aggregation. The findings from the analysis of the tensile fracture morphology align well with the results of the tensile experiments. Of all the alloys examined, the Al-5Ni-0.1Sm alloy exhibits the most advantageous tensile characteristics.

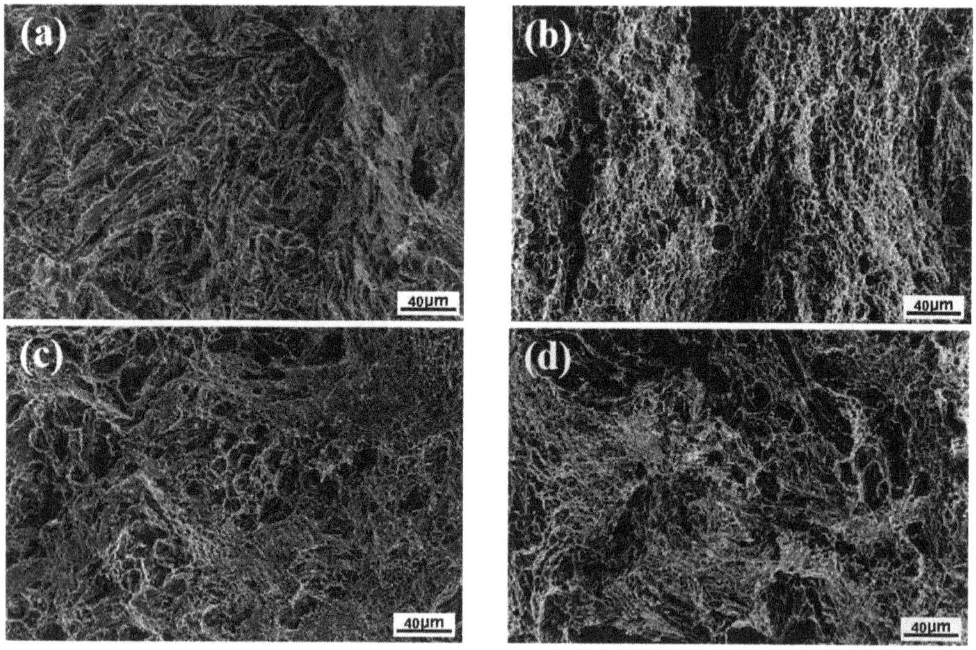

Figure 6. Tensile fracture patterns of Al-5Ni alloys with different Sm contents: (**a**) 0 wt.%Sm, (**b**) 0.1 wt.%Sm, (**c**) 0.3 wt.%Sm, (**d**) 0.5 wt.%Sm.

The mechanical performance of near-eutectic Al-5Ni-xSm alloys is governed by the morphological characteristics and spatial distribution of the (α-Al + Al_3Ni) eutectic architecture. In Sm-free alloys (Figure 1a), the eutectic structure exhibits a coarse, irregular morphology with uneven dispersion, creating stress concentration sites that detrimentally impact mechanical integrity. This interconnected network of brittle Al_3Ni phases facilitates preferential crack propagation pathways under applied loads, as evidenced by the progressive elongation reduction at elevated Sm concentrations. Sm addition modifies the eutectic structure through morphological refinement and homogenized dispersion (Figure 1b–d), enhancing load transfer efficiency while mitigating stress concentrations. The resultant microstructure optimization, achieved through Sm-induced eutectic modification, demonstrates a strengthening–toughening synergy that significantly improves tensile performance despite minor ductility trade-offs at higher Sm levels. This microstructural engineering approach effectively decouples conventional strength–ductility conflicts in near-eutectic Al-Ni systems through controlled interface design.

3.4. Effect of Sm Addition on the Thermal Conductivity

Figure 7a presents the thermophysical properties of Al-5Ni-xSm alloys, illustrating a linear decrease in specific heat capacity from 0.878 J/(g·K) (0 wt.% Sm) to 0.874 J/(g·K) (0.5 wt.% Sm), attributed to the comparatively low heat capacity of Sm J/(g·K). Concurrently, density increased from 3.009 g/cm^3 to 3.033 g/cm^3 due to Sm's higher atomic mass. Figure 7b depicts the properties of thermal diffusivity, electrical conductivity, and thermal conductivity. The thermal conductivities are calculated based on thermal diffusivity, heat capacity, and density, in accordance with Equation (1). It is observable that the changes in electrical and thermal conductivity of Al-5Ni-xSm alloys, when varying in Sm content, are similar to the fluctuations in thermal diffusivities. Initially, there is an increase in the electrical and thermal conductivities, alongside a rise in thermal diffusivity, followed by a gradual decline. The alloy without Sm exhibits the minimal levels of thermal diffusivity and electrical and thermal conductivity, recorded at 82.55 $\pm$ 0.38 mm^2/s, 31.33 $\pm$ 0.15 MS/m, and 218.1 $\pm$ 1.0 W/(m·K), respectively. The alloy of Al-5Ni-0.1Sm exhibits maximum thermal diffusivity, electrical and thermal conductivity, achieving 82.93 $\pm$ 0.29 mm^2/s, 31.50 $\pm$ 0.11 MS/m, and 219.3 $\pm$ 0.8 W/(m·K). The data indicates a rough rise of 0.5%, 0.5%, and 0.6% in comparison to the original Al-5Ni alloy. With an increase in Sm content to 0.5%, there is a gradual reduction in thermal diffusivity, electrical conductivity, and thermal conductivity, dropping to 81.56 $\pm$ 0.39 mm^2/s, 31.08 $\pm$ 0.15 MS/m, and 216.3 $\pm$ 1.0 W/(m·K).

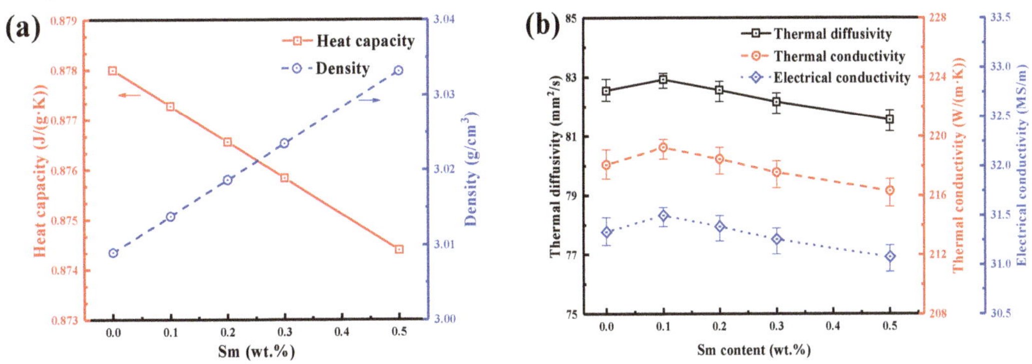

Figure 7. (**a**) The heat capacity and density of Al-5Ni-xSm alloys, (**b**) thermal diffusivities and thermal/electrical conductivities of Al-5Ni-xSm alloys.

Research on Al-5Ni-xSm alloys indicates that their increased heat transfer rate is due to the poor solubility of Ni and Sm within the α-Al matrix. Sm exhibits moderate capability for microstructural modification, and the variation in thermal conductivity is intricately correlated with the evolution of the alloy's microstructure. The trace additions of Sm refine (α-Al + Al_3Ni) eutectic structures, concomitantly increasing electron conduction pathways. However, excessive Sm incorporation (0.5 wt.%) gives rise to eutectic coarsening (as shown in Figure 1e), thereby deteriorating thermal conductivity. Previous studies [27] confirm that rare earth elements purify the α-Al matrix through impurity segregation effects. However, detailed interfacial chemistry analysis (to be presented subsequently) is required to fully elucidate these conductivity enhancement mechanisms.

3.5. Promotion Mechanism of Sm Addition on Al-5Ni Alloys

The SEM results, as provided in Figure 8, disclose the effect of samarium addition on the compositions and morphologies of secondary phases in the Al-5Ni alloy. The morphologies of secondary phases in the Al-5Ni alloy are displayed in Figure 8a. As seen, two kinds of morphologies of secondary phases can be found in the alloy. The secondary phase near the boundary of the α-Al grain, marked as phase 'A' and termed as 'Eutectic boundary phase', is flake-like. The needle-like phases, labeled as phase 'B' and termed as 'Eutectic internal phase', are far from the boundary of the α-Al grain. These secondary phases with different morphologies are extensively generated in the Al-Ni hypoeutectic alloy with different preparation methods [28,29].

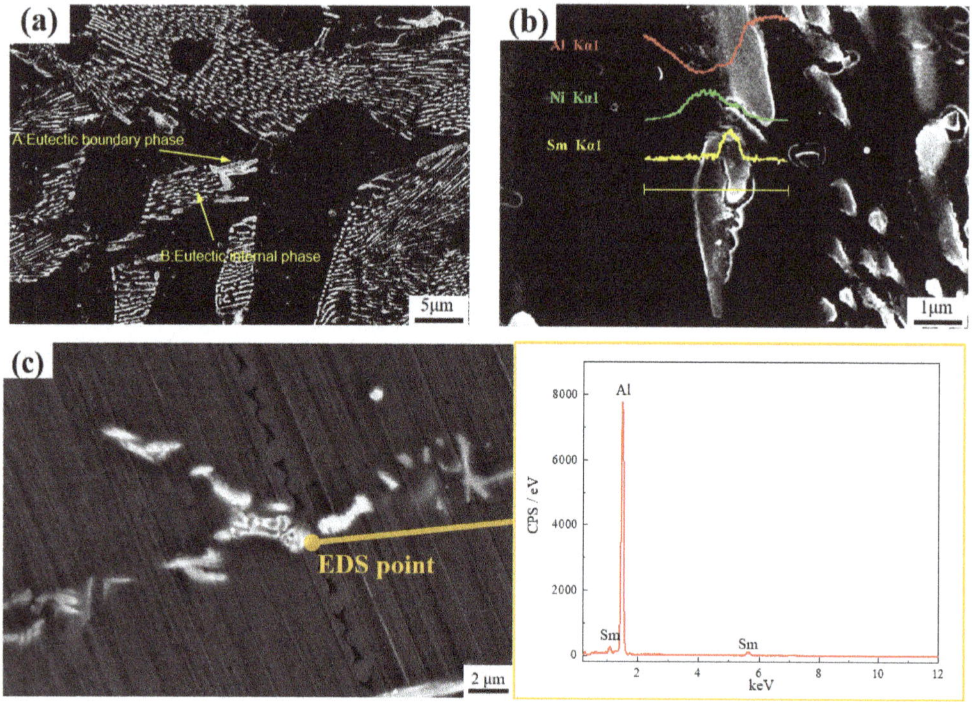

Figure 8. The SEM observation (**a**) Al-5Ni, (**b**) Al–5Ni-0.1Sm alloys with corresponding results of line scanning, (**c**) the precipitation of Al-Sm phase in the Al-5Ni-0.5Sm alloy.

The morphological difference between the eutectic internal phase and eutectic boundary phase is closely connected to the solidification process. Previous studies [30,31] establish 27.6 vol.% as the critical secondary phase fraction for morphological transitions. For this study, the theoretical volume fraction of the Al_3Ni phase, about 8.3 vol.%, is computed by the thermodynamic software JmatPro 7.0. During the early phase of solidification, the Ni element's concentration remains fairly consistent, indicating that the Al_3Ni phase's phase volume fraction is less than 27.6 vol.%. Observing the rod or needle-shaped Al_3Ni phase within the eutectic grain is straightforward. During the last phase of solidification, the residual liquid exhibits a high Ni element concentration due to the redistribution of solutes and non-equilibrium solidification. This occurrence reveals that the Al_3Ni phase's related phase volume fraction exceeds 27.6 vol.%. Hence, the eutectic boundary phase manifests a morphology resembling flakes.

Introducing the Sm element might lead to the creation of a flake-like eutectic boundary phase due to adsorption poisoning. Highly active Sm element will be enriched at the boundary of the liquid/solid interface, inhibiting the migration and diffusion of Ni atoms. Therefore, there are more remaining liquids with high Ni concentration, which facilitates the formation of flake-like eutectic boundary phase. The relative line scanning results for the secondary phase in the Al-5Ni-0.1Sm alloy are provided in Figure 8b. The experimental results demonstrate that the Sm element mainly distributes alongside the flake-like eutectic boundary phase. Combined with the XRD analysis, the Sm-rich layer may be the Al_3Sm phase. In other words, the flake-like Al_3Ni phases are coated with Al_3Sm phases. Figure 8c displays the morphology of the Al-Sm phase. By referring to the XRD patterns, the Al-Sm phase can be determined as Al_3Sm, indicating that the excessive addition of samarium will lead to the precipitation of Al_3Sm in the Al-5Ni matrix.

Figure 9 presents EPMA elemental mapping of the Al-5Ni alloy. Illustrated in Figure 9c,d, the minimal presence of Fe and Si impurities renders these elements almost undetectable in the initial alloy, a fact supported by the experimental findings. Figure 10 depicts the EPMA mapping outcomes for the Al-5Ni-0.1Sm alloy. Importantly, Figure 10c,d distinctly show clusters of Fe and Si located at the edges of grains. The finding suggests that introducing a minor quantity of the Sm element markedly reduces the solubility of Fe and Si impurities in the α-Al matrix. Furthermore, an examination of Figure 10e makes it evident that Sm predominantly accumulates in the region surrounding the eutectic Al_3Ni phase.

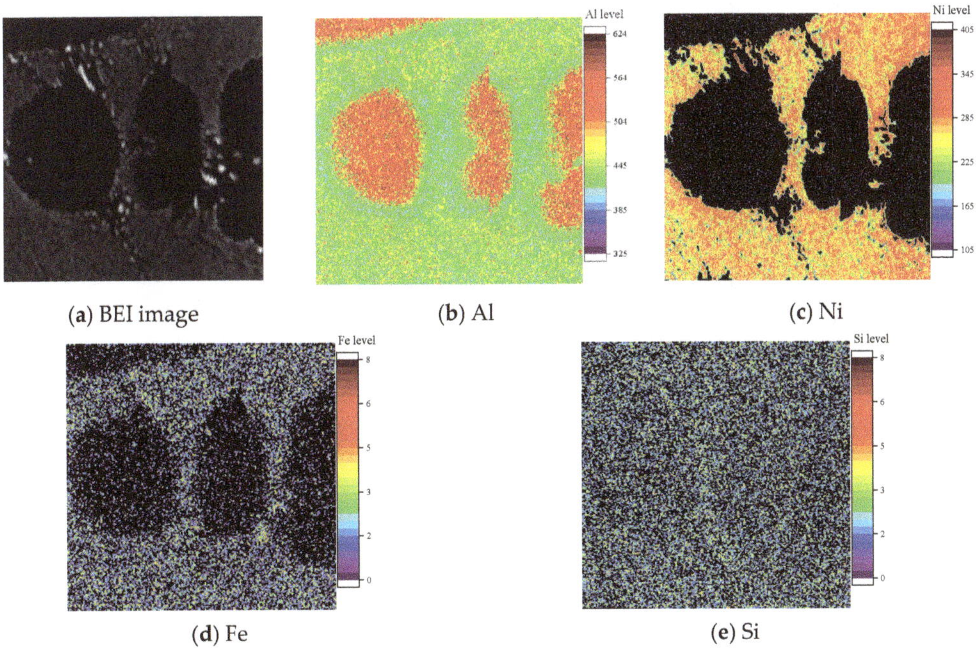

Figure 9. EPMA mapping of Al-5.0Ni alloy. (**a**) BEI image, (**b**) Al, (**c**) Ni, (**d**) Fe, and (**e**) Si.

In metallic systems, thermal transport arises from coupled electron and phonon contributions, with total thermal conductivity (λ) comprising electronic (λ_e) and lattice (λ_g) components as per established models [22]. Among these, the electronic thermal conductivity plays a more dominant role compared to the lattice component. According to the Wiedemann–Franz rule [32,33], a proportional correlation is observed between electronic thermal and electrical conductivity. Concurrently, the Mattissen–Flemming rule [34,35] suggests that the main contributors to electrical resistance are solid solute atoms, contaminants, and imperfections. The associated expression is shown as follows:

$$\rho = \rho^{Alpure} + \rho^{precipitates} + S_{GB}\Delta\rho^{GB} + L_{dislo}\Delta\rho^{dislo} + C_v\Delta\rho^{vac} + \sum_i C^i_{solu}\Delta\rho^i_{solu}. \quad (3)$$

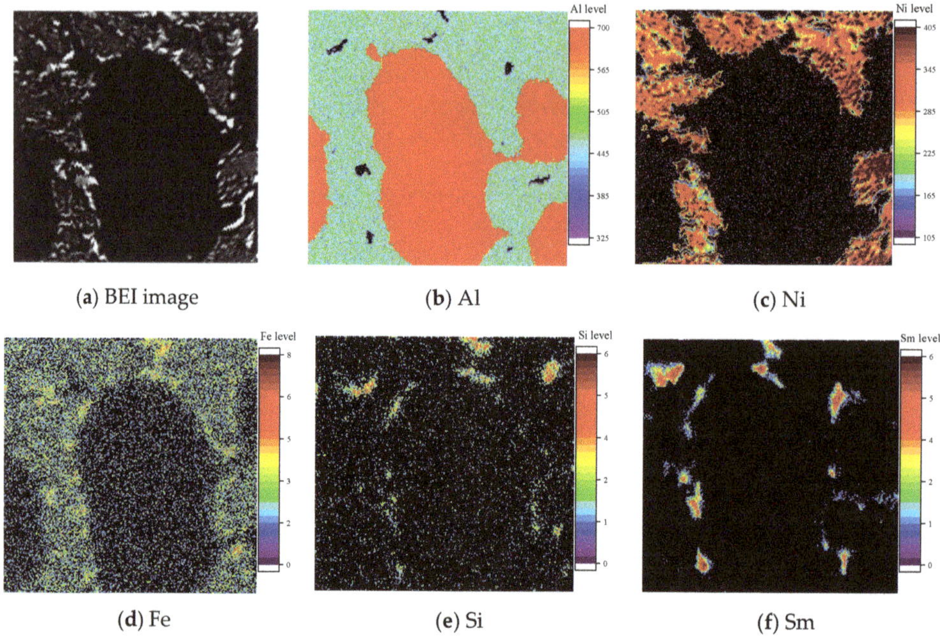

Figure 10. EPMA mapping of Al-5.0Ni-0.1Sm alloy. (**a**) BEI image, (**b**) Al, (**c**) Ni, (**d**) Fe, (**e**) Si, and (**f**) Sm.

The value of ρ^{Alpure} is 2.665 × 10^{-8} Ω·m, as reported in reference [34]. $\rho^{precipitates}$ represents the electrical resistivity caused by precipitates. The negligible effect of minor Sm additions on precipitate volume fraction permits the exclusion of precipitation-mediated resistivity contributions. The coefficients ($\Delta\rho^{GB}$, $\Delta\rho^{dislo}$, $\Delta\rho^{vac}$, and $\Delta\rho^i_{solu}$), represent the individual contributions of grain boundaries, dislocations, vacancies, and solid solute atoms to the electrical resistance. S_{GB} is the grain boundary fraction per unit volume, L_{dislo} is the dislocation density, C_v is the vacancy concentration and C^i_{solu} is the concentration of solid solute atom *i*. According to Miyajama's study [36], for aluminum, $\Delta\rho^{GB}$ is approximately 2.6 × 10^{-15} Ω·m^2, $\Delta\rho^{dislo}$ is around 2.7 × 10^{-25} Ω·m^3, and $\Delta\rho^{vac}$ is about 2.6 × 10^{-8} Ω·m/at.%. Considering the minimal magnitudes of $\Delta\rho^{GB}$, $\Delta\rho^{dislo}$, and $\Delta\rho^{vac}$, the impact of grain boundaries, dislocations, and vacancies on electrical resistance is trivial. As a result, the electrical resistance is greatly influenced by the levels of atoms in solid solutions. Mulazimoglu [37] and Cooper [38] quantified solute-specific resistivity coefficients through systematic alloy characterization, deriving empirical relationships between aluminum's electrical resistivity and solute concentrations. These correlations are mathematically expressed as

$$\Delta\rho^{Si}_{solu} = 7 \times 10^{-9}\ \Omega\cdot\text{m}\cdot(\text{wt.\%})^{-1} \quad (4)$$

$$\Delta\rho^{Fe}_{solu} = 1.682 \times 10^{-8}\ \Omega\cdot\text{m}\cdot(\text{wt.\%})^{-1} \quad (5)$$

According to the aforementioned formulas, each 1 wt.% rise in Si or Fe content results in a roughly 26.4% or 63.4% reduction in the electrical resistance of pure Al, respectively. In reality, trace amounts of Fe and Si elements are invariably present in commercially pure aluminum. Moreover, as Jiang et al. [8] and Zhang et al. [27] emphasized that rare earth elements present in molten aluminum possess pronounced interfacial adsorption capabilities. Such a trait grants them exceptional purification capabilities for the α-Al matrix. In this study, the same observation (as illustrated in Figure 9c,d) reveals that Sm can also extract Fe and Si elements from the Al-5Ni melt. This extraction process effectively reduces the solid solubility of these impurities within the alloy. Moreover, as illustrated in Figure 1, the incorporation of Sm results in notable refinement of both α-Al grains and the eutectic (α-Al + Al_3Ni) structure. This microstructural refinement contributes significantly to improved thermal conductivity. The mechanism driving this phenomenon has been thoroughly detailed in our earlier work [13]. The reduction in grain size and optimization of the eutectic arrangement increases intergranular contact surface, thereby expanding pathways for electron flow. As a result, Sm-induced refinement directly promotes higher thermal conductivity in the near-eutectic Al-5Ni alloy. However, the excessive addition of samarium in the Al-5Ni alloy results in the precipitation of Al_3Sm within the Al matrix (as illustrated in Figure 8c). As an intermetallic phase in the Al-5Ni alloy, Al_3Sm acts as a scattering source that hinders the transmission of free electrons. Consequently, as illustrated in Figure 7b, the TC property of the Al-5Ni alloy decreases with increasing samarium content.

4. Conclusions

This study systematically investigates the influence of samarium incorporation on the microstructure, solidification behavior, mechanical characteristics, and thermophysical behavior of Al-5Ni alloys, offering insight into the underlying rare earth modification mechanisms. The principal findings are outlined as follows:

(1) The introduction of Sm leads to simultaneous refinement of α-Al grains and eutectic (α-Al + Al_3Ni) phases, driven by two synergistic mechanisms—enhanced nucleation undercooling of the α-Al matrix and suppressed Al_3Ni phase growth via interfacial adsorption effects.

(2) The controlled Sm alloying significantly enhances the tensile properties and thermal conductivity (TC). This is attributed to the addition of moderate Sm (0.1 wt.%), which lowers the nucleation temperature of dendritic particles and promotes solute redistribution, thereby refining the size of eutectic structures and optimizing their morphology and distribution to improve mechanical properties. Meanwhile, the thermal conductivity is enhanced by inhibiting the solid solubility of Fe/Si elements in the α-Al matrix.

Author Contributions: Conceptualization, K.W.; data curation, Y.H.; formal analysis, Y.H.; funding acquisition, W.L.; investigation, Y.H., Z.Z. and W.C.; methodology, K.W.; project administration, W.L.; resources, W.L.; supervision, W.L.; validation, Z.Z.; writing—original draft, Y.H.; writing—review and editing, Y.H. All authors have read and agreed to the published version of the manuscript.

Funding: The Guangdong Major Project of Basic and Applied Basic Research (2021B1515130010), Guangdong Major Project of Basic and Applied Basic Research (2020B0301030001), Science and Technology Service Network Initiative of Chinese Academy of Sciences (Dongguan Special Project: 20211600200082), Guangdong University's Innovative Team Project (2021KCXTD022), Dongguan Social Development Science and Technology Project (NO.20231800938572), and The Fund Project of Laboratory Management Professional Committee of Guangdong Higher Education Association (GDJ20240022) have all provided funding for this study.

Data Availability Statement: The original contributions presented in the study are included in the article; further inquiries can be directed at the corresponding author.

Acknowledgments: The writers express their gratitude for the support provided by Dongguan University of Technology's Analytical and Testing Center in conducting the SEM observation. In the meantime, we recognize the assistance provided by the Guangdong Research Center for High Performance Light Alloys Forming Technology, along with the Novel Light Alloy and its Process Technology Key Laboratory in Dongguan City.

Conflicts of Interest: The authors declare no conflicts of interest.

References

1. Qi, M.; Kang, Y.; Li, J.; Shang, B. Improvement in Mechanical, Thermal Conductivity and Corrosion Performances of a New High-Thermally Conductive Al-Si-Fe Alloy through a Novel R-HPDC Process. *J. Mater. Process. Technol.* **2020**, *279*, 116586. [CrossRef]
2. Luo, G.; Zhou, X.; Li, C.; Du, J.; Huang, Z. Design and Preparation of Al-Fe-Ce Ternary Aluminum Alloys with High Thermal Conductivity. *Trans. Nonferrous Met. Soc. China* **2022**, *32*, 1781–1794. [CrossRef]
3. Wen, C.; Gan, J.; Li, C.; Huang, Y.; Du, J. Comparative Study on Relationship Between Modification of Si Phase and Thermal Conductivity of Al-7Si Alloy Modified by Sr/RE/B/Sb Elements. *Int. J. Met.* **2021**, *15*, 194–205. [CrossRef]
4. Song, Z.-X.; Li, Y.-D.; Liu, W.-J.; Yang, H.-K.; Cao, Y.-J.; Bi, G.-L. Effect of La and Sc Co-Addition on the Mechanical Properties and Thermal Conductivity of As-Cast Al-4.8% Cu Alloys. *Metals* **2021**, *11*, 1866. [CrossRef]
5. Liu, J.; Wu, S.; Zheng, J. *Microstructure and Properties of Semi-Solid Die Casting of High Silicon Aluminum Alloy with High Thermal Conductivity*; Springer: Singapore, 2020; pp. 463–470.
6. Shin, J.-S.; Ko, S.-H.; Kim, K.-T. Development and Characterization of Low-Silicon Cast Aluminum Alloys for Thermal Dissipation. *J. Alloys Compd.* **2015**, *644*, 673–686. [CrossRef]
7. Kaya, H.; Boyuk, U.; Cadirli, E.; Marasli, N. Measurements of the Microhardness, Electrical and Thermal Properties of the Al-Ni Eutectic Alloy. *Mater. Des.* **2012**, *34*, 707–712. [CrossRef]
8. Jiang, H.; Li, S.; Zheng, Q.; Zhang, L.; He, J.; Song, Y.; Deng, C.; Zhao, J. Effect of Minor Lanthanum on the Microstructures, Tensile and Electrical Properties of Al-Fe Alloys. *Mater. Des.* **2020**, *195*, 108991. [CrossRef]
9. Luo, G.; Huang, Y.; Li, C.; Huang, Z.; Du, J. Microstructures and Mechanical Properties of Al-2Fe-xCo Ternary Alloys with High Thermal Conductivity. *Materials* **2020**, *13*, 3728. [CrossRef]
10. Yousefi, F.; Taghiabadi, R.; Baghshahi, S. Improving Mechanical Properties of Mn-Added Hypoeutectic Al-4Ni Alloy by Friction Stir Processing. *Trans. Nonferrous Met. Soc. China* **2019**, *29*, 460–472. [CrossRef]
11. Jones, H. Some Effects of Solidification Kinetics on Microstructure Formation in Aluminium-Base Alloys. *Mater. Sci. Eng. A* **2005**, *413*, 165–173. [CrossRef]
12. Srivastava, N.; Chaudhari, G.P. Effect of Ultrasonic Treatment on the Mechanical Behaviour of Al-Ni Alloys. *Mater. Sci. Technol.* **2019**, *35*, 1239–1247. [CrossRef]
13. Wang, K.; Li, W.; Xu, W.; Hou, S.; Hu, S. Simultaneous Improvement of Thermal Conductivity and Strength for Commercial A356 Alloy Using Strontium Modification Process. *Met. Mater. Int.* **2021**, *27*, 4742–4756. [CrossRef]
14. Xia, X.; Zhao, Q.; Peng, Y.; Zhang, P.; Liu, L.; Ding, J.; Luo, X.; Wang, L.; Huang, L.; Zhang, H.; et al. Precipitation Behavior and Mechanical Performances of A356.2 Alloy Treated by Al-Sr-La Composite Refinement-Modification Agent. *J. Alloys Compd.* **2020**, *818*, 153370. [CrossRef]
15. Qiu, H.; Yan, H.; Hu, Z. Modification of Near-Eutectic Al-Si Alloys with Rare Earth Element Samarium. *J. Mater. Res.* **2014**, *29*, 1270–1277. [CrossRef]
16. Yuansheng, R.; Hong, Y.; Zhi, H. Modification of Eutectic Silicon and β-Al5FeSi Phases in as-Cast ADC12 Alloys by Using Samarium Addition. *J. Rare Earths* **2013**, *31*, 916–922.
17. Chihlangu, C.; Gorlov, L.E.; Barkov, R.Y.; Bykov, A.S.; Medvedeva, S.V.; Glavatskikh, M.V.; Pozdniakov, A.V. The Effect of Ni Addition on the Phase Composition and Properties of Quasi-Binary Al-Cu-Y(Er,Yb,Gd) Alloys. *J. Mater. Eng. Perform.* **2025**. [CrossRef]
18. Suwanpreecha, C.; Toinin, J.P.; Michi, R.A.; Pandee, P.; Dunand, D.C.; Limmaneevichitr, C. Strengthening Mechanisms in Al-Ni-Sc Alloys Containing Al3Ni Microfibers and Al3Sc Nanoprecipitates. *Acta Mater.* **2019**, *164*, 334–346. [CrossRef]
19. Okamoto, H. *Phase Diagrams for Binary Alloys*, 2nd ed.; ASM International: Materials Park, OH, USA, 2010.
20. Hongxu, Q.; Hong, Y.; Zhi, H. Effect of Samarium (Sm) Addition on the Microstructures and Mechanical Properties of Al-7Si-0.7Mg Alloys. *J. Alloys Compd.* **2013**, *567*, 77–81.
21. Rudajevova, A.; Buch, F.; Mordike, B. ChemInform Abstract: Thermal Diffusivity and Thermal Conductivity of MgSc Alloys. *ChemInform* **2010**, 31. [CrossRef]
22. Pan, H.; Pan, F.; Yang, R.; Peng, J.; Zhao, C.; She, J.; Gao, Z.; Tang, A. Thermal and Electrical Conductivity of Binary Magnesium Alloys. *J. Mater. Sci.* **2014**, *49*, 3107–3124. [CrossRef]
23. Yang, L.; Li, W.; Du, J.; Wang, K.; Tang, P. Effect of Si and Ni Contents on the Fluidity of Al-Ni-Si Alloys Evaluated by Using Thermal Analysis. *Thermochim. Acta* **2016**, *645*, 7–15. [CrossRef]

24. Farahany, S.; Ourdjini, A.; Abu Bakar, T.A.; Idris, M.H. A New Approach to Assess the Effects of Sr and Bi Interaction in ADC12 Al–Si Die Casting Alloy. *Thermochim. Acta* **2014**, *575*, 179–187. [CrossRef]
25. Wang, T.; Zhao, Y.; Chen, Z.; Zheng, Y.; Kang, H. The Bimodal Effect of La on the Microstructures and Mechanical Properties of In-Situ A356-TiB2 Composites. *Mater. Des.* **2015**, *85*, 724–732. [CrossRef]
26. Alexopoulos, N.D.; Pantelakis, S.G. Quality Evaluation of A357 Cast Aluminum Alloy Specimens Subjected to Different Artificial Aging Treatment. *Mater. Des.* **2004**, *25*, 419–430. [CrossRef]
27. Zhang, Y.; Wei, F.; Mao, J.; Niu, G. The Difference of La and Ce as Additives of Electrical Conductivity Aluminum Alloys. *Mater. Charact.* **2019**, *158*, 109963. [CrossRef]
28. Boussinot, G.; Döring, M.; Hemes, S.; Stryzhyboroda, O.; Apel, M.; Schmidt, M. Laser powder bed fusion of eutectic Al-Ni alloys: Experimental and phase-field studies. *Mater. Des.* **2021**, *198*, 109299. [CrossRef]
29. Suwanpreecha, C.; Pandee, P.; Patakham, U.; Limmaneevichitr, C. New generation of eutectic Al-Ni casting alloys for elevated temperature services. *Mater. Sci. Eng. A* **2018**, *709*, 46–54. [CrossRef]
30. Bian, Z.; Liu, Y.; Dai, S.; Chen, Z.; Wang, M.; Chen, D.; Wang, H. Regulating Microstructures and Mechanical Properties of Al-Fe-Ni Alloys. *Prog. Nat. Sci. Mater. Int.* **2020**, *30*, 54–62. [CrossRef]
31. Bian, Z.; Xiao, Y.; Geng, J.; Hu, L.; Chen, Z.; Wang, M.; Chen, D.; Wang, H. Optimizing Zr Addition Method to Improve the Comprehensive High Temperature Performance of Al-Fe-Ni-Sc Eutectic Alloy. *J. Alloys Compd.* **2021**, *866*, 158883. [CrossRef]
32. Wang, M.; Pan, N. Predictions of Effective Physical Properties of Complex Multiphase Materials. *Mater. Sci. Eng. R Rep.* **2008**, *63*, 1–30. [CrossRef]
33. Chen, J.K.; Hung, H.Y.; Wang, C.F.; Tang, N.K. Thermal and Electrical Conductivity in Al-Si/Cu/Fe/Mg Binary and Ternary Al Alloys. *J. Mater. Sci.* **2015**, *50*, 5630–5639. [CrossRef]
34. Sauvage, X.; Bobruk, E.V.; Murashkin, M.Y.; Nasedkina, Y.; Enikeev, N.A.; Valiev, R.Z. Optimization of Electrical Conductivity and Strength Combination by Structure Design at the Nanoscale in Al-Mg-Si Alloys. *Acta Mater.* **2015**, *98*, 355–366. [CrossRef]
35. Klemens, P.G. Deviations from Matthiessen's Rule and the Electronic Thermal Conductivity of Alloys. *Therm. Conduct.* **1978**, *15*, 203–207.
36. Miyajima, Y.; Komatsu, S.-Y.; Mitsuhara, M.; Hata, S.; Nakashima, H.; Tsuji, N. Change in Electrical Resistivity of Commercial Purity Aluminium Severely Plastic Deformed. *Philos. Mag.* **2010**, *90*, 4475–4488. [CrossRef]
37. Mulazimoglu, M.H.; Drew, R.A.L.; Gruzelski, J.E. Electrical Conductivity of Aluminium-Rich Al-Si-Mg Alloys. *J. Mater. Sci. Lett.* **1989**, *8*, 297–300. [CrossRef]
38. Cooper, P.S.; Kearns, M.A. Removal of Transition Metal Impurities in Aluminium Melts by Boron Additives. *Mater. Sci. Forum* **1996**, *217–222*, 141–146. [CrossRef]

crystals

MDPI

Article

Understanding the Interface Characteristics Between TiB_2(0001) and $L1_2$-Al_3Zr(001): A First-Principles Investigation

Xingzhi Pang [1], Loujiang Yang [2], Hang Nong [1,*], Mingjun Pang [3], Gaobao Wang [4], Jian Li [1], Zhenchao Chen [1], Wei Zeng [1], Zhihang Xiao [1], Zengxiang Yang [1] and Hongqun Tang [1,*]

1 State Key Laboratory of Featured Metal Materials and Life-Cycle Safety for Composite Structures, MOE Key Laboratory of New Processing Technology for Nonferrous Metals and Materials, Center of Ecological Collaborative Innovation for Aluminum Industry in Guangxi, and School of Resources, Environment and Materials, Guangxi University, Nanning 530004, China; pxzh2006@126.com (X.P.); 2139200618@st.gxu.edu.cn (J.L.); 2139200603@st.gxu.edu.cn (Z.C.); 2139200613@st.gxu.edu.cn (W.Z.); 2139200645@st.gxu.edu.cn (Z.X.); 2139200647@st.gxu.edu.cn (Z.Y.)

2 Ningbo Boway Alloy Material Co., Ltd., Ningbo 315000, China; ylj2260@163.com

3 SAIC-GM-Wuling Automobile Co., Ltd., Liuzhou 545007, China; mjpang@163.com

4 Guangxi Youhe Metal Materials Technology Co., Ltd., Pingguo 531400, China; wanggaobao063@163.com

* Correspondence: nh1998728@163.com (H.N.); hqtang@gxu.edu.cn (H.T.)

Abstract: This study employs first-principles calculation methods to explore the characteristics of the TiB_2(0001)/$L1_2$-Al_3Zr(001) interface, including the atomic structure, adhesion work, interfacial energy, and electronic structure of various interface models. Considering four different terminations and three different stacking positions, twelve potential interface models were investigated. Surface tests revealed that a stable interface could be formed when a 9-layer TiB_2(0001) surface is combined with a 7-layer ZrAl-terminated and a 9-layer Al-terminated Al_3Zr(001) surface. Among these interfaces, the bridge-site stacking at the T/Al termination (TAB), hollow-site stacking at the Ti/ZrAl termination (TZH), top-site stacking at the B/Al termination (BAT), and hollow-site stacking at the B/ZrAl termination (BZH) were identified as the optimal structures. Particularly, the TAB interface exhibits the strongest adhesion strength and the lowest surface energy, indicating the highest stability. A Detailed analysis of the electronic structure further reveals that most interfaces predominantly exhibit covalent bonding, with the TAB, TZH, and BZH interfaces primarily featuring covalent bonds, while the BAT interface displays a combination of ionic and covalent bonds. The study ultimately ranks the stability of the interfaces from highest to lowest as TAB, BZH, TZH, and BAT.

Keywords: first-principles calculations; interface stability; TiB_2/Al_3Zr interfaces; electronic structure analysis

1. Introduction

Aluminum–silicon alloys, characterized by low density, high strength, high wear resistance, and good machinability, are extensively used in mechanical manufacturing, automotive, maritime, and aerospace industries [1,2]. The performance of the alloy largely depends on the microstructure of its grains. However, in aluminum–silicon alloys, the coarse eutectic silicon phase can lead to grain coarsening, thereby diminishing the material's performance [3]. Studies have shown that the addition of Zr to aluminum alloys can significantly enhance their performance. In these Zr-doped aluminum alloys, nanoscale Al_3Zr precipitates have been observed and are considered to contribute to grain refinement [4].

Al_3Zr in aluminum alloys exists in two structures, $D0_{23}$ and $L1_2$ [5,6]. $D0_{23}$-Al_3Zr is a stable structure formed by tetragonal lattice distortion and intra-cell atomic migration, and it exhibits higher brittleness [7]. $L1_2$-Al_3Zr is a metastable cubic structure, primarily precipitated from supersaturated aluminum solid solutions during rapid cooling processes [8]. $L1_2$-Al_3Zr is completely coherent with the α-Al matrix, and it significantly contributes to inhibiting grain recrystallization and enhancing alloy strength through its shearing suppression effect on dislocation movement [9,10]. Studies have indicated that $L1_2$-Al_3Zr serves as an effective heterogeneous nucleation substrate for α-Al at the atomic level [11,12], and first-principles calculations have confirmed that liquid Al atoms tend to nucleate on the $L1_2$-$Al_3Zr(001)$ surface at the ZrAl terminated [13].

Adding Al-Ti-B grain refiners to aluminum melt is also a widely applied method for improving the microstructure of grains, which contain strong nucleating particles like TiB_2 [14,15]. TiB_2 serves as an effective nucleating particle and plays a crucial role in the refinement of α-Al grains [16]. It has been reported that the refining mechanism of Al-Ti-B grain refiners is the formation of a two-dimensional Al_3Ti compound on the (0001) TiB_2 surface, which effectively promotes heterogeneous nucleation of α-Al [17]. Reports indicate that the distribution of TiB_2 particles follows a log-normal distribution, with most nanoparticles being smaller than 100 nanometers [18]. Nanoscale TiB_2 particles can pin dislocations at grain boundaries and inhibit the growth of recrystallized grains. Additionally, TiB_2 is a boride with a hexagonal structure, known for its excellent stiffness, hardness, and wear resistance, making it an outstanding reinforcing material [19,20].

When both Al-Ti-B grain refiners and excess Zr and Si elements are present in the aluminum melt, the grain refining capability is significantly reduced, leading to the so-called "poisoning" phenomenon of the grain refiner [21,22]. The prevailing view on the mechanism of the "poisoning" phenomenon is that Zr atoms in the Al melt cause the dissolution of Al_3Ti on the $TiB_2(0001)$ surface, thereby reducing the nucleation capability of the refiner [15,23–25]. Some studies also speculate that TiB_2 loses its ability to refine Al grains due to the formation of interfacial nucleation with Al_3Zr [26]. Therefore, to further elucidate the mechanism of the "poisoning" phenomenon, it is necessary to study the relationship between the two nucleating particles, TiB_2 and Al_3Zr. Current research on the interface relationship between TiB2 and $D0_{23}$-Al_3Zr is already quite extensive. Researchers have observed the precipitation of $D0_{23}$-Al_3Zr at the TiB_2 interface using scanning electron microscopy, with the orientation relationships identified as $(0001)_{TiB2}//(1(\_)03(\_))_{D023\text{-}Al3Zr}$, $(101(\_)0)_{TiB2}//(105(\_))_{D023\text{-}Al3Zr}$, and $[121(\_)0]_{TiB2}//[010]_{D023\text{-}Al3Zr}$ [27,28]. However, research on the interface relationship between TiB_2 and $L1_2$-Al_3Zr remains quite scarce. During the solidification process of the alloy melt, the precipitation of the $D0_{23}$-Al_3Zr phase at the TiB_2 interface nearly depletes Zr from the matrix, leading to scant precipitation of $L1_2$-Al_3Zr around the TiB_2 clusters. This situation makes it difficult to observe the interface relationship between TiB_2 and $L1_2$-Al_3Zr through experimental means [29]. Therefore, simulations based on Density Functional Theory can provide a convenient means to study the potential interface relationship between these two phases.

This article employs first-principles calculation methods to systematically investigate the structural and electronic properties of the $TiB_2(0001)/L1_2$-$Al_3Zr(001)$ interface at the atomic level. To gain a more comprehensive understanding of the bonding behavior at the $TiB_2(0001)/L1_2$-$Al_3Zr(001)$ interface, we also predicted the ideal adhesive work and interfacial energy. Finally, the electronic structure of $TiB_2(0001)/L1_2$-$Al_3Zr(001)$ was analyzed through the differential charge density and a density of states analysis.

2. Calculation Method

First-principles calculations were performed using the Cambridge Serial Total Energy Package (CASTEP) code [30]. The interactions between ionic cores and valence electrons were described using ultrasoft pseudopotentials [31]. Within the framework of the Generalized Gradient Approximation (GGA), the Perdew–Burke–Ernzerhof (PBE) function was employed as the exchange-correlation potential [32]. Theoretical calculations were performed with a plane-wave cutoff energy of 380 eV. For the calculations of bulk and surfaces, Brillouin zone sampling was carried out using the Monkhorst–Pack method, with K-points set to $10 \times 10 \times 10$ for bulk and $6 \times 6 \times 1$ for surfaces [33]. During the relaxation of all atomic models, the Broyden–Fletcher–Goldfarb–Shanno (BFGS) minimization scheme was used for ensemble optimization steps, with all atomic structures and volumes being fully relaxed [34]. The structural optimizations of these systems were relaxed until the total energy was less than 1×10^{-5} eV/atom. The convergence tolerances for the forces between the atoms, maximum stress, and maximum ionic displacement were set to 0.01 eV/Å, 0.02 GPa, and 5×10^{-4} Å, respectively. Additionally, for the interfaces, a vacuum layer of 14 Å was chosen to isolate the free surfaces, ensuring sufficient space to prevent their interactions.

3. Results and Discussion

3.1. Crystal Properties

The schematic diagram of the crystal structure model used in the calculations is shown in Figure 1. From Figure 1a, it can be seen that the crystal structure model of the bulk TiB_2 was of the AlB_2 type, with the space group P6/mmm. As shown in Figure 1b, the crystal structure model of the bulk Al_3Zr was of the Cu_3Au type, with the space group Pm-3m.

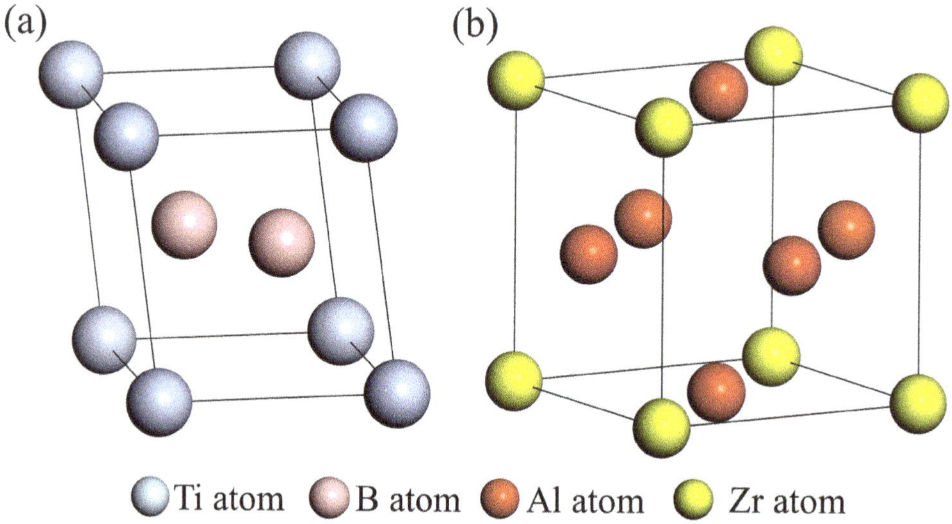

Figure 1. Crystal structures of TiB_2 and Al_3Zr: (**a**) TiB_2; (**b**) Al_3Zr.

To ensure the accuracy of the calculations in this paper, especially the accuracy of the pseudopotentials used, a series of calculations were performed on the lattice constants and formation enthalpies of TiB_2 and Al_3Zr using the Perdew–Burke–Ernzerhof (PBE) within the Generalized Gradient Approximation. The lattice constants were obtained directly from the calculations, and the formation enthalpies of TiB_2 and Al_3Zr were derived using Equations (1) and (2) [35]:

$$\Delta\gamma H_{TiB_2} = \frac{E_{bulk}^{TiB_2} - N_{bulk}^{Ti}(TiB_2) \times \frac{E_{bulk}^{Ti}}{N_{bulk}^{Ti}(Ti)} - N_{bulk}^{B}(TiB_2) \times \frac{E_{bulk}^{B}}{N_{bulk}^{B}(B)}}{N_{bulk}^{Ti}(TiB_2) + N_{bulk}^{B}(TiB_2)} \quad (1)$$

$$\Delta\gamma H_{Al_3Zr} = \frac{E_{bulk}^{Al_3Zr} - N_{bulk}^{Al}(Al_3Zr) \times \frac{E_{bulk}^{Al}}{N_{bulk}^{Al}(Al)} - N_{bulk}^{Zr}(Al_3Zr) \times \frac{E_{bulk}^{Zr}}{N_{bulk}^{Zr}(Zr)}}{N_{bulk}^{Al}(Al_3Zr) + N_{bulk}^{Zr}(Al_3Zr)} \quad (2)$$

where $\Delta\gamma H_{TiB_2}$ and $E_{bulk}^{TiB_2}$ represent the formation enthalpy and the total energy of bulk TiB_2, respectively; E_{bulk}^{Ti} denotes the energy of a Ti atom in the bulk, and E_{bulk}^{B} is the energy of boron in its conventional cell; $N_{bulk}^{Ti}(TiB_2)$ and $N_{bulk}^{B}(TiB_2)$ are the numbers of Ti and B atoms in the bulk model of TiB_2, respectively; and $N_{bulk}^{Ti}(Ti)$ and $N_{bulk}^{B}(B)$ represent the numbers of atoms in the conventional cells of titanium and boron, respectively. Equation (2) follows the same rationale and will not be reiterated here.

Table 1 presents the lattice constants and formation enthalpies of TiB_2 and $L1_2$-Al_3Zr, comparing them with other computational and experimental results. In this work, the lattice constants of TiB_2 differ by less than 0.004 Å from the theoretical values reported in the literature [36,37], and the deviation from the experimental values reported in the literature does not exceed 0.006 Å [38]. The calculated formation enthalpy of TiB_2 had a maximum deviation of 0.013 eV/atom from other theoretical values obtained from calculations and a difference of 0.062 eV/atom compared with the experimental formation enthalpies. For the $L1_2$-Al_3Zr unit cell structure, the maximum discrepancy in lattice constants compared to theoretical values from the other literature was 0.035 Å, with a formation enthalpy error within 0.021 eV/atom [11,39], and the discrepancy from the experimental lattice constant reference values was 0.008 Å [40]. These results indicate that our current computational methods and settings are reliable, and the calculation accuracy was sufficient.

Table 1. Calculated and experimental lattice constants and formation energies of bulk TiB_2 and bulk Al_3Zr.

Phases	Method	Lattice Constant (Å)			$\Delta_r H$(eV/atom)
		a	b	c	
TiB_2	Present	3.026	3.026	3.229	−1.072
	Calculation [a]	3.034	3.034	3.224	−1.079
	Calculation [b]	3.030	3.030	3.221	−1.085
	Experiment [c]	3.032	3.032	3.229	−1.134
Al_3Zr	Present	4.082	4.082	4.082	−0.487
	Calculation [d]	4.117	4.117	4.117	−0.471
	Calculation [e]	4.096	4.096	4.096	−0.466
	Experiment [f]	4.090	4.090	4.090	-

[a] Ref. [36] GGA-PBE. [b] Ref. [37] GGA-PBE. [c] Ref. [38]. [d] Ref. [11] GGA-PBE. [e] Ref. [39] GGA-PW91. [f] Ref. [40].

3.2. Surface Convergence and Surface Energy

Due to the significant difference in lattice constants between the primitive cells of $TiB_2(0001)$ and the $Al_3Zr(001)$ surfaces, direct construction of the interface model was not feasible. To satisfy the periodicity conditions and lattice relationships, it was necessary to find matching lattice constants for the $TiB_2(0001)$ and $Al_3Zr(001)$ surfaces to build the interface model. After testing, it was found that the lattice constants of the $TiB_2(0001)$-$(\sqrt{3} \times 2)$R30° surface unit and the $Al_3Zr(001)$-$(\sqrt{2} \times \sqrt{2})$R45° surface unit are numerically similar, and the mismatch rate at the interface can be calculated as follows:

$$\xi = \frac{2A}{A_{Al_3Zr} + A_{TiB_2}} - 1 \quad (3)$$

where in the above equation, A represents the interfacial area after the $TiB_2(0001)$ and $Al_3Zr(001)$ surfaces form an interface, while A_{Al_3Zr} and A_{TiB_2}, respectively, refer to the areas of the $Al_3Zr(001)$ and $TiB_2(0001)$ surfaces before forming the interface. Using Equation (1), the mismatch rate of this interface was calculated to be less than 1%. Therefore, the $TiB_2(0001)/Al_3Zr(001)$ interface model was considered reasonable.

Before establishing the interface model, it was necessary to perform convergence tests on the atomic thickness of the phases on both sides of the interface to ensure that the atoms deep within the surface exhibited characteristics similar to those of bulk phase atoms. Typically, the convergence of the surface energy for both the $TiB_2(0001)$ and $Al_3Zr(001)$ surfaces was used to determine this. Since both the $TiB_2(0001)$ and $Al_3Zr(001)$ surfaces are non-polar, the convergence could be assessed by measuring the percentage change in relaxation between the atomic layers (Δ_{ij}, where i, j represent the atomic layers in the surface model).

The change in interlayer distance (Δij) is defined as follows:

$$\Delta_{ij} = \frac{d_{ij} - d_{ij}^0}{d_{ij}^0} \times 100\% \quad (4)$$

where Δ_{ij} is the initial distance between layers i and j in the ideal crystal structure and is the vertical distance between the adjacent i and j layers after relaxation.

After the geometric optimization of the atoms, the changes in the interlayer distances of the TiB_2(0001) and Al_3Zr(001) surfaces are shown in Table 2. For TiB_2, when the number of atomic layers *nn* in the Ti-terminated and B-terminated TiB_2(0001) surfaces was ≥9, the changes in the interlayer distances in TiB_2(0001) were essentially converged. A comparison of the percentage changes in the interlayer distances of the same number of atomic layers shows that the relaxation of the outermost interlayer distance on the B-terminated surface was much greater than that on the Ti-terminated surface. The results indicate that the outermost interlayer distances on the Ti-terminated and B-terminated surfaces with nine atomic layers decreased by 7.7% and 9.6% of their bulk volumes, respectively. From Table 2, it can be seen that for Al_3Zr, the interlayer distances on the Al-terminated side essentially converged when *nn* was ≥9, while on the ZrAl-terminated side, a good convergence effect was achieved when *nn* was ≥7.

Table 2. Interlayer relaxation percentages for TiB_2(0001) and Al_3Zr(001) surfaces at different slab thicknesses.

Surface	Termination	Interlayer	Slab Thickness, n				
			3	5	7	9	11
TiB_2(0001)	Ti	Δ_{12}	−6.96	−7.69	−7.54	−7.73	−7.63
		Δ_{23}		−1.61	−2.09	−2.00	−1.86
		Δ_{34}			−1.98	−2.51	−2.29
		Δ_{45}				−2.72	−2.14
		Δ_{56}					−2.73
	B	Δ_{12}	−8.15	−10.20	−9.53	−9.61	−9.71
		Δ_{23}		0.47	0.11	−0.11	0.02
		Δ_{34}			−2.31	−2.24	−2.40
		Δ_{45}				−3.16	−2.69
		Δ_{56}					−2.71
Al_3Zr(001)	Al	Δ_{12}	−1.45	3.02	1.33	0.63	0.91
		Δ_{23}		−1.52	−1.20	1.68	1.05
		Δ_{34}			0.65	2.89	2.62
		Δ_{45}				−0.81	−0.46
		Δ_{56}					0.67
	Zr-Al	Δ_{12}	−15.33	−18.17	−16.38	−16.68	−16.92
		Δ_{23}		−4.95	−5.10	−4.43	−4.73
		Δ_{34}			4.28	2.51	3.21
		Δ_{45}				0.09	−0.39
		Δ_{56}					0.89

For simulations related to surfaces, the more atomic layers the surface model contains, the more precise the results become. However, as the number of atoms increases, the computational resources required also increase. Therefore, to further determine the minimum thickness of the TiB_2(0001) and Al_3Zr(001) surfaces that satisfy the properties of the bulk phase, and to reduce the error in determining the number of layers, we further discussed the impact of different layer thicknesses on the convergence of surface energy. This approach makes the selection of convergence layer numbers more reliable [41].

Surface energy is an important parameter for measuring surface stability. For non-stoichiometric TiB_2 surface models, the chemical potentials of both Ti and B should be considered. Therefore, the surface energy of the $TiB_2(0001)$ surface was calculated using a generalized definition of surface energy [42,43]:

$$\gamma_{TiB_2(0001)} = \frac{E_{surf}^{TiB_2(0001)} - N_{TiB_2(0001)}^{Ti} \times \mu_{bulk}^{TiB_2} + \left(2N_{TiB_2(0001)}^{Ti} - N_{TiB_2(0001)}^{B}\right) \times \mu_{slab}^{B}}{2 \times A_{surf}^{TiB_2(0001)}} \quad (5)$$

In the formula, $\gamma_{TiB_2(0001)}$ represents the surface energy of the $TiB_2(0001)$ surface. $E_{surf}^{TiB_2(0001)}$ refers to the total energy of the $TiB_2(0001)$ surface used in the calculation. $N_{TiB_2(0001)}^{Ti}$ is the number of Ti atoms on the $TiB_2(0001)$ surface. $N_{TiB_2(0001)}^{B}$ represents the number of B atoms on the $TiB_2(0001)$ surface. $\mu_{bulk}^{TiB_2}$ is the chemical potential of bulk TiB_2. μ_{slab}^{B} represents the chemical potential of B in the surface. $A_{surf}^{TiB_2(0001)}$ is the area of the $TiB_2(0001)$ surface model.

It is well known that the chemical potentials of various elements in the surface composition are always lower than the corresponding bulk chemical potentials because the surface optimizes energy through a specific ordered arrangement [44]. This implies that $\mu_{slab}^{B} \leq \mu_{bulk}^{B}$ and $\mu_{slab}^{Ti} \leq \mu_{bulk}^{Ti}$. Therefore, the range of chemical potential for B is as follows:

$$\mu_{bulk}^{TiB_2} = \mu_{bulk}^{Ti} + 2\mu_{bulk}^{B} + \Delta\gamma H_{TiB_2}, \frac{1}{2}\Delta\gamma H_{TiB_2} \leq \mu_{slab}^{B} - \mu_{bulk}^{B} \leq 0 \quad (6)$$

In the formula mentioned above, $\Delta\gamma H_{TiB_2}$ represents the formation enthalpy of TiB_2 previously discussed. μ_{bulk}^{Ti} is the chemical potential of Ti in the bulk phase. μ_{bulk}^{B} is the chemical potential of B in the α-B_{12} bulk phase. $\mu_{bulk}^{TiB_2}$ is the chemical potential of the TiB_2 bulk phase.

The $Al_3Zr(001)$ surface model used in the calculations was also non-stoichiometric, hence the surface energy of the $Al_3Zr(001)$ surface could be determined using Equation (7):

$$\gamma_{ZrAl_3(001)} = \frac{E_{surf}^{ZrAl_3(001)} - N_{ZrAl_3(001)}^{Zr} \times \mu_{bulk}^{ZrAl_3} + \left(3N_{ZrAl_3(001)}^{Zr} - N_{ZrAl_3(001)}^{Al}\right) \times \mu_{slab}^{Al}}{2 \times A_{surf}^{ZrAl_3(001)}} \quad (7)$$

Similarly, the range for the chemical potential of aluminum is as follows:

$$\mu_{bulk}^{Al_3Zr} = \mu_{bulk}^{Zr} + 3\mu_{bulk}^{Al} + \Delta\gamma H_{Al_3Zr}, \frac{1}{3}\Delta\gamma H_{Al_3Zr} \leq \mu_{slab}^{Al} - \mu_{bulk}^{Al} \leq 0 \quad (8)$$

Table 3 lists the surface energy data for different structures of the $TiB_2(0001)$ surface and the $Al_3Zr(001)$ surface under conditions rich in B and Al, respectively. The surface energies of the Ti-terminated and B-terminated ends of $TiB_2(0001)$ converged as the number of atomic layers increased, reaching convergence when the atomic layers were ≥9. Similarly, the Al-terminated end of $Al_3Zr(001)$ converged at the atomic layers ≥7; the ZrAl-terminated end converged when the atomic layers were ≥9. The smaller the surface energy value, the better the stability of the surface, thus the stability of the surface was consistent with the assessment of the interlayer distances. This implies that choosing a 9-layer $TiB_2(0001)$ in conjunction with a 7-layer ZrAl-terminated end and a 9-layer Al-terminated end of $Al_3Zr(001)$ to construct the TiB_2/Al_3Zr interface is feasible for subsequent research.

Table 3. The surface energy (J/m^2) calculation results of $TiB_2(0001)$ and $Al_3Zr(001)$ surfaces.

Layer	$TiB_2(0001)$		$Al_3Zr(001)$	
	Ti-Terminated	B-Terminated	ZrAl-Terminated	Al-Terminated
3	5.442	3.014	1.768	1.193
5	5.447	2.987	1.721	1.113
7	5.392	2.914	1.707	1.073
9	5.354	2.876	1.705	1.053
11	5.305	2.836	1.637	1.017
13	5.264	2.779	1.686	1.062

The surface energy of non-stoichiometric surfaces is influenced by the chemical potential difference between the surface atoms and the bulk atoms. Therefore, we calculated the surface energy for a 9-layer $TiB_2(0001)$ with different B chemical potential differences ($\mu^{B}_{slab} - \mu^{B}_{bulk}$), and also for a 7-layer ZrAl-terminated and a 9-layer Al-terminated $Al_3Zr(001)$ with different Al chemical potential differences $\left(\mu^{Al}_{slab} - \mu^{Al}_{bulk}\right)$. The relationship between the calculated surface energies and the different chemical potential differences is shown in Figure 2. From Figure 2a, it can be seen that the surface energy range for the B-terminated TiB2(0001) was from 6.124 to 2.876 J/m^2; for the Ti-terminated $TiB_2(0001)$, the range was from 2.106 to 5.354 J/m^2. Under Ti-rich conditions, the surface energy of the Ti-terminated side was lower than that of the B-terminated side, indicating that the Ti-terminated $TiB_2(0001)$ surface was more stable at this time. However, as the B chemical potential difference increased, the surface energy of the Ti-terminated side gradually increased, while that of the B-terminated side gradually decreased. When the B chemical potential difference was greater than −0.613 eV, the B-terminated $TiB_2(0001)$ surface became more stable. This is consistent with the results reported in the literature [45]. From Figure 2b, it is evident that the surface energy of $L1_2$-$Al_3Zr(001)$ also showed a linear relationship with the chemical potential difference in Al. The surface energy range for the Al-terminated $Al_3Zr(001)$ was from 1.545 to 1.053 J/m^2; for the ZrAl-terminated $Al_3Zr(001)$, the range was from 1.213 to 1.705 J/m^2. The results indicate that as the chemical potential difference shifted from Zr-rich conditions to Al-rich conditions, the surface energy of the Al-terminated end showed a decreasing trend, while the surface energy of the ZrAl end showed an increasing trend. Within the range of −0.487 eV to −0.322 eV, the surface energy of the Al end was approximately lower than that of the ZrAl end, and in the range from −0.322 eV to 0 eV, the opposite was true. This suggests that the Al-terminated end of $Al_3Zr(001)$ is more stable in an aluminum-rich environment, while the ZrAl-terminated end is more stable in a zirconium-rich environment.

3.3. *Interface Characteristics of $TiB_2(0001)/Al_3Zr(001)$*

3.3.1. Atomic Structure of the Interface

Based on the results of the convergence tests, the TiB_2/Al_3Zr interface was constructed as follows: nine layers of $TiB_2(0001)$ were stacked on top of the $Al_3Zr(001)$ surface, with nine layers chosen for the Al-terminated surface of $Al_3Zr(001)$ and seven layers for the ZrAl-terminated surface. The mismatch degree of this interface was less than 1%, indicating that it was a typical coherent interface. Considering that the $TiB_2(0001)$ and $Al_3Zr(001)$ surfaces each have two different terminations (Ti-terminated, B-terminated and Al-terminated, ZrAl-terminated) and three different stacking sequences—top site (TS), bridge site (BS), and hollow site (HS)—a total of 12 interface models were established. These models are illustrated in Figure 3. The "interface name" was designated based on the first letter of the terminations and stacking order of $TiB_2(0001)$ and $Al_3Zr(001)$.

3.3.2. Interface Adhesion Work

The strength of the bonding at the interface can be described by the adhesion work. The adhesion work between the $TiB_2(0001)$ surface and the $Al_3Zr(001)$ surface can be determined using the following definition [46]:

$$W_{ad} = \frac{E^{slab}_{Al_3Zr} + E^{slab}_{TiB_2} - E^{total}_{\frac{Al_3Zr}{TiB_2}}}{A} \tag{9}$$

where $E^{total}_{Al_3Zr/TiB_2}$ and A represent the total energy and the interface area of the $TiB_2(0001)/Al_3Zr(001)$ interface, respectively. $E^{slab}_{TiB_2}$ and $E^{slab}_{Al_3Zr}$ are the energies of the relaxed $TiB_2(0001)$ and $Al_3Zr(001)$ slabs, respectively.

(**a**)

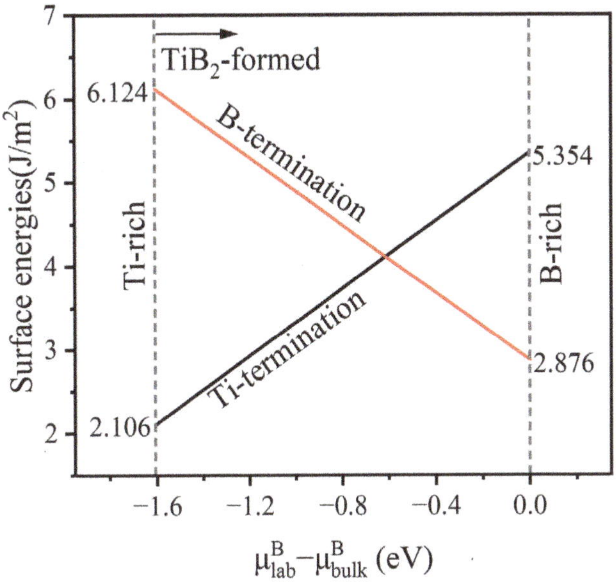

(**b**)

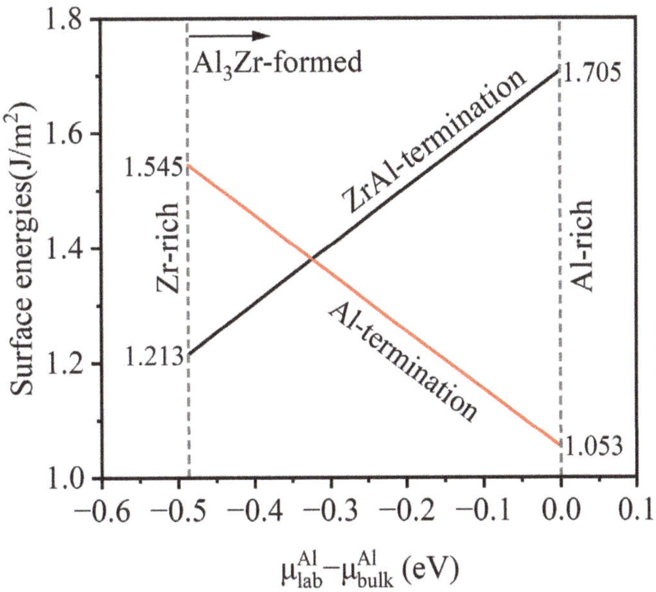

Figure 2. Relationship between the surface energy and chemical potential: (**a**) the function of surface energy with boron chemical potential difference for the Ti-terminated and B-terminated TiB_2 surfaces; (**b**) the function of surface energy with aluminum chemical potential difference for the Al-terminated and ZrAl-terminated Al_3Zr surfaces. Vertical dashed lines indicate the stability range of the compounds.

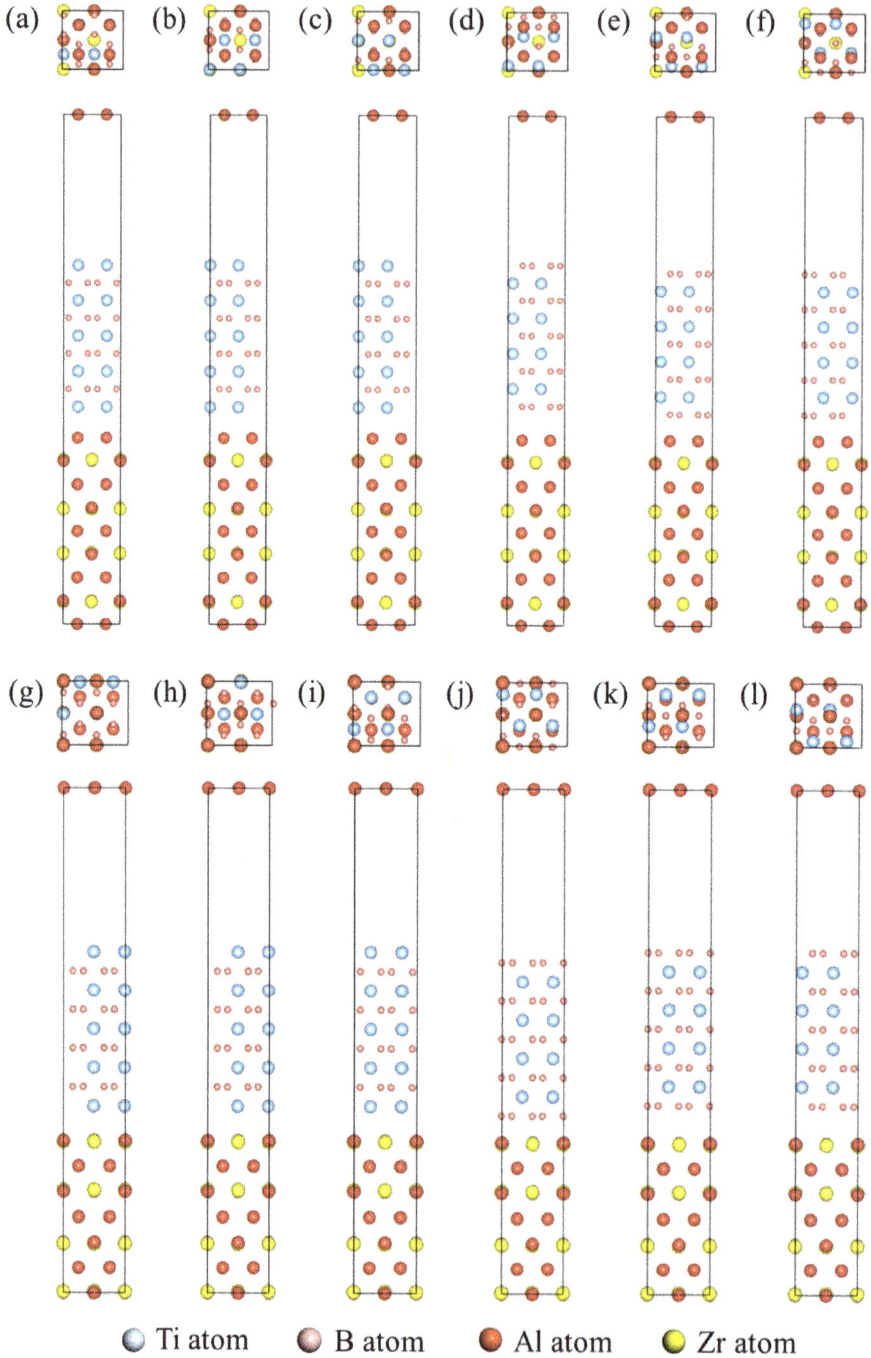

Figure 3. Twelve models for the TiB_2/Al_3Zr interface: (**a**) TAT interface; (**b**) TAB interface; (**c**) TAH interface; (**d**) BAT interface; (**e**) BAB interface; (**f**) BAH interface; (**g**) TZT interface; (**h**) TZB interface; (**i**) TZH interface; (**j**) BZT interface; (**k**) BZB interface; (**l**) BZH interface.

Table 4 displays the interfacial distances and corresponding adhesion work for 12 different interfaces. From Table 4, it is evident that the stacking position significantly impacted both the equilibrium distance and the adhesion work of the interfaces. Within the same stacking order, the adhesion work of the interfaces between the Ti-terminated and Al-terminated sides of the TiB_2/Al_3Zr was higher than that of other terminated interfaces, indicating better stability for interfaces involving the Ti termination compared with those with the Al termination.

Table 4. The comparison of interfacial distance (d_0) and adhesion work (W_{ad}) for 12 different interfaces.

Terminations of $TiB_2(0001)$	Terminations of $Al_3Zr(001)$	Stacking Sequence	Interface Name	d_0 (Å)	W_{ad} (J/m^2)
Ti	Al	TS	TAT	2.32	2.37
		BS	TAB	2.34	2.41
		HS	TAH	2.18	2.40
Ti	Zr-Al	TS	TZT	2.60	1.26
		BS	TZB	2.87	1.26
		HS	TZH	2.41	1.97
B	Al	TS	BAT	2.07	1.92
		BS	BAB	1.98	1.87
		HS	BAH	1.92	1.88
B	Zr-Al	TS	BZT	2.27	2.22
		BS	BZB	2.16	2.22
		HS	BZH	2.18	2.29

For interfaces between the B-terminated and ZrAl-terminated sides, the adhesion work for the BZT, BZB, and BZH interfaces were very close, at 2.22, 2.22, and 2.29 J/m^2, respectively, suggesting that the stacking order had a minimal impact on the bonding strength of these interfaces. However, for interfaces between the Ti-terminated and ZrAl-terminated sides, the effect of the stacking order was relatively more significant. The TZH interface, for example, had an adhesion work of 1.97 J/m^2, which was substantially higher than that of the TZT and TZB interfaces, yet their interfacial distances were the opposite. This is because, in the TZH interface, the Ti atoms interacted with the surrounding two Zr atoms and two Al atoms, affecting the interfacial distance. Generally, the higher the W_{ad}, the stronger the bonding force between the interface atoms, hence the TAB interface was the most stable among the 12 interface structures, possessing the greatest bonding strength.

3.3.3. Interfacial Energy

Interfacial energy is a key thermodynamic quantity used to measure the stability of an interface. Generally, the lower the interfacial energy, the more stable the interface structure. The formula for calculating the interfacial energy of the $TiB_2(0001)/Al_3Zr$ interface is as follows [47,48]:

$$\gamma_{int} = \gamma_{TiB_2(0001)} + \gamma_{ZrAl_3(001)} - W_{ad} \tag{10}$$

In the formula, $\gamma_{TiB_2(0001)}$ represents the interfacial energy; $\gamma_{TiB_2(0001)}$ and $\gamma_{ZrAl_3(001)}$, respectively, denote the surface energies of the $TiB_2(0001)$ and $Al_3Zr(001)$ surfaces; and W_{ad} represents the adhesion work of the corresponding interface.

Based on the previous analysis, we have identified interface structures with the lowest interfacial energies for further study: the TAB, TZH, BAT, and BZH interfaces. Figure 4 illustrates the variation in interfacial energy for these four interfaces as a function of the chemical potential difference between boron in the slab and bulk $\left(\mu^B_{slab} - \mu^B_{bulk}\right)$. Throughout the range of $\left(\mu^B_{slab} - \mu^B_{bulk}\right)$, the interfacial energies for TAB, TZH, BAT, and BZH were, respectively, 1.730 to 4.430 J/m^2, 1.848 to 5.509 J/m^2, 2.455 to 6.118 J/m^2, and 2.895 to 5.595 J/m^2. Evidently, for interfaces on the titanium side, TAB exhibits the lowest interfacial energy and the highest interface stability, consistent with the results obtained from the adhesion work analysis. For interfaces on the boron side, under titanium-rich conditions, BZH displayed relatively lower interfacial energy, indicating a more stable interface structure. In contrast, under boron-rich conditions, BAT showed better thermodynamic stability.

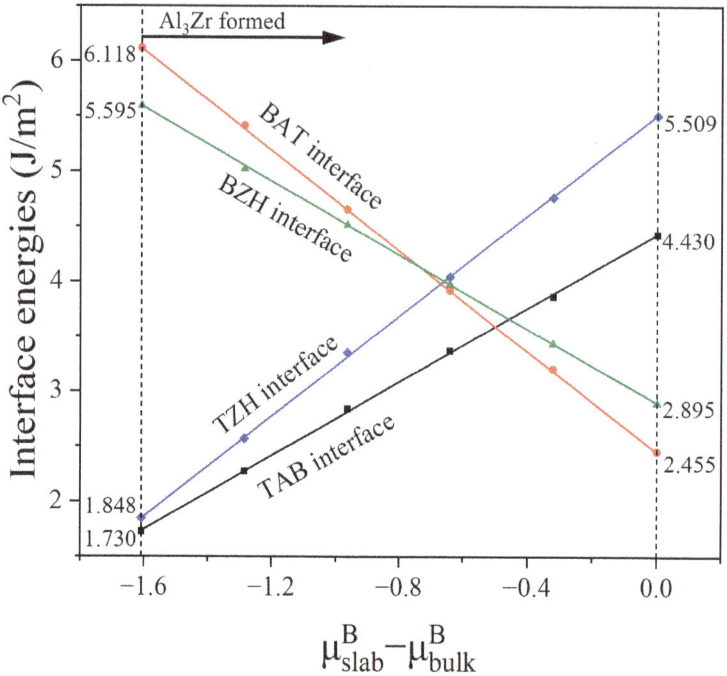

Figure 4. The relationship between the twelve types of $TiB_2(0001)/Al_3Zr(001)$ interfaces and varying chemical potentials of boron.

3.4. Electron Structure

The bonding at interfaces fundamentally involves interactions among the electrons located at the interface. To further analyze the interactions within the $TiB_2(0001)/Al_3Zr(001)$ interface, we selected representative models from four different terminated interfaces and calculated their differential charge density maps. These interfaces were the TAB, BAT, TZH, and BZH interfaces. Each of these interface models exhibited the highest adhesion work and the lowest interfacial energy within their respective terminated interface models. The differential charge density for the $TiB_2(0001)/Al_3Zr(001)$ interface can be expressed as follows [47]:

$$\Delta\rho = \rho_{toal} - \rho_{TiB2(0001)} - \rho_{Al3Zr(001)} \quad (11)$$

In this formula, ρ_{toal} is the total charge density of the $TiB_2(0001)/Al_3Zr(001)$ interface; and $\rho_{TiB2(0001)}$ and $\rho_{Al3Zr(001)}$ represent the independent charge densities of the $TiB_2(0001)$ surface and the $Al_3Zr(001)$ surface that make up the interface structure, respectively.

Figure 5 shows the differential charge density results for the four interface models. Red and blue represent the accumulation and loss of charge, respectively. From Figure 5a, it can be seen that there was a significant charge accumulation between the Ti and Al atoms at the TAB interface, which was due to the loss of charge from the Al and Ti atoms. This indicates the formation of strong Ti−Al covalent bonds at the $L1_2$-$Al_3Ti(111)$ surface. Additionally, some charge accumulated between the Al and Zr atoms, suggesting the formation of Al−Zr covalent bonds. As shown in Figure 5b, electron transfer occurred between the Al and B atoms, indicating the presence of ionic bonds; furthermore, there was a small accumulation of charge between the Al and B atoms, and the presence of B-Al covalent bonds. Figure 5c describes the differential charge density at the TZH interface, clearly showing Ti-Zr covalent bonding. On the BZH interface, strong covalent bonds were formed between the Zr and B atoms. It is noteworthy that Zr formed covalent bonds with both Al atoms in the same layer and those in the second layer, but the covalent bond between Zr and Al in the same layer was stronger than that between Zr and the second-layer Al, which was related to the distances between the atoms.

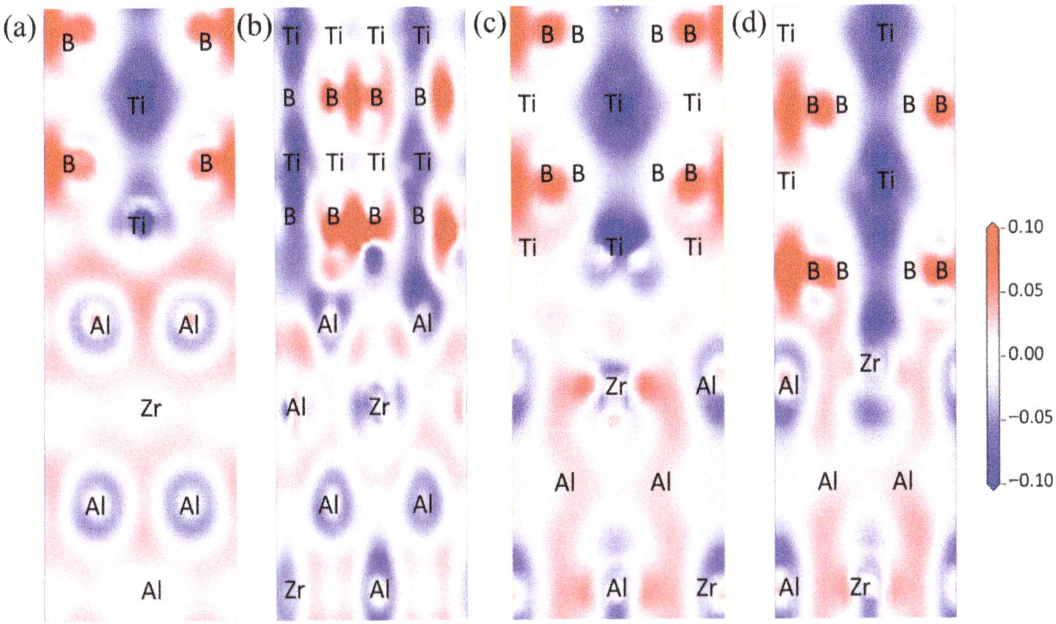

Figure 5. Differential charge density of the $TiB_2(0001)/Al_3Zr(001)$ interfaces: (**a**) TAB interface; (**b**) BAT interface; (**c**) TZH interface; (**d**) BZH interface.

To further investigate the electronic structure and analyze the bonding nature of the atoms at the $TiB_2(0001)/Al_3Zr(001)$ interface, we further analyzed the density of states (DOSs) of the stable interfaces, with results shown in Figure 6. Notable distinctions were observed in the sharpness of the peaks associated with the same elements in the $TiB_2(0001)$ slab near the interface junction, and similarly in the $Al_3Zr(001)$ slab. This indicates that the bonding process in these four interface models was primarily contributed by the outer electrons of the first layer of atoms at the interface end.

Continuing the analysis with a focus on the TAB interface model, we aimed to understand the bonding characteristics of the interface. From Figure 6a, it is evident that the first layer of Ti at the $TiB_2(0001)$ end and the first layer of Al at the $Al_3Zr(001)$ end exhibited a peak between 0 and 5 eV, primarily contributed by the Ti 3d and Al 3p orbitals. Moreover, hybridization between Ti 3d and Al 3p also occurred around 10 eV. Notably, there was intense hybridization near 8 eV between the first layer of Ti at the $TiB_2(0001)$ end and the second layer of Zr at the $Al_3Zr(001)$ end, mainly contributed by the Ti 3d and Zr 4d orbitals. The peak intensities indicated strong hybridization at the TAB interface.

In Figure 6b, the hybridization between the first layer of B on the $TiB_2(0001)$ side and the first layer of Al on the $Al_3Zr(001)$ side was also evident. A distinct peak appeared near the Fermi level, mainly due to the contributions of the B 2p and Al 3s orbitals. Additionally, the 2p orbital of the first-layer B atoms showed a prominent peak around 5 eV, and there were indications of increased orbital density around 5 eV in the 3p orbital of the first-layer Al atoms. This suggests a possible overlap between the B 2p and Al 3p orbitals, potentially pointing to the presence of hybridization. Such hybridization could lead to the formation of strong B-Al covalent bonds at the interface.

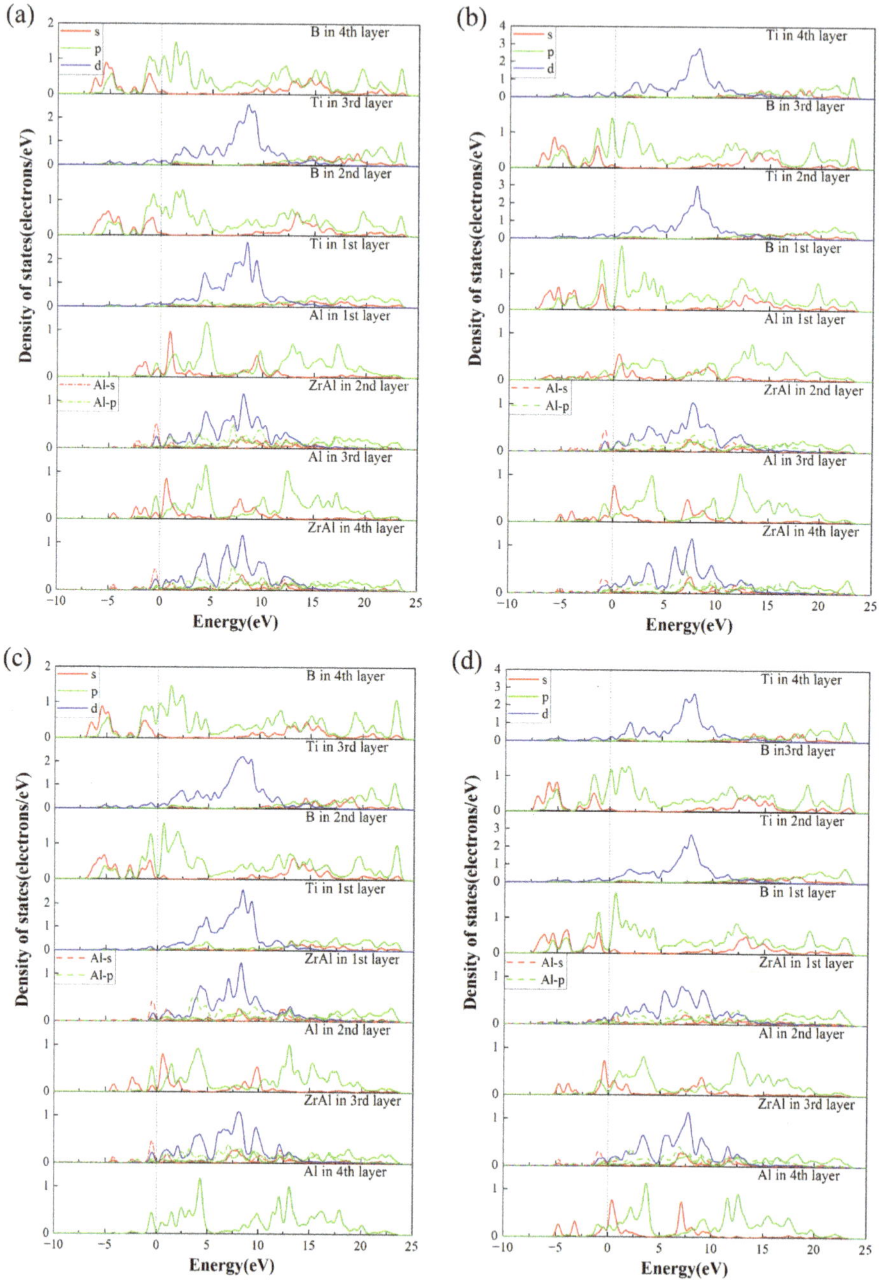

Figure 6. Partial density of states for the $TiB_2(0001)/Al_3Zr(001)$ interfaces: (**a**) TAB interface; (**b**) BAT interface; (**c**) TZH interface; (**d**) BZH interface. Vertical dashed lines indicate the Fermi level.

Figure 6c shows significant hybridization between Ti 3d and Zr 4d in the 5 to 10 eV range, forming Ti-Zr covalent bonds. The interface composition in Figure 6d, which consisted of the first layer of B at the $TiB_2(0001)$ end and the first layer at the ZrAl end of $Al_3Zr(001)$, however, showed no significant common peaks, with only slight hybridization near the Fermi level.

The density of states analysis in Figure 6 revealed variations in the electronic states of Al atoms when adjacent to Zr compared with their presence in layers without Zr. This observation suggests differences in orbital interactions, likely due to hybridization, between Al and neighboring atoms under different compositional contexts. Notably, the analysis showed that the hybridization patterns were more pronounced in the TAB interface, where peak intensities suggested a stronger covalent character in the bonding interactions, particularly involving Al 3p and Zr 4d orbitals. This finding underscores the impact of atomic positioning and local chemical environment on the electronic properties at the interface. Therefore, it can be concluded that the hybridization at the TAB interface exhibited a higher intensity and complexity; based on the analysis of the density of states and differential charge density, the order of interface stability is as follows: TAB interface > BZH interface > TZH interface > BAT interface. This is consistent with the previous analyses of adhesion work and interfacial energy.

4. Conclusions

To understand the interfacial bonding characteristics of the $TiB_2(0001)/Al_3Zr(001)$ interface, first-principles studies were conducted focusing on the adhesion work, interfacial energy, and electronic structure. This study considered 12 different TiB_2/Al_3Zr interfaces featuring four different terminations and three stacking sequences, with a detailed investigation on four of the most stable structures. The main findings of the study are summarized as follows:

(1) Simulation tests have demonstrated that a stable interface structure can be formed when a 9-layer $TiB_2(0001)$ surface is combined with a 7-layer ZrAl-terminated and a 9-layer Al-terminated $Al_3Zr(001)$ surface.
(2) For interfaces with the same termination in TiB_2/Al_3Zr, the study identifies the bridge-site stacking (TAB) at the T/Al termination, hollow-site stacking (TZH) at the Ti/ZrAl termination, top-site stacking (BAT) at the B/A termination, and hollow-site stacking at the B/ZrAl termination as the optimal structures. Among these interfaces, the TAB interface has the strongest adhesion strength due to the highest adhesion work of 2.41 J/m^2.
(3) The interfacial energies of TAB, TZH, BAT, and BZH across the entire range of $\left(\mu^{B}_{slab} - \mu^{B}_{bulk}\right)$ are, respectively, 1.730 to 4.430 J/m^2, 1.848 to 5.509 J/m^2, 6.118 to 2.455 J/m^2, and 5.595 to 2.895 J/m^2. Evidently, TAB exhibits the lowest surface energy and the highest interface stability.
(4) Electronic structure analysis indicates that the TAB, TZH, and BZH interfaces primarily feature covalent bonding, while the BAT interface shows both ionic and covalent bonds. According to the analysis of the electronic structure, the stability order of the $TiB_2(0001)$ and $Al_3Zr(001)$ interfaces is as follows: TAB interface > BZH interface > TZH interface > BAT interface.

Author Contributions: Conceptualization, H.N.; Methodology, H.N.; Software, J.L. and Z.X.; Validation, G.W.; Formal analysis, Z.C.; Investigation, W.Z.; Data curation, Z.Y.; Writing—original draft preparation, X.P. and L.Y.; Writing—review and editing, M.P.; Visualization, H.N.; Supervision, X.P.; Project administration, H.T.; Funding acquisition, H.T. All authors have read and agreed to the published version of the manuscript.

Funding: This research work is supported by the National Natural Science Foundation of China-Guangxi Joint Fund (U20A20276), the Guangxi Natural Science Foundation (2021GXNSFAA220088, 2021GXNSFBA220080), the National Natural Science Foundation of China (51761002, 51961008), the Special Funds for Local Scientific and Technological Development Funds Guided by the Central Government in 2021 (GuiKeZY21195030), the Guangxi Science and Technology Base and Talent Project in 2022 (GuiKeAD21238010), the Key research and development project of Guangxi (GuiKeAB22080061), and the Research Project of Guangxi Key Laboratory of Processing for Non-ferrous Metallic and Featured Materials (2021GXMPSF06).

Data Availability Statement: The original contributions presented in the study are included in the article, further inquiries can be directed to the corresponding author.

Conflicts of Interest: Author Liujiang Yang was employed by the company Jiangsu Runyang Yueda Century Photovoltaic Technology. Author Mingjun Pang was employed by the company SAIC-GM-Wuling Automobile. Author Gaobao Wang was employed by the company Guangxi Youhe Metal Materials Technology Co., Ltd. The remaining authors declare that the research was conducted in the absence of any commercial or financial relationships that could be construed as a potential conflict of interest.

References

1. Cheng, W.; Liu, C.; Ge, Z. Optimizing the mechanical properties of Al–Si alloys through friction stir processing and rolling. *Mater. Sci. Eng. A* **2021**, *804*, 140786. [CrossRef]
2. Jin, S.; Luo, Z.; An, X.; Liao, X.; Li, J.; Sha, G. Composition-dependent dynamic precipitation and grain refinement in Al-Si system under high-pressure torsion. *J. Mater. Sci. Technol.* **2021**, *68*, 199–208. [CrossRef]
3. Yin, J.; Li, Y.; Niu, Y.; Zhang, Z.; Song, D.; Chen, Y.; Liu, Y.; Zhong, Y. Effect of particle size on the microstructure and tensile property of the Al-28.5 Si alloy prepared by continuous powder extrusion. *Mater. Res. Express* **2021**, *8*, 026524. [CrossRef]
4. Wang, F.; Qiu, D.; Liu, Z.-L.; Taylor, J.A.; Easton, M.A.; Zhang, M.-X. The grain refinement mechanism of cast aluminium by zirconium. *Acta Mater.* **2013**, *61*, 5636–5645. [CrossRef]
5. Xu, J.-h.; Freeman, A. Bandfilling and structural stability of trialuminides: YAl_3, $ZrAl_3$, and $NbAl_3$. *J. Mater. Res.* **1991**, *6*, 1188–1199. [CrossRef]
6. Moon, K.I.; Kim, S.C.; Lee, K.S. A study on the microstructure of $D0_{23}$ Al_3Zr and $L1_2$ (Al+ 12.5 at.% Cu) 3Zr intermetallic compounds synthesized by PBM and SPS. *Intermetallics* **2002**, *10*, 185–194. [CrossRef]
7. Xu, J.-H.; Freeman, A. Phase stability and electronic structure of $ScAl_3$ and $ZrAl_3$ and of Sc-stabilized cubic $ZrAl_3$ precipitates. *Phys. Rev. B* **1990**, *41*, 12553. [CrossRef]
8. Yang, Q.-B.; Deng, Y.-J.; Mou, Y.; Zhang, Z.-Q.; Li, W.-G.; Qing, L. Effect of Al_3Zr particles on hot-compression behavior and processing map for Al-Cu-Li based alloys at elevated temperatures. *Trans. Nonferrous Met. Soc. China* **2020**, *30*, 872–882. [CrossRef]
9. Li, Y.; Lu, B.; Yu, W.; Fu, J.; Xu, G.; Wang, Z. Two-stage homogenization of Al-Zn-Mg-Cu-Zr alloy processed by twin-roll casting to improve $L1_2$ Al_3Zr precipitation, recrystallization resistance, and performance. *J. Alloys Compd.* **2021**, *882*, 160789. [CrossRef]
10. Wang, Y.; Wu, X.; Cao, L.; Tong, X.; Zou, Y.; Zhu, Q.; Tang, S.; Song, H.; Guo, M. Effect of Ag on aging precipitation behavior and mechanical properties of aluminum alloy 7075. *Mater. Sci. Eng. A* **2021**, *804*, 140515. [CrossRef]
11. Yang, T.; Han, X.; Ding, Z.; Wang, Y.; Li, J. Electronic and structural properties of low-index $L1_2$–Al_3Zr surfaces by first-principle calculations. *Calphad* **2019**, *66*, 101645. [CrossRef]
12. Yang, T.; Wei, M.; Ding, Z.; Han, X.; Li, J. First-principle calculations on the Al/$L1_2$-Al_3Zr heterogeneous nucleation interface. *Calphad* **2020**, *69*, 101768. [CrossRef]
13. Yang, T.; Han, X.; Li, W.; Chen, X.; Liu, P. Angular dependent potential for Al-Zr binary system to study the initial heterogeneous nucleation behavior of liquid Al on $L1_2$-Al_3Zr. *Comput. Mater. Sci.* **2023**, *230*, 112480. [CrossRef]
14. Birol, Y. Effect of silicon content in grain refining hypoeutectic Al–Si foundry alloys with boron and titanium additions. *Mater. Sci. Technol.* **2012**, *28*, 385–389. [CrossRef]
15. Wang, Y.; Fang, C.; Zhou, L.; Hashimoto, T.; Zhou, X.; Ramasse, Q.; Fan, Z. Mechanism for Zr poisoning of Al-Ti-B based grain refiners. *Acta Mater.* **2019**, *164*, 428–439. [CrossRef]
16. Mao, G.; Tong, G.; Gao, W.; Liu, S.; Zhong, L. The poisoning effect of Sc or Zr in grain refinement of Al-Si-Mg alloy with Al-Ti-B. *Mater. Lett.* **2021**, *302*, 130428. [CrossRef]
17. Fan, Z.; Wang, Y.; Zhang, Y.; Qin, T.; Zhou, X.; Thompson, G.; Pennycook, T.; Hashimoto, T. Grain refining mechanism in the Al/Al–Ti–B system. *Acta Mater.* **2015**, *84*, 292–304. [CrossRef]
18. Tang, Y.; Chen, Z.; Borbely, A.; Ji, G.; Zhong, S.; Schryvers, D.; Ji, V.; Wang, H. Quantitative study of particle size distribution in an in-situ grown Al–TiB_2 composite by synchrotron X-ray diffraction and electron microscopy. *Mater. Charact.* **2015**, *102*, 131–136. [CrossRef]
19. Wang, Q.; Li, Y.; Chen, S.; Liu, X.; Chen, Z.; Wang, M.; Zhu, H.; Wang, H. Interface alloying design to improve the dispersion of TiB_2 nanoparticles in Al composites: A first-principles study. *J. Phys. Chem. C* **2021**, *125*, 5937–5946. [CrossRef]
20. Liu, Z.; Dong, Z.; Cheng, X.; Zheng, Q.; Zhao, J.; Han, Q. On the supplementation of magnesium and usage of ultrasound stirring for fabricating in situ TiB 2/A356 composites with improved mechanical properties. *Metall. Mater. Trans. A* **2018**, *49*, 5585–5598. [CrossRef]
21. Li, Y.; Hu, B.; Liu, B.; Nie, A.; Gu, Q.; Wang, J.; Li, Q. Insight into Si poisoning on grain refinement of Al-Si/Al-5Ti-B system. *Acta Mater.* **2020**, *187*, 51–65. [CrossRef]

22. Li, Y.; Jiang, Y.; Hu, B.; Li, Q. Novel Al-Ti-Nb-B grain refiners with superior efficiency for Al-Si alloys. *Scr. Mater.* **2020**, *187*, 262–267. [CrossRef]
23. Xiao, H.-Y.; Li, Y.-G.; Geng, J.-W.; Li, H.-P.; Wang, M.-L.; Dong, C.; Li, Z.-G.; Wang, H.-W. Effects of nano-sized TiB_2 particles and Al_3Zr dispersoids on microstructure and mechanical properties of Al–Zn–Mg–Cu based materials. *Trans. Nonferrous Met. Soc. China* **2021**, *31*, 2189–2207. [CrossRef]
24. Liu, S.-D.; Zhang, X.-M.; Chen, M.-A.; You, J.-H.; Zhang, X.-Y. Effect of Zr content on quench sensitivity of AIZnMgCu alloys. *Trans. Nonferrous Met. Soc. China* **2007**, *17*, 787–792. [CrossRef]
25. Pang, X.; Yang, L.; Yang, J.; Pang, M.; Xu, Z.; Li, A.; Wei, B.; Tang, H. Understanding the poisoning mechanisms of Si and Zr atoms on $L1_2$ Al_3Ti (111) surface: A first-principles investigation. *Vacuum* **2023**, *210*, 111891. [CrossRef]
26. Wu, J.; Ruan, Q.; Chen, S.; Meng, C.; Xu, Z.; Wei, C.; Tang, H.; Wang, J. Insights into poisoning mechanism of Zr by first principle calculation on adhesion work and adsorption energy between TiB_2, Al_3Ti, and Al_3Zr. *Metals* **2022**, *12*, 286. [CrossRef]
27. Huang, J.; Xiang, Z.; Tang, Y.; Li, J.; Shen, G.; Shi, G.; Ma, X. Precipitation behavior of $D0_{23}$, $L1_2$ type Al_3zr and interfacial characteristics in Tib_2 reinforced Al-Zn-Mg-Cu-Zr composite. SSRN. Available online: https://papers.ssrn.com/abstract_id=4100289 (accessed on 17 October 2024).
28. Huang, J.; Xiang, Z.; Tang, Y.; Li, J.; Shen, G.; Shi, G.; Ma, X.; Chen, Y.; Chen, Z. Interfacial precipitation of the in-situ TiB_2-reinforced Al–Zn–Mg–Cu–Zr composite. *Mater. Chem. Phys.* **2023**, *301*, 127561. [CrossRef]
29. Hu, H.; Zhao, M.; Wu, X.; Jia, Z.; Wang, R.; Li, W.; Liu, Q. The structural stability, mechanical properties and stacking fault energy of Al_3Zr precipitates in Al-Cu-Zr alloys: HRTEM observations and first-principles calculations. *J. Alloys Compd.* **2016**, *681*, 96–108. [CrossRef]
30. Segall, M.; Lindan, P.J.; Probert, M.a.; Pickard, C.J.; Hasnip, P.J.; Clark, S.; Payne, M. First-principles simulation: Ideas, illustrations and the CASTEP code. *J. Phys. Condens. Matter* **2002**, *14*, 2717. [CrossRef]
31. Vanderbilt, D. Soft self-consistent pseudopotentials in a generalized eigenvalue formalism. *Phys. Rev. B* **1990**, *41*, 7892. [CrossRef]
32. Wu, Z.; Cohen, R.E. More accurate generalized gradient approximation for solids. *Phys. Rev. B* **2006**, *73*, 235116. [CrossRef]
33. Monkhorst, H.J.; Pack, J.D. Special points for Brillouin-zone integrations. *Phys. Rev. B* **1976**, *13*, 5188. [CrossRef]
34. Head, J.D.; Zerner, M.C. A Broyden—Fletcher—Goldfarb—Shanno optimization procedure for molecular geometries. *Chem. Phys. Lett.* **1985**, *122*, 264–270. [CrossRef]
35. Zhang, H.; Xiong, H.; Wang, D.; Wang, W. Mechanism of NbC heterogeneous nucleation on TiN in microalloyed steel: A first-principles study. *Comput. Mater. Sci.* **2018**, *146*, 126–133. [CrossRef]
36. Wang, Q.; Liu, C.; Yao, R.; Zhu, H.; Liu, X.; Wang, M.; Chen, Z.; Wang, H. First-principles study on the stability and work function of low-index surfaces of TiB_2. *Comput. Mater. Sci.* **2020**, *172*, 109356. [CrossRef]
37. Liu, R.; Yin, X.; Feng, K.; Xu, R. First-principles calculations on Mg/TiB_2 interfaces. *Comput. Mater. Sci.* **2018**, *149*, 373–378. [CrossRef]
38. Topor, L.; Kleppa, O. Enthalpies of formation of first-row transition-metal diborides by a new calorimetric method. *J. Chem. Thermodyn.* **1985**, *17*, 1003–1016. [CrossRef]
39. Ghosh, G.; Vaynman, S.; Asta, M.; Fine, M. Stability and elastic properties of $L1_2$-(Al, Cu) 3 (Ti, Zr) phases: Ab initio calculations and experiments. *Intermetallics* **2007**, *15*, 44–54. [CrossRef]
40. Desch, P.; Schwarz, R.; Nash, P. Formation of metastable $L1_2$ phases in Al_3Zr and Al-12.5% X-25% Zr (X ≡ Li, Cr, Fe, Ni, Cu). *J. Less Common Met.* **1991**, *168*, 69–80. [CrossRef]
41. Li, Y.; Gao, Y.; Xiao, B.; Min, T.; Fan, Z.; Ma, S.; Yi, D. Theoretical study on the electronic properties and stabilities of low-index surfaces of WC polymorphs. *Comput. Mater. Sci.* **2011**, *50*, 939–948. [CrossRef]
42. Rapcewicz, K.; Chen, B.; Yakobson, B.; Bernholc, J. Consistent methodology for calculating surface and interface energies. *Phys. Rev. B* **1998**, *57*, 7281. [CrossRef]
43. Batyrev, I.; Alavi, A.; Finnis, M.W. Ab initio calculations on the Al_2O_3 (0001) surface. *Faraday Discuss.* **1999**, *114*, 33–43. [CrossRef]
44. Song, M.; Huang, B.; Zhang, M.; Li, J. Study of formation behavior of TiC ceramic obtained by self-propagating high-temperature synthesis from Al–Ti–C elemental powders. *Int. J. Refract. Met. Hard Mater.* **2009**, *27*, 584–589. [CrossRef]
45. Han, Y.; Dai, Y.; Shu, D.; Wang, J.; Sun, B. First-principles study of TiB_2 (0001) surfaces. *J. Phys. Condens. Matter* **2006**, *18*, 4197. [CrossRef] [PubMed]
46. Jin, W.; Li, L.; Zhang, S.; Yang, H.; Gao, K.; Pang, X.; Volinsky, A.A. First principles calculations of interfacial properties and electronic structure of the AlN (0 0 0 1)/Ti (0 0 0 1) interface. *Chem. Phys. Lett.* **2018**, *713*, 153–159. [CrossRef]
47. Pang, X.; Yang, J.; Pang, M.; He, J.; Yang, W.; Qin, H.; Zhan, Y. Theoretical understanding of atomic and electronic structures of the ZrC (111)/Cu (111) interface. *J. Alloys Compd.* **2019**, *791*, 431–437. [CrossRef]
48. Xiong, H.; Liu, Z.; Zhang, H.; Du, Z.; Chen, C. First principles calculation of interfacial stability, energy and electronic properties of SiC/ZrB_2 interface. *J. Phys. Chem. Solids* **2017**, *107*, 162–169. [CrossRef]

crystals

MDPI

Article

Toughening Mechanism of $CaAl_{12}O_{19}$ in Red Mud–Al_2O_3 Composite Ceramics

Shiwei Jiang [1], Anmin Li [2,]* and Zhengliang Wang [2]

[1] School of Resources, Environment and Materials, Guangxi University, Nanning 530004, China; 2215391037@st.gxu.edu.cn

[2] State Key Laboratory of Featured Metal Materials and Life-Cycle Safety for Composite Structures, Guangxi University, Nanning 530004, China; 2315391099@st.gxu.edu.cn

* Correspondence: lianmin@gxu.edu.cn

Abstract: The utilization of red mud in the production of ceramic products represents an efficient approach for harnessing red mud resources. Composite ceramics were prepared from Al_2O_3, red mud, and Cr_2O_3 by atmospheric pressure sintering, and the phase composition and microscopic morphology of the composite ceramics were investigated by XRD, SEM, and EDS. The flexural strength and fracture toughness of composite ceramics were measured by three-point bending and SENB methods. The results showed that the composite ceramics sintered at 1500 °C with the addition of 1.5 wt.% Cr_2O_3 had a flexural strength of 297.03 MPa, a hardness of 17.44 GPa, and a densification of 97.75% and fracture toughness of 6.57 MPa·$m^{1/2}$. The addition of Cr_2O_3 helps to improve the low strength of red mud composite ceramic samples. The $CaAl_{12}O_{19}$ phase can form a similar "endo-crystalline" structure with Al_2O_3 grains, which changes the fracture mode of the ceramics and thus significantly improves the fracture toughness. The wettability tests conducted on Cu and RM–Al_2O_3 composite ceramic materials revealed that the composites exhibited non-wetting behavior towards Cu at elevated temperatures, while no interfacial reactions or elemental diffusion were observed. Composites have higher surface energy than Al_2O_3 ceramic at high temperatures. The present study provides a crucial foundation for enhancing the comprehensive utilization value of red mud and the application of red mud ceramics in the field of electronic packaging.

Keywords: red mud; alumina composite ceramics; mechanical properties; surface energy

1. Introduction

"Red mud" (RM) refers to solid waste that is discharged during the process of generating Al_2O_3. Comprised primarily of various complex mineral phases, red mud possesses a highly alkaline nature. The current situation lacks an effective solution for red mud treatment, primarily relying on the construction of multiple containment dams to store the red mud [1,2]. With prolonged stockpiling, the harmful elements in red mud can contaminate nearby water sources and cause soil alkalization. At the same time, red mud is highly adsorbent, and dams may also be at risk of breaking during the rainy season. This seriously jeopardizes the safety of people's lives and property. Wang et al. [3] took red mud and kaolin as raw materials, changed the reaction path of mullite by adding $(NH_4)_6Mo_7O_{24}$, and sintered the sample at 1180 °C to obtain a sample with a flexion strength of 185.6 MPa, a bulk weight of 1.45 g/cm^3, and a water absorption rate of 5.5%. He et al. [4] used red mud, clay and raw materials, and a 20% RM ceramic body was prepared by sintering at 1050 °C for 2 h. The ceramic body had a bulk weight of 1.69 g/cm^3, loss on ignition of 10.32%, sintering shrinkage of 5.71%, water absorption of 19.91%, and compressive strength of 36.5 MPa. Pérez-Villarejo et al. [5] mixed red mud and clay in a 1:1 ratio and sintered at 950 °C for 1 h. The samples obtained had a linear shrinkage of 0.46%, a water absorption of 21.00%, a weight loss of 12.6% after sintering, and a compressive strength of

52.54 MPa. Triviño et al. [6] mixed red mud and granite waste in a ratio of 6:4 and sintered the sample at 1150 °C, and the sample obtained had the highest hardness (545.6 kgf/mm^2) and a flexural strength of 27 MPa. The addition of red mud should be carefully controlled in order to enhance the properties of ceramic products, as red mud itself contains high levels of Na_2O and CaO, which are detrimental to the properties of cement and ceramics. Furthermore, the high content of red mud in the sample also makes it easier for the harmful elements in the red mud to leak into the environment, which limits the application of red mud ceramic products.

Cr_2O_3 is often used as one of the additives to improve the physical properties of alumina, Cr_2O_3 and Al_2O_3 have isostructural corundum crystal structures (hexagonal system with the same space group R-3c), and it is well known that Cr_2O_3 forms a substitutional solid solution in the Al_2O_3 lattice by ion exchange with Al^{3+} and does not form any eutectic in the whole compositional range [7]. Several studies have shown the addition of Cr_2O_3 to Al_2O_3 in different ratios to improve the mechanical properties of ceramics by means of different solidification techniques. Kafkaslıoğlu Yıldız et al. [8] investigated the effect of Cr_2O_3 additions with different volume ratios (0.5, 1, 5 vol.%) on the microstructure and mechanical properties of Al_2O_3. When the Cr_2O_3 addition was 0.5 vol.%, the flexural strength of the material was 286 MPa, which increased by 44% compared to pure Al_2O_3. This is due to the localized compressive stress caused by the grain boundary alteration of Cr^{3+} ions. The crack extension behavior changes from a predominantly along-crystal extension to mixed along-crystal and through-crystal extensions with the addition of localized compressive stresses from Cr_2O_3, leading to grain boundary strengthening. Azhar et al. [9] reported that the microstructures of ZTA composite were significantly affected by the addition of a small amount of Cr_2O_3 (~0.6 wt.%), with the grains becoming larger and acquiring a plate-like shape. The fracture toughness was increased from 4.41 MPa·m$^{1/2}$ to 4.73 MPa·m$^{1/2}$. Manshor et al. [10] reported that with addition of 0.6 wt.% of Cr_2O_3 to the composite, an optimum density of 4.0630 g/cm^3, hardness of 1681 HV, and fracture toughness of 7.15 MPa·m$^{1/2}$ were obtained. The addition of 0.6 wt.% Cr_2O_3 led to the formation of large plate-like grains, enhancing the interaction between the crack and the matrix grains. The inter-granular fracture mode caused the cracks to spread along the borders of the grains, eventually increasing the strength due to grain deflection.

The work in this paper is motivated by the exhaust gas recirculation (EGR) system utilized in diesel engines, which introduces an appropriate quantity of exhaust gas into the cylinder to participate in combustion. As a result, it effectively reduces the maximum temperature within the cylinder and subsequently mitigates NOx emissions. Similarly, adding an appropriate amount of red mud to Al_2O_3 ceramics can be used as a sintering aid for the sintering of Al_2O_3 ceramics, so as to realize the resource utilization of red mud. In this paper, the effects of different sintering temperatures and Cr_2O_3 contents on the properties and microstructures of RM–Al_2O_3 composite ceramics have been investigated by adding 10% RM and different contents of Cr_2O_3 to alumina, and the optimal sintering temperatures and addition amounts have been determined. The wetting behavior of Cu on the samples at elevated temperatures have been investigated by using the seated drop method, while the surface energy of the composite ceramics at high temperature has been estimated based on the data obtained from high-temperature contact angle measurements, so as to explore the possibility of applying RM–Al_2O_3 composite ceramics in the field of electronic packaging.

2. Materials and Methods

2.1. Raw Materials

The Al_2O_3 powder (200 nm, $Al_2O_3 \geq 99.9\%$) used in this study was produced by Beasley New Materials Co. (Suzhou, China). Cr_2O_3 powder (100 nm, $Cr_2O_3 \geq 99.9\%$) was produced by Hebei Yuanying New Materials Co. (Xingtai, China). Red mud was provided by Guangxi Huayin Aluminium Co., Ltd. (Baise, China)., and Fe_3O_4 in red mud was separated by strong magnetic separation. The red mud was dried in a drying oven at

120 °C for 24 h. The dried mass was crushed using a pulverizer. The composition of the red mud analyzed by XRF is shown in Table 1. Stearic acid, polypropylene, and paraffin were produced by Shanghai McLean Biochemical Technology Co. (Shanghai, China) Cu (≥99.95%) used was produced by CMT New Materials Co. (Attleboro, MA, USA) and processed into 3 mm × 3 mm × 3 mm cubes by wire cutting.

Table 1. The chemical composition of red mud (wt.%).

Fe_2O_3	Al_2O_3	SiO_2	CaO	Na_2O	TiO_2	Loss
29.60	18.90	16.90	15.60	11.10	5.78	2.12

2.2. Preparation of Composite Ceramics

The powder ingredients were proportioned in accordance with the composition presented in Table 2. In addition, 1 g of polypropylene, 1 g of paraffin, and 1 g of stearic acid were additionally added to the mixed powder (30 g total weight). A high-energy planetary ball mill was employed to grind the mixed powder for a duration of 30 h at a rotational speed of 300 revolutions per minute, utilizing anhydrous ethanol as the grinding medium. The mixed slurry was subjected to vacuum drying at 70 °C for 24 h in a drying oven. Subsequently, the dried mass was finely ground using an onyx mortar and then sieved through an 80-mesh sieve. The powder was pressed into 40 mm × 4 mm × 10 mm slabs using a hydraulic press at 15 MPa for 3 min. The samples were sintered using a $MoSi_2$ electric furnace (KSL-1700-2A). The temperature was ramped up to 1200 °C at a rate of 5 °C/min, followed by a further ramp-up to the target temperature at a rate of 2 °C/min. The target temperature was maintained for 2 h, and then the samples were cooled with the furnace. Additionally, each sample required a debinding process where it was held at 800 °C for 2 h.

Table 2. Chemical compositions (wt.%) of the samples.

Sample	Al_2O_3	RM	Cr_2O_3
A	100	0	0
ARC0	90	10	0
ARC1	89.1	9.9	1
ARC2	88.65	9.85	1.5
ARC3	88.2	9.8	2
ARC4	87.75	9.75	2.5
ARC5	87.3	9.7	3

2.3. Characterization of Composite Ceramics

The test samples underwent pre-treatment prior to conducting the tests. The ceramic samples were individually polished using 2.5 w and 1 w abrasive pastes, respectively. Subsequently, the ceramic samples with observed surface morphology were subjected to etching using an etching solution (3% HF, 5% HN_3) for a duration of 1 min in order to eliminate the glassy phase. The elemental composition of the red mud was analyzed using an X-ray fluorescence spectrometer (S8 TIGER, BRUKER, Berlin, Germany). Following this, ultrasonic cleaning was performed on the ceramic samples in anhydrous ethanol for a period of 5 min. X-ray diffraction analyses were carried out using an X-ray diffractometer (MiniFlex600-C, Rigaku Co., Ltd., Akishima, Japan) at 40 kV/15 mA, with a scanning range of 10~85 at a 2θ angle and a scanning speed of 5 °/min. A scanning electron microscope (SU8020, Hitachi High-Technologies, Tokyo, Japan) was used to observe the surface morphology and fracture morphology of the samples. The apparent density of the composite ceramics was measured using the Archimedes drainage method, and the densification was expressed as the ratio of the apparent density to the theoretical density. The phase compositions of the samples were analyzed semi-quantitatively by EDS (energy-dispersive X-ray spectroscopy). Vickers hardness tester was used to measure the hardness of the samples (loading force 5 Kg, holding pressure 15 s), and in each group of 4 specimens, each specimen was tested in 10 different positions, and then the average value was taken.

The samples were machined into 40 mm × 3 mm × 6 mm blocks by diamond wire cutting, the fracture toughness of the samples was measured using the single edge notched beam (SENB) method as shown in Figure 1a, and the loading speed was 0.5 mm/min during the test. The fracture toughness K_{IC} is calculated from Equation (1):

$$K_{IC} = Y\frac{3PL}{2bh^2}\sqrt{a} \tag{1}$$

where Y is the shape factor, P is the fracture load, L is the support span, a is the depth of the cut, b is the width, w is the thickness, and c is the width of the cut. When span length/thickness/width = 8:2:1, the shape factor Y is as follows:

$$Y = 1.96 - 2.75\left(\frac{a}{h}\right) + 13.66\left(\frac{a}{h}\right)^2 - 23.98\left(\frac{a}{h}\right)^3 + 25.22\left(\frac{a}{h}\right)^4 \tag{2}$$

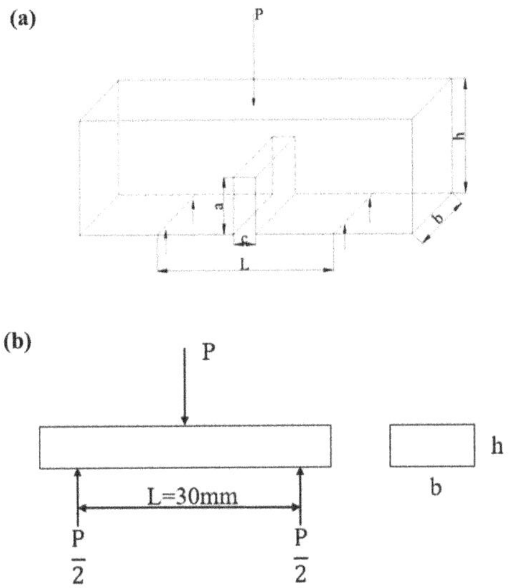

Figure 1. (**a**) Single edge notched beam schematic. (**b**) Flexural strength test schematic.

The flexural strength was measured by the three-point bending method as shown in Figure 1b. The size of the specimen was 40 mm × 3 mm × 12 mm, and the loading speed during the test was 0.5 mm/min. The bending strength can be calculated by Equation (3):

$$\sigma = \frac{3PL}{2Bh^2} \tag{3}$$

where P is the fracture load, L is the support span, b is the width, and h is the thickness.

High-temperature wettability tests were performed in a vacuum tube furnace (KJ-GSL-1800X, Hefei, China). The oxide film on the surface of the copper used for the test was polished with sandpaper and ultrasonically cleaned in anhydrous ethanol. The copper particles were placed on the ceramic samples and heated together in a vacuum tube furnace. The temperature was ramped up to the target temperature at a rate of 10 °C/min and held for ten minutes, and when the copper droplets were stabilized, a high-definition camera was used to take pictures. Finally, the contact angle was calculated using ImageJ (1.54v) software. The contact angles of Cu droplets with ceramics were measured at 1133, 1183, 1233, 1283, and 1333 °C, respectively.

3. Results and Discussion

3.1. Phase Composition

Figure 2a shows the XRD patterns of ARC2 samples sintered at different temperatures for 2 h. The main phases in the sample are corundum and $CaAl_{12}O_{19}$ (CA6), with small amounts of calcium silicate, quartz, pyroxene, and NaFe $(SiO_3)_2$. As the sintering temperature rises, most of the silicate is gradually burnt out. The XRD patterns of samples sintered at 1500 °C for 2 h, ranging from ARC1 to ARC5, are depicted in Figure 2b. The composition of the phases remains unchanged. It is worth noting that an increase in Cr_2O_3 content leads to a slight shift of the diffraction peak towards smaller angles. This can be attributed to that the formation of a complete replacement solid solution between Cr^{3+}(0.64 m) and Al^{3+} (0.57 nm), resulting in the lattice constant changing, which results in a small shift in the diffraction peak [11].

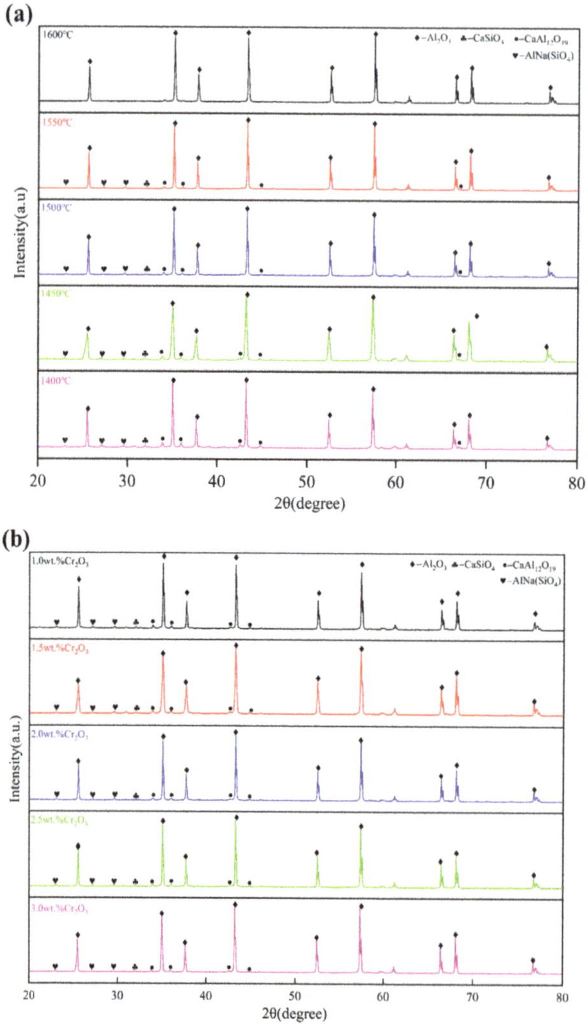

Figure 2. XRD patterns of the samples: (**a**) ARC2 samples sintered at different temperatures for 2 h; (**b**) ARC1~ARC5 samples sintered at 1500 °C.

Figure 3a shows the EDS data of the corroded ARC4 sample. Based on the compositional data of the point energy spectra, it can be judged that the columnar grains in the matrix are alumina crystals, and the hexagonal flake particles are $CaAl_{12}O_{19}$. From the surface spectral analysis, it is observed that the Ca element is concentrated in the hexagonal flake particles, and the Cr element is uniformly distributed in the matrix. Figure 3b shows the EDS data of the uncorroded AR sample. From the results of the surface spectrum analysis, Na and Si elements are mainly concentrated at the grain boundaries, and Fe elements are uniformly distributed in the matrix, with a slightly higher concentration at the grain boundaries. From the results of the point energy spectra, the silicate phase mainly shows an amorphous pattern distributed between the alumina grains and fills the grain-to-grain voids.

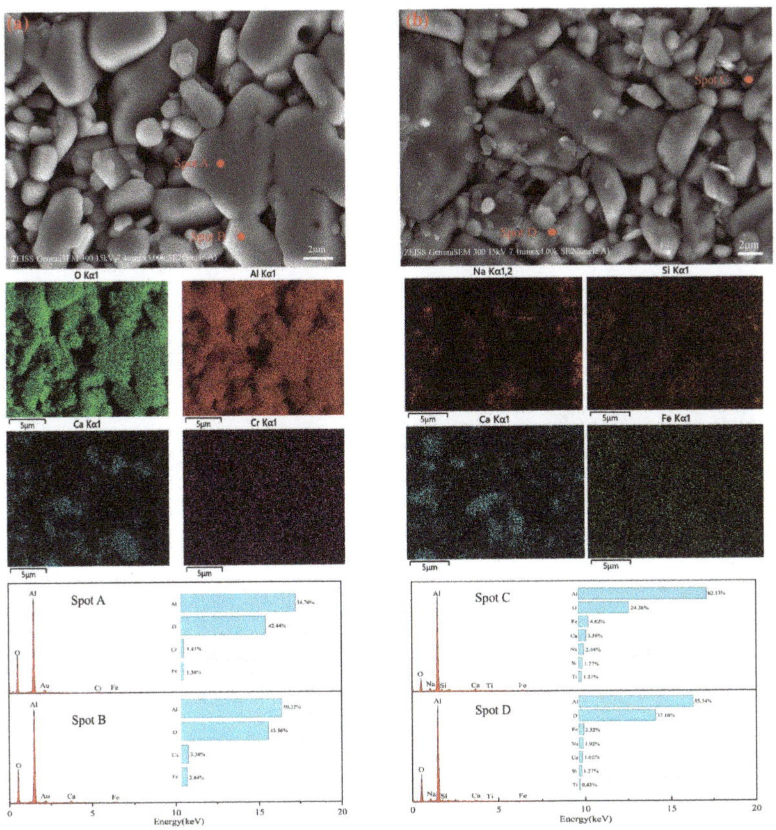

Figure 3. EDS results of composite ceramics: (**a**) ARC4 samples after 1 min in corrosion solution; (**b**) ARC0 samples not corroded by corrosion solution.

3.2. Microstructural Characterization

Figure 4b–f shows SEM images of ARC2 sintered at different temperatures for 2 h. The samples sintered at 1400 °C have been incompletely sintered, with more holes inside, and the microstructure consists of fine equiaxed grains and a few hexagonal lamellar particles. The CA6 grain exhibits preferential growth along the basal plane, and the coupled diffusion of Ca^{2+} and O^{2-} from the Ca-rich phase to the Al-rich phase controls the formation of CA6. This anisotropy of growth rate leads to the formation and growth of CA6, preferring the orientation of the base plane which is perpendicular to the reaction front, which leads flaky CA6 being obtained after reaction and sintering [12,13]. The samples sintered at

1450 °C exhibit abnormally grown grains with surface pores. This phenomenon can be attributed to the rapid decrease in the viscosity of the glassy phase, which facilitates gas diffusion in the liquid phase and subsequently leads to pore formation on the ceramic body's surface. A significant number of plate-like grains are observed in the samples sintered at 1500 °C. This phenomenon can be attributed to the formation of a small quantity of liquid phase within the lower-melting-point oxides present in the red mud. The uneven distribution of this liquid phase leads to the coexistence of dry and wet interfaces, resulting in variations in interfacial energy between the base surface of Al_2O_3 grains and other interfaces. Consequently, different growth rates occur on different surfaces, leading to anisotropic growth in certain grains [14].

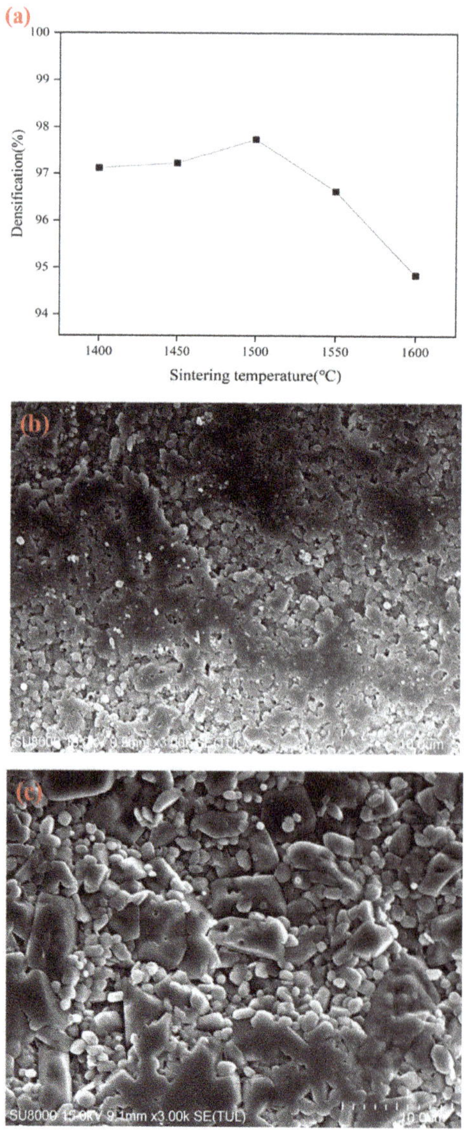

Figure 4. *Cont.*

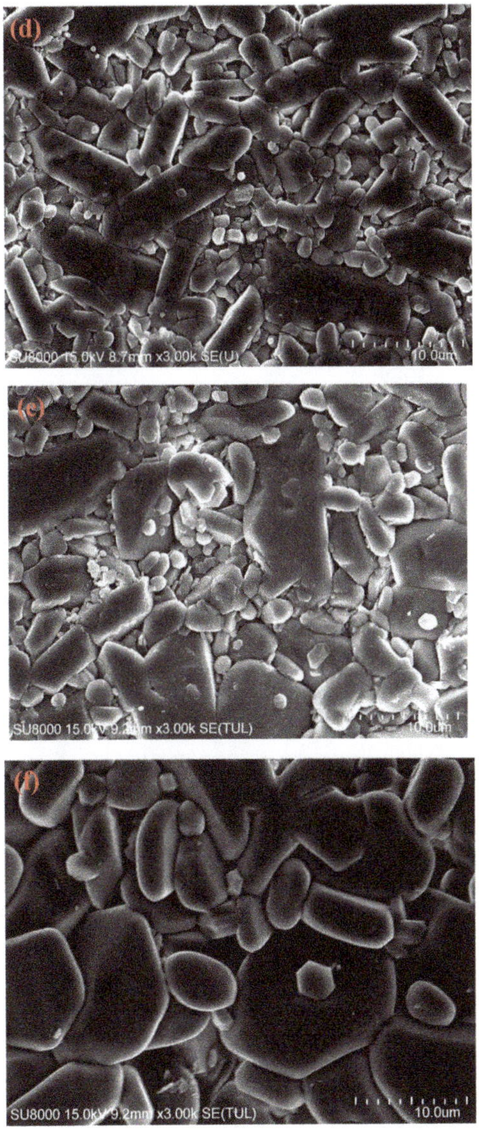

Figure 4. SEM micrographs of ARC2 sintered at different temperatures for 2 h: (**a**) densification of ARC2; (**b**) 1400 °C; (**c**) 1450 °C; (**d**) 1500 °C; (**e**) 1550 °C; (**f**) 1600 °C.

The densifications profiles of ARC2 samples sintered at various temperatures for 2 h are depicted in Figure 4a. It can be observed that the densifications of the composite ceramic exhibit a gradual increase followed by a rapid decrease with increasing temperature. In the initial stage of ceramic sintering, the bonding between the particles is weak, and there are more pores in the material. As the temperature increases, sintering and solid-state diffusion begin to occur between the particles, the pores are gradually filled, the material becomes denser, and the density gradually increases. The maximum densification achieved is 97.75% for samples sintered at 1500 °C. Obvious lath-like grains can be observed from Figure 4d. This is in agreement with the findings of Cahoon et al. [15]. They observed relatively uniform-sized anisotropic growth grains in pure alumina ceramics, which appeared abruptly as the matrix approached final densification. The densities of the composite ceramics decreased rapidly when the sintering temperature exceeded 1500 °C. The density of the composite ceramics sintered at 1600 °C for 2 h was 94.85%. According to the classical equation for grain growth kinetics:

$$G^n - G_0^n = K \cdot t \tag{4}$$

where G is the grain size, G_0 is the initial grain size, n is the growth index, K is a temperature-dependent constant, and t is the sintering time. This equation shows that the grain size increases with time and temperature. According to the Arrhenius equation, the grain growth rate constant K is related to the temperature T as follows:

$$K = K_0 \exp\left(\frac{-Q}{RT}\right) \tag{5}$$

where K_0 is a constant, Q is the activation energy for grain boundary migration, R is the gas constant, and T is the absolute temperature. At high temperatures, the K value increases significantly, and the grain growth rate is accelerated, especially in the late stage of high-temperature sintering, and the abnormal grain growth phenomenon is more obvious. Samples sintered at 1600 °C have very coarse grains, as shown in Figure 4f. In the case of abnormal grain growth, the larger grains occupy more volume, while the smaller grains are more difficult to move and cannot completely eliminate the inter-grain porosity. Especially, larger residual pores may form between large grains. During normal grain growth, the grain boundaries migrate uniformly and can effectively reduce the pores. However, at the interface between abnormally grown grains and small grains, the grain boundaries migrate unevenly, making it difficult to fill the pores in these regions. As a result, the densification of the material decreases and the porosity increases.

Figure 5b–f shows the surface morphology of ARC1~ARC5. Overall, the effect of Cr_2O_3 content on crystal morphology is not obvious. When the Cr_2O_3 content increased to 2.5 wt.%, obvious holes were observed in the matrix.

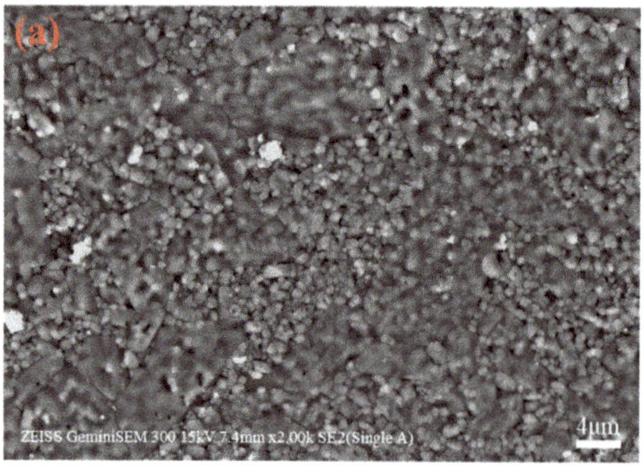

Figure 5. *Cont.*

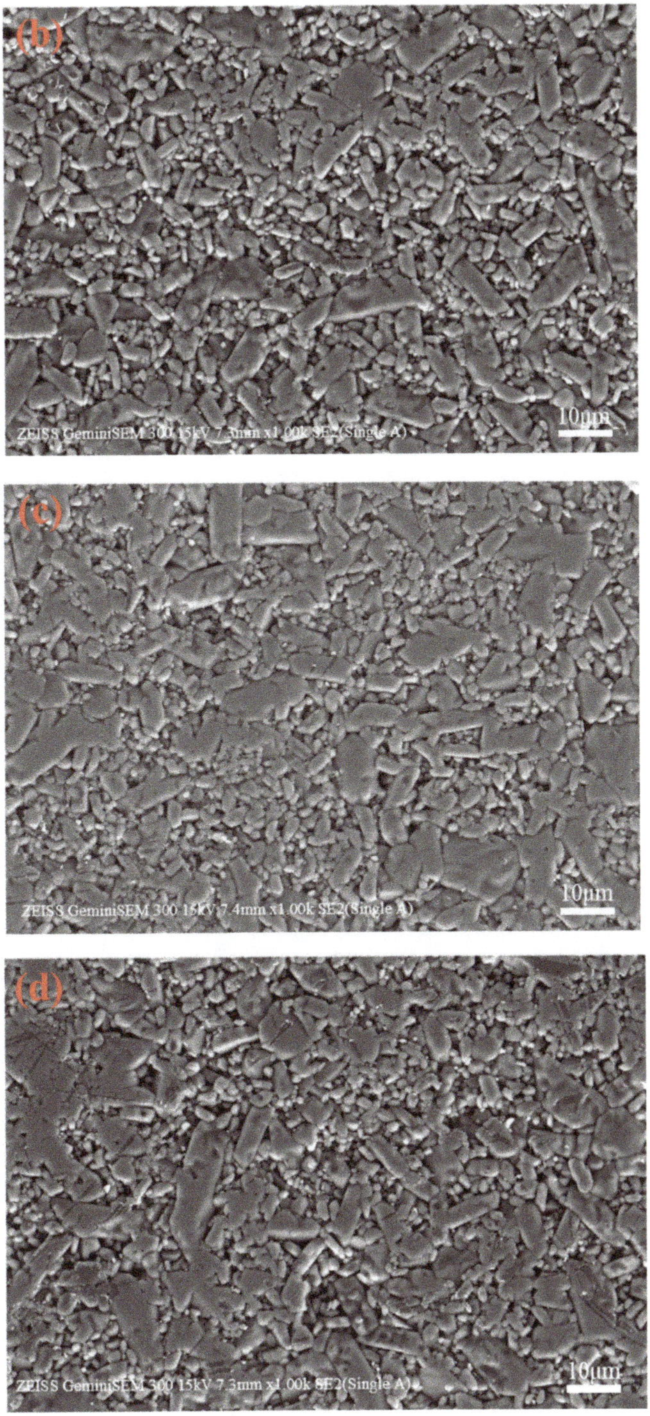

Figure 5. *Cont.*

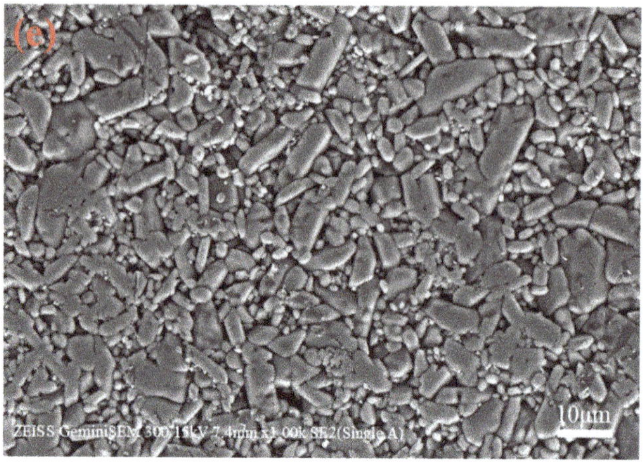

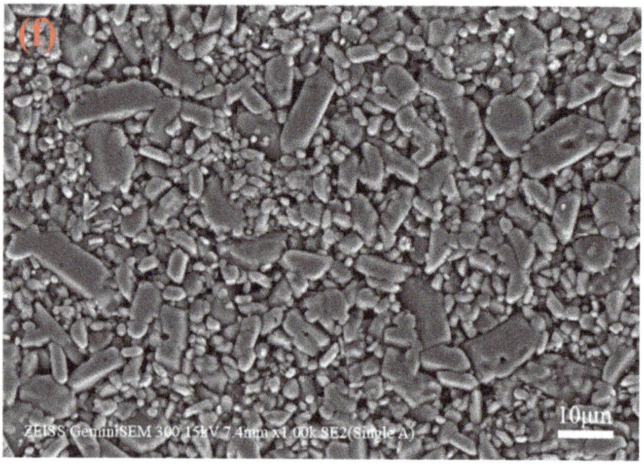

Figure 5. SEM micrographs of different samples sintered at 1500 °C: (**a**) A; (**b**) ARC1; (**c**) ARC2; (**d**) ARC3; (**e**) ARC4; (**f**) ARC5.

The densification of pure alumina specimens sintered at 1500 °C for 2 h was 92.14%. Figure 5a shows an incompletely densified structure with uneven grain size distribution and a large number of holes in the matrix. In the case of alumina, ordinary sintering at atmospheric pressure is usually above 1650 °C. The addition of red mud makes a small amount of liquid phase appear in the matrix, which accelerates the mass transfer rate during the sintering process. At the same time, under the action of capillary force generated by the liquid phase, it causes pressure and sliding between particles, which prompts the original particles to rearrange and improves the particle packing density. Based on this, this study determined that the optimum sintering temperature of ARC composite ceramics is 1500 °C, while the addition of red mud can effectively reduce the sintering temperature of Al_2O_3 ceramics.

3.3. Mechanical Properties

Figure 6a,b show the flexural strength and hardness of ARC2 sintered at different temperatures for 2 h. With the increase in sintering temperature, the flexural strength and Vickers hardness of the samples increase first and then decrease. The maximum flexural strength of the sample sintered at 1500 °C is 297.03 MPa, and the hardness value is 17.44 GPa. This is because, when sintered below 1500 °C, the sample presents an incomplete sintering state, retains more original particles, has poor interfacial binding force, and has more pores, which is mainly inter-granular fracture. When the sintering temperature is higher than 1500 °C, the grains in the matrix grow rapidly, and along with the liquid phase volatilization at a low boiling point in the grain boundary, the density of the sample decreases, resulting in a decrease in strength. The flexural strength of the sample sintered at 1600 °C is higher than that of the sample sintered at 1550 °C, because as the grain grows, it is more difficult for the crack to bypass the large grain in the process of expansion, resulting in a significantly higher proportion of trans-granular fracture (as shown in Figure 7a).

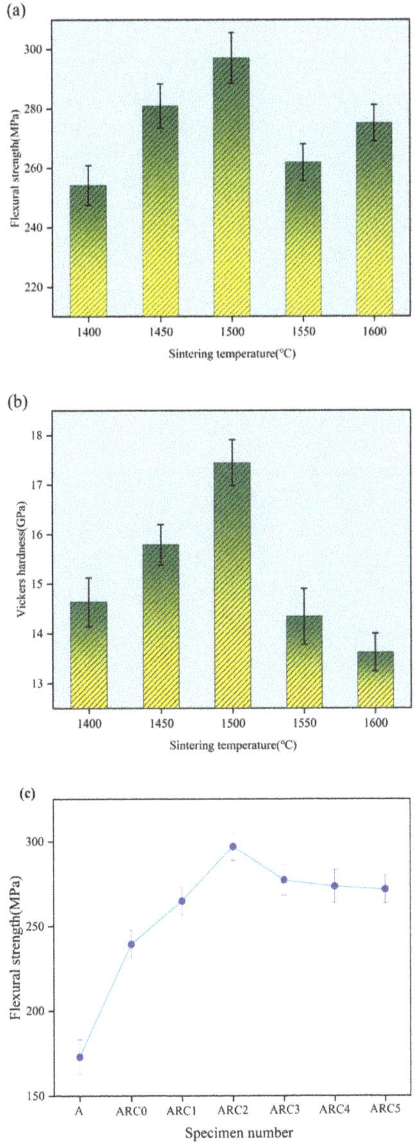

Figure 6. *Cont.*

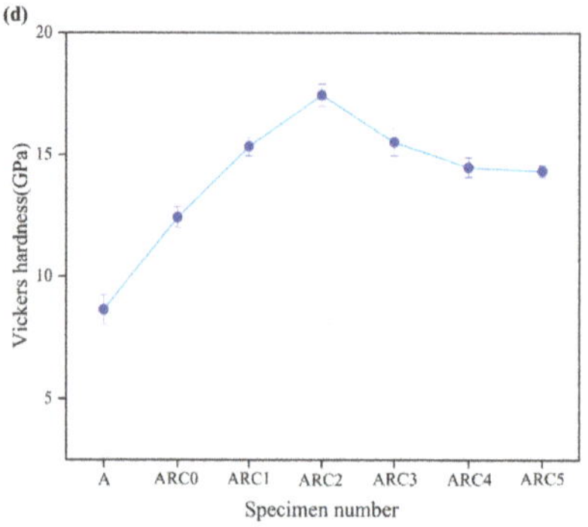

Figure 6. Mechanical properties of samples: (**a**) flexural strength of ARC2 sintered at different temperatures; (**b**) Vickers hardness of ARC2 sintered at different temperatures; (**c**) flexural strength of A~ARC5; (**d**) Vickers hardness of A~ARC5.

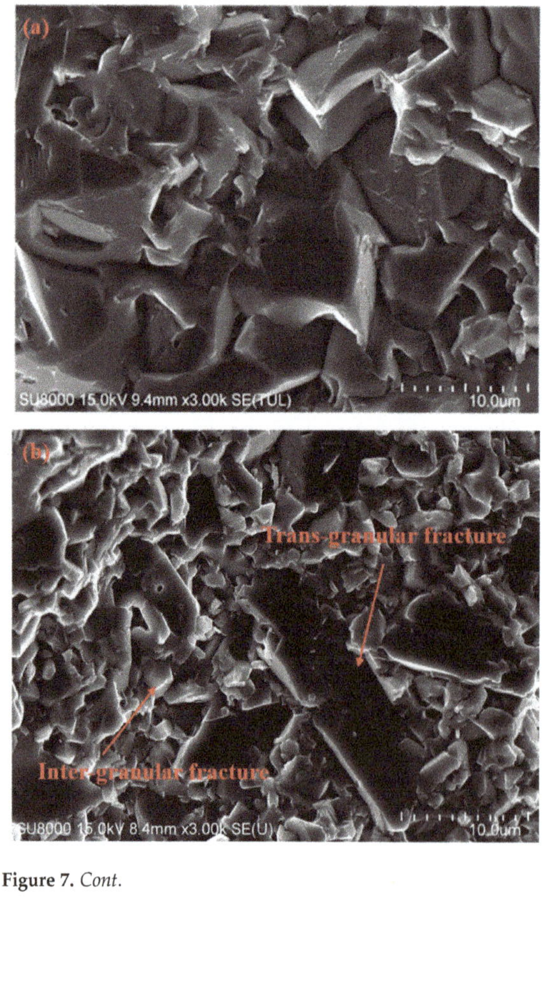

Figure 7. *Cont.*

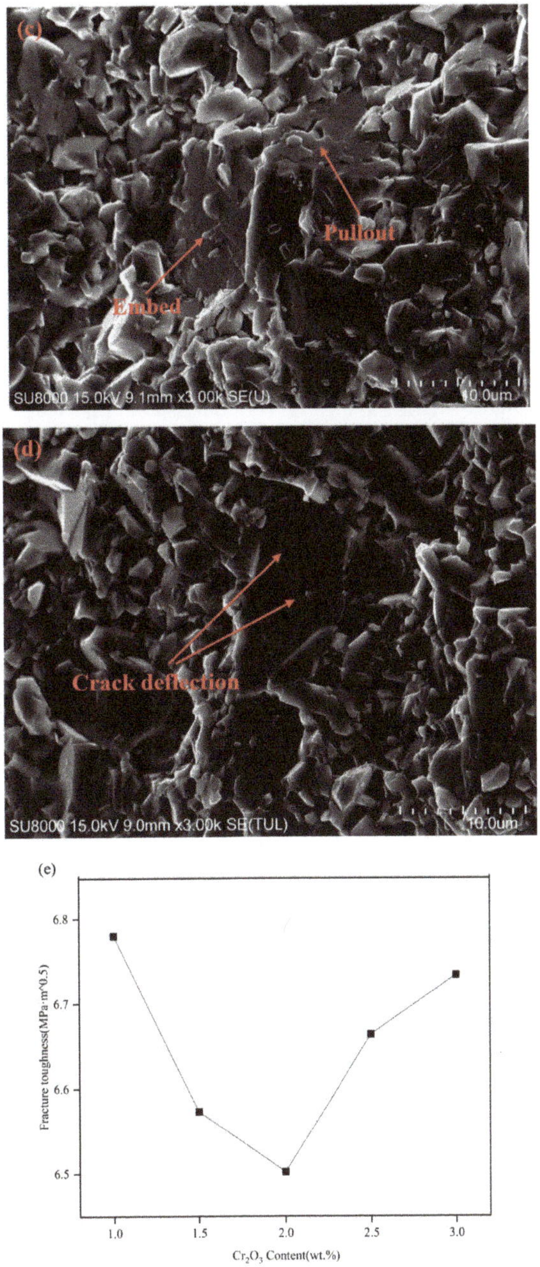

Figure 7. Fracture morphologies of some samples: (**a**) ARC2 sintered at 1600 °C; (**b**) ARC1 sintered at 1500 °C; (**c**) ARC4 sintered at 1500 C; (**d**) ARC2 sintered at 1500 °C; (**e**) toughness of ARC1~ARC5 sintered at 1500 °C.

Figure 6c,d show the flexural strength and hardness of A~ARC5 sintered at 1500 °C. Since the pure alumina specimen was not fully sintered, its flexural strength was only 173.5 MPa, and its hardness was 8.65 GPa. Notably, the sample containing 1.5 wt.% Cr_2O_3

exhibits the highest flexural strength. In the early stage of sintering, Cr^{3+} is easily enriched on the surface of Al_2O_3 grains [16,17]. Due to the solid solution of Cr ions, the distortion energy of the Al_2O_3 grain interface is increased, and the growth rate of the chromium-rich interface is accelerated, which promotes the sintering of ceramic samples. On the other hand, the increase in grain-boundary distortion energy has a positive effect on hindering the dislocation motion, thus enhancing the flexural strength of the composite ceramics. The mechanical properties of ceramics are very sensitive to defects such as porosity and cracks. Cr_2O_3 is unstable under atmospheric pressure sintering and is easily converted to CrO_3 through a gasification–condensation mechanism, leading to a decrease in the densification of the ceramic samples. Consequently, the flexural strength and hardness of the ceramic samples increase and then decrease with the increase in Cr_2O_3 content.

The fracture morphologies of some samples are depicted in Figure 7, exhibiting a combination of through-crystal and inter-granular fractures. The growth of CA6 exhibits a strong preferred crystallographic orientation, and the reaction primarily occurs by the diffusion of Ca^{2+} into Al_2O_3. The distribution of CA6 phase in the matrix can be observed in three different forms. Firstly, they distribute at the grain boundaries and fill the voids between grains, which enhances material densification and reduces porosity and pore size, thereby improving the strength of the composite material [18]. Secondly, CA6 grows along the surface of Al_2O_3 grains, which prolong the path of crack growth and increase the energy required for crack growth during inter-granular fracture and thus improve the fracture toughness of the material [19]. However, due to the introduction of less Ca element, CA6 growing in the Al_2O_3 grain is very thin, so it can provide a limited strengthening effect, as shown in Figure 7c.

It is worth noting that in the third case, CA6 is embedded between the Al_2O_3 grains at different angles and forms an "endo-crystalline" structure with the Al_2O_3 grains [20,21], as shown in Figure 7c. CA6 and Al_2O_3 have similar coefficients of thermal expansion (8.0×10^{-6} °C^{-1}, 8.6×10^{-6} °C^{-1}, respectively) [22], and it is generally believed that CA6 and Al_2O_3 do not undergo thermal mismatch and have good compatibility in the matrix. However, this is limited to the independent existence of CA6 grains and Al_2O_3 grains, because there are gaps between grains to make up for this difference. And from the perspective of the crystal structure of CA6 and Al_2O_3, the thermal expansion behavior of the two phases is anisotropic, and the occurrence of thermal mismatch is predictable when CA6 grows at different angles within the Al_2O_3 grains. This kind of structure will produce micro-cracks at the sub-interface, weaken the effect of the main grain boundary, inhibit the dislocation movement during the fracture process, play the role of crack pinning and change the expansion direction of the main crack, and may change its fracture mode, i.e., from the original inter-granular fracture to trans-granular fracture, as shown in Figure 7d.

3.4. Wettability at High Temperature

Al_2O_3 ceramics have been extensively utilized in the fabrication of ceramic substrates and ceramic cleavers [23]. With increasing demand for lower costs in semiconductor packaging, the adoption of low-cost bonding wire becomes imperative. Consequently, copper wire will inevitably replace gold wire as the primary alternative for bonding wire in the future. By investigating the wetting behavior and interface reactions of Cu liquid on RM–Al_2O_3 composite ceramic materials at elevated temperatures, this research establishes a fundamental framework for the potential application of RM–Al_2O_3 composite ceramic materials in the field of electronic packaging.

The contact angle data of Cu and RM–Al_2O_3 composite ceramics at different temperatures are presented in Figure 8. It can be observed that the contact angle gradually decreases with increasing temperature. Moreover, all contact angles measured at temperatures above the melting point are greater than 90°, indicating the non-wetting behavior of Cu and ARC composite ceramic. The SEM images in Figure 9 illustrate the wetted interface of ARC2 with Cu droplets, revealing a clean and flat interface without any dissolution or presence of reactants. It shows that at high temperature, the elements of red mud such as iron, nickel, sodium, calcium, and silicon in RM–Al_2O_3 composite ceramics have no diffusion phenomenon, and there is no pollution of copper liquid.

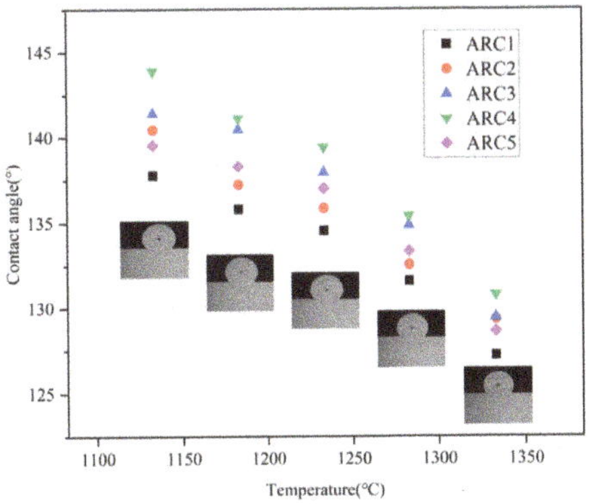

Figure 8. Contact angle of samples' surfaces with Cu.

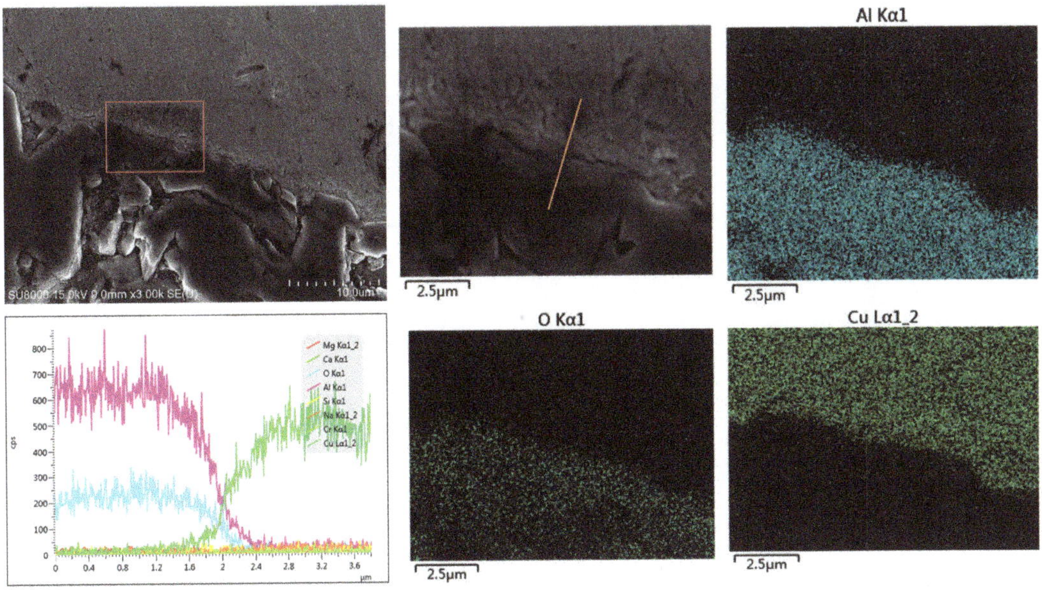

Figure 9. SEM and EDS images of ARC2–Cu interface. The red square are face-scanning area and the lines are line-scanning area.

Thomas Young proposed the contact angle of a liquid as the mechanical equilibrium of the drop resting on a plane solid surface at the three-phase boundary.

$$\gamma_{lv}\cos\theta = \gamma_{sv} - \gamma_{sl} \tag{6}$$

The London dispersion theory is employed by Berthelot [24] to calculate interfacial tension based on the surface tensions of two phases, using the geometric mean combination rule:

$$\gamma_{sl} = \gamma_s + \gamma_l - 2\sqrt{\gamma_s\gamma_l} \tag{7}$$

The surface tension of Cu can be calculated from the following equation [25]:

$$\gamma_l = 1.32 - 2.8 \times 10^{-4}(T - 1358) \tag{8}$$

The experimental values obtained by measuring the contact angle of copper and ceramics at different temperatures are plotted as cosθ vs. γ_l. The Zismen approximation was employed to establish a linear correlation between cosθ and γ_l:

$$\cos\theta = a\gamma_l + b \tag{9}$$

The above equations are ultimately interconnected to establish the correlation between γ_s and T:

$$\gamma_s = AT^3 + BT^2 + CT + D \tag{10}$$

The fitted parameters are presented in Table 3. The surface energy data from ARC1~ARC5 have been compared with the surface energy results for aluminum oxide reported by Eustathopoulos et al. [26] (γ_s = 20.77 − 0.007083 T), as illustrated in Figure 10. The surface energy of ARC composite ceramics decreases more slowly as the temperature decreases than that of Al_2O_3, and ARC3 and ARC4 have higher surface energy than Al_2O_3 at high temperature. As the Cr content increases, the wettability of Cu on the composite ceramic becomes worse. This can be explained in two ways. With the increase in Cr oxide, the solid solution of Cr ions in Al_2O_3 increases, leading to an increase in lattice distortion, which leads to an increase in the surface energy of the composite ceramics. Secondly, the excess addition of Cr led to an increase in the internal pores of the composite ceramics, which also led to an increase in the surface roughness of the material. According to the Wenzel model, on a hydrophobic surface, the larger the roughness is, the more hydrophobic it is.

Table 3. Values of fitting parameters and corresponding calculation.

Sample	a	b	R^2	A	B	C	D
ARC1	−2.32	2.28	0.94	-2.95×10^{-11}	1.82×10^{-8}	−0.0027	13.90
ARC2	−2.37	2.34	0.98	-3.08×10^{-11}	2.00×10^{-8}	−0.0028	14.43
ARC3	−2.56	2.55	0.92	-3.60×10^{-11}	2.91×10^{-8}	−0.0032	16.34
ARC4	−2.68	2.69	0.96	-3.94×10^{-11}	3.47×10^{-8}	−0.0035	17.68
ARC5	−2.38	2.33	0.92	-3.11×10^{-11}	2.24×10^{-8}	−0.0028	14.36

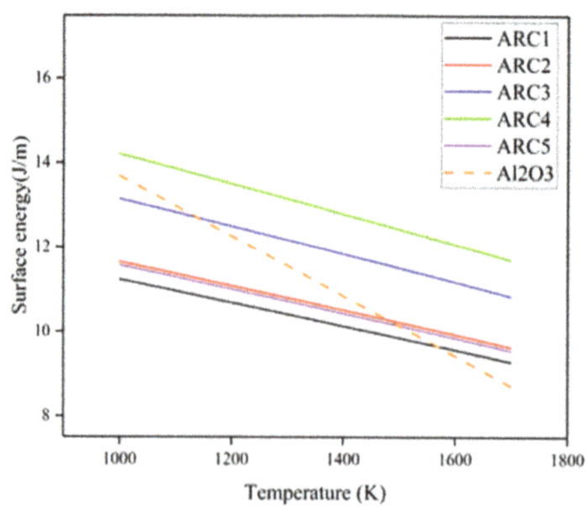

Figure 10. Surface energy of ARC1~ARC2 at high temperature.

4. Conclusions

Using red mud, Al_2O_3, and Cr_2O_3 as raw materials, high-performance red mud alumina composite ceramics have been prepared by press-free sintering. The experimental results demonstrate the following:

(1) Red mud can significantly lower the sintering temperature of Al_2O_3 ceramics. The samples sintered at 1500 °C with the addition of 1.5 wt.% Cr_2O_3 have better properties, with a bending strength of 297.03 MPa, a fracture toughness of 6.50 $MPa/m^{1/2}$, a hardness value of 17.44 GPa and a densification of 97.75%.

(2) There are three different distributions of CA6 in ARC composite ceramics. CA6 distributing at grain boundaries can fill the voids between grains and improve the densification of the ceramics. CA6 distributing parallel to the Al_2O_3 grains extends the crack propagation path. CA6 and alumina grains form an "endo-crystalline" structure, which can change the direction of crack propagation and transform from an inter-granular fracture to trans-granular fracture.

(3) Cu liquid does not wet the ARC composite ceramics at high temperatures, no chemical reaction occurs at the interface, and there is no elemental diffusion into Cu liquid. Moreover, the ARC composite ceramics have a higher surface energy than Al_2O_3 at high temperatures and decrease more slowly with temperature.

Author Contributions: Conceptualization, S.J. and A.L.; methodology, S.J.; investigation, S.J.; formal analysis, S.J.; writing—original draft, S.J. and Z.W.; funding acquisition, A.L.; resources, A.L.; supervision, A.L.; writing—review and editing, A.L.; data curation, Z.W. All authors have read and agreed to the published version of the manuscript.

Funding: This research was funded by the Specific Research Project of Guangxi for Research Bases and Talents, grant number GuiKeAD21238010. 2021 Special funds for central guidance of local scientific and technological development funds, grant number GuiKeZY21195030. Guangxi Key Research and Development Program, grant number GuiKeAB22080015.

Data Availability Statement: Data are contained within the article.

Conflicts of Interest: The authors declare no conflicts of interest.

References

1. Liu, R.-X.; Poon, C.-S. Utilization of red mud derived from bauxite in self-compacting concrete. *J. Clean. Prod.* **2016**, *112*, 384–391. [CrossRef]
2. Zhang, N.; Li, H.; Liu, X. Hydration mechanism and leaching behavior of bauxite-calcination-method red mud-coal gangue based cementitious materials. *J. Hazard. Mater.* **2016**, *314*, 172–180. [CrossRef] [PubMed]
3. Wang, W.; Chen, W.; Liu, H.; Han, C. Recycling of waste red mud for production of ceramic floor tile with high strength and lightweight. *J. Alloys Compd.* **2018**, *748*, 876–881. [CrossRef]
4. He, H.; Yue, Q.; Qi, Y.; Gao, B.; Zhao, Y.; Yu, H.; Li, J.; Li, Q.; Wang, Y. The effect of incorporation of red mud on the properties of clay ceramic bodies. *Appl. Clay Sci.* **2012**, *70*, 67–73. [CrossRef]
5. Pérez-Villarejo, L.; Corpas-Iglesias, F.A.; Martínez-Martínez, S.; Artiaga, R.; Pascual-Cosp, J. Manufacturing new ceramic materials from clay and red mud derived from the aluminium industry. *Constr. Build. Mater.* **2012**, *35*, 656–665. [CrossRef]
6. Gonzalez Triviño, I.; Pascual-Cosp, J.; Moreno, B.; Benítez-Guerrero, M. Manufacture of ceramics with high mechanical properties from red mud and granite waste. *Mater. Construcción* **2019**, *69*, 180. [CrossRef]
7. Nath, M.; Kumar, P.; Maldhure, A.V.; Sinhamahapatra, S.; Dana, K.; Ghosh, A.; Tripathi, H.S. Anomalous densification behavior of Al_2O_3–Cr_2O_3 system. *Mater. Charact.* **2016**, *111*, 8–13. [CrossRef]
8. Kafkaslıoğlu Yıldız, B.; Yılmaz, H.; Tür, Y.K. Evaluation of mechanical properties of Al_2O_3–Cr_2O_3 ceramic system prepared in different Cr_2O_3 ratios for ceramic armour components. *Ceram. Int.* **2019**, *45*, 20575–20582. [CrossRef]
9. Azhar, A.Z.A.; Choong, L.C.; Mohamed, H.; Ratnam, M.M.; Ahmad, Z.A. Effects of Cr_2O_3 addition on the mechanical properties, microstructure and wear performance of zirconia-toughened-alumina (ZTA) cutting inserts. *J. Alloys Compd.* **2012**, *513*, 91–96. [CrossRef]
10. Manshor, H.; Azhar, A.Z.A.; Rashid, R.A.; Sulaiman, S.; Abdullah, E.C.; Ahmad, Z.A. Effects of Cr_2O_3 addition on the phase, mechanical properties, and microstructure of zirconia-toughened alumina added with TiO_2 (ZTA–TiO_2) ceramic composite. *Int. J. Refract. Met. Hard Mater.* **2016**, *61*, 40–45. [CrossRef]
11. Zhao, P.; Zhao, H.; Yu, J.; Zhang, H.; Gao, H.; Chen, Q. Crystal structure and properties of Al_2O_3–Cr_2O_3 solid solutions with different Cr_2O_3 contents. *Ceram. Int.* **2018**, *44*, 1356–1361. [CrossRef]

12. An, L.; Chan, H.M.; Soni, K.K. Control of calcium hexaluminate grain morphology in in-situ toughened ceramic composites. *J. Mater. Sci.* **1996**, *31*, 3223–3229. [CrossRef]
13. Domínguez, C.; Chevalier, J.; Torrecillas, R.; Fantozzi, G. Microstructure development in calcium hexaluminate. *J. Eur. Ceram. Soc.* **2001**, *21*, 381–387. [CrossRef]
14. Song, H.; Coble, R.L. Origin and Growth Kinetics of Platelike Abnormal Grains in Liquid-Phase-Sintered Alumina. *J. Am. Ceram. Soc.* **1990**, *73*, 2077–2085. [CrossRef]
15. Cahoon, H.P.; Christensen, C.J. Sintering and Grain Growth of Alpha-Alumina. *J. Am. Ceram. Soc.* **1956**, *39*, 337–344. [CrossRef]
16. Cho, S.-A.; Arenas, F.J.; Ochoa, J. Densification and hardness of Al_2O_3–Cr_2O_3 system with and without Ti addition. *Ceram. Int.* **1990**, *16*, 301–309. [CrossRef]
17. Paek, Y.K.; Lee, H.Y.; Kang, S.J.L. Direction of Liquid Film Migration Induced by Chromic Oxide in Alumina-Anorthite. *J. Am. Ceram. Soc.* **1996**, *79*, 3029–3032. [CrossRef]
18. Kunkun, C.; Tao, F.; Yingyi, Z.; Jie, W.; Haobo, M.; Tianbiao, T. Microstructure and mechanical properties of $CaAl_{12}O_{19}$ reinforced Al_2O_3–Cr_2O_3 composites. *J. Eur. Ceram. Soc.* **2021**, *41*, 7935–7945. [CrossRef]
19. An, L.; Chan, H.M. R-Curve Behavior of In-Situ-Toughened Al_2O_3:$CaAl_{12}O_{19}$ Ceramic Composites. *J. Am. Ceram. Soc.* **1996**, *79*, 3142–3148. [CrossRef]
20. Yin, Z.; Huang, C.; Zou, B.; Liu, H.; Zhu, H.; Wang, J. Study of the mechanical properties, strengthening and toughening mechanisms of Al_2O_3/TiC micro-nano-composite ceramic tool material. *Mater. Sci. Eng. A* **2013**, *577*, 9–15. [CrossRef]
21. Bai, X.; Huang, C.; Wang, J.; Zou, B.; Liu, H. Fabrication and characterization of Si3N4 reinforced Al_2O_3-based ceramic tool materials. *Ceram. Int.* **2015**, *41*, 12798–12804. [CrossRef]
22. Xi, Z.; Jinsheng, L.; Junguo, L.; Yanan, Z.; Suju, H.; Pengyao, L.; Hao, N. The properties characterization and strengthening-toughening mechanism of Al_2O_3-CA6-MA-Ni multi-phase composites prepared by adding calcined dolomite. *Mater. Charact.* **2022**, *186*, 111810. [CrossRef]
23. Kim, I.J.; Kim, H.S.; Seo, M.Y.; Gauckler, L.J. Advanced ceramics in wire bonding capillaries for semiconductor package technology. *Mater. Sci. Eng. A* **2008**, *498*, 129–134. [CrossRef]
24. Sharma, P.K.; Rao, K.H. Analysis of different approaches for evaluation of surface energy of microbial cells by contact angle goniometry. *Adv. Colloid Interface Sci.* **2002**, *98*, 341–463. [CrossRef] [PubMed]
25. Chawla, K.K.; Withers, P.J. International Materials Reviews in 2013. *Int. Mater. Rev.* **2013**, *58*, 1. [CrossRef]
26. Bao, S.; Tang, K.; Kvithyld, A.; Engh, T.; Tangstad, M. Wetting of pure aluminium on graphite, SiC and Al_2O_3 in aluminium filtration. *Trans. Nonferrous Met. Soc. China* **2012**, *22*, 1930–1938. [CrossRef]

crystals

MDPI

Article

Spectroscopic Ellipsometry and Wave Optics: A Dual Approach to Characterizing TiN/AlN Composite Dielectrics

Mohamed El Hachemi [1,*,†], Nikhar Khanna [1,2,†] and Emanuele Barborini [1,†]

[1] Materials Research and Technology Department (MRT), Luxembourg Institute of Science and Technology, 5, Avenue des Hauts Fournaux, Belval, L-4362 Esch-sur-Alzette, Luxembourg; emanuele.barborini@list.lu (E.B.)
[2] Department of Physics & Materials Science (PHyMS), Luxembourg University, 2, Avenue de l'Université, L-4365 Esch-sur-Alzette, Luxembourg
* Correspondence: mohamed.elhachemi@list.lu
† These authors contributed equally to this work.

Abstract: In this paper, we present a method for retrieving the optical properties of a nano-designed TiN/AlN composite dielectric, using spectroscopic ellipsometry for experimental measurements and wave optics simulations for numerical analysis. Composite cermets have gained attention for solar–thermal energy conversion, but their fundamental optical properties are not well understood. While characterizing uniformly deposited layers is generally straightforward, the process becomes more complex for nanoparticulate composites. The refractive index is essential for investigating and tuning the optical characteristics of the composite. Our method employs COMSOL Multiphysics software, validated by experimental spectroscopic ellipsometry studies. The strong agreement between experimental and numerical results supports this approach as a rational way to design material models for optical property studies across a broad spectrum.

Keywords: composite dielectric absorber; cermet; spectroscopic ellipsometry analysis; numerical homogenization; wave optics modelling

1. Introduction

The refractive index (n) of a material controls the reflection, transmission, and absorption of electromagnetic (EM) radiation in the range of the optical spectrum. Tuning the optical properties of materials has attracted noticeable attention for applications in opto-electronics, nano-optics, plasmonics, and many other technological fields, particularly in the area of nanostructured films. Optical properties, in terms of refractive index, describe the interaction between light and media and hence are of prime importance to materials sciences. The refractive index of a material is a property that is necessary in a wide range of applications and studies but is challenging to measure (Knöner et al. [1]), especially when the design and the structure of the material are complex. The refractive index depends on the wavelength and is defined as the ratio of light speed in vacuum to that inside the material. It associates light scattering to the size, shape, and chemical composition of a material/nanostructure (Brown [2], Bohren and Huffman [3]) and defines the magnitude of the optical force that an EM field applies on nanoparticles (Righini et al. [4], Bendix and Oddershede [5]). As the effective refractive index of a composite depends on the chemical composition of the material phases as well as the size and concentration of the nanoparticles, its determination can also be employed to differentiate between different material phases in a sample.

The absorption coefficient (a) determines how far in a material the light of a particular wavelength penetrates before it gets absorbed. It depends on the material and also on the wavelength. The (a) value of the material is directly dependent on the extinction coefficient of the material (κ). Properties like reflectance (R), absorptance (A), and transmittance (T) are optical characteristics of the coating film that determine the performance of a device; hence, precise and exact measurements of the effective optical properties become of utmost importance for the improvement of optical technology and its applications. The composite absorber coating under investigation comprises a uniform distribution of 20 nm diameter titanium nitride (TiN) nanoparticles embedded in an aluminium nitride (AlN) matrix, forming a composite dielectric. This nanostructure design aims to achieve high absorbance in the visible region of the spectrum while minimizing reflectance in the infrared region, as detailed in Khanna et al. [6]. Accurate characterization of this composite is crucial for validating the proposed TiN/AlN nanostructure design.

In this work, we used Scanning Electron Microscope (SEM) and X-ray Photoelectron Spectroscopy (XPS) characterization data to inform about the nanostructure and molecular composition of the composite. This allowed us to accurately build a material model and investigate the effective refractive index $n(\lambda)$ and extinction coefficient $\kappa(\lambda)$ of the nano-designed TiN/AlN composite dielectric, both numerically and experimentally. For the numerical study, we calculated the effective parameters from finite element simulations using COMSOL Multiphysics V6.0software. For the experimental study, we retrieved the effective parameters from spectroscopic ellipsometry (SE) measurements. Determining the effective refractive index of rough nanoparticulate composite thin coatings using SE presents several challenges. The calculated effective refractive index often does not physically correspond to the homogenized material properties. Consequently, additional characterization techniques, such as SEM and XPS, are required to quantify surface roughness and layer composition and to incorporate these factors into SE analysis. Furthermore, resolving ambiguities in the refractive index necessitates iterative constrained fitting with complex optical models.

Following this introduction, the remainder of this paper is structured as follows: Section 2 presents the SEM and XPS characterization, along with a detailed analysis of the nanostructure composition of the TiN/AlN composite samples. Section 3 provides an overview of the numerical modelling approaches, outlining the equations governing the interaction between electromagnetic waves and the composite dielectric material, as well as the methods used to determine the effective refractive index from both numerical simulations and SE measurements. Section 4 details the nanoscale model simulations conducted using COMSOL Wave Optics software to obtain scattering parameters and retrieve the effective refractive index. It also describes the optical models used to derive the effective refractive index from SE measurements. A detailed comparison of the effective optical properties obtained through numerical and experimental studies is then discussed. Finally, Section 5 concludes the paper by summarizing the key findings and providing a qualitative review of the proposed approaches.

2. Structure of the Composite

2.1. SEM Characterization

To synthesize the AlN/TiN thin-film coating, a layer-by-layer process was employed. This involved a wet-deposition technique to distribute TiN nanoparticles, followed by an atomic layer deposition (ALD) technique for AlN layers, as sketched in Figure 1. The simultaneous formation of both phases at the same time could not be achieved in a controlled manner (Khanna et al. [6]). Instead, the DC magnetron sputtering method proposed by Bilokur et al. [7], Du et al. [8] has enabled the deposition of titanium aluminium nitride

(TiAlN) crystals. In contrast, our work aims to combine the distinct properties of TiN and AlN within a single composite. The composite was formed by depositing multiple layers of TiN nanoparticles embedded in an AlN matrix, stacked on top of each other, resulting in a total thickness of a few hundred nanometres. Each layer was created by first dispersing TiN nanoparticles onto the previous layer using a liquid suspension, followed by embedding them in the AlN matrix. A conformal deposition of AlN onto the TiN NP dispersion is performed. The synthesis process is detailed in Khanna [9]. Figure 1 shows a schematic of the composite, showing details of deposition in cross-sectional views. The layer's number, from 1 to 8, is indicated on the right-hand side of the image, where odd numbers correspond to TiN deposition cycles, and even numbers correspond to alternating AlN layer depositions. Two specimens of the TiN NP suspension were considered. The first specimen was made from dry powder of TiN, while the second specimen was a ready-to-use suspension. The development of the composite was guided by a material design approach based on numerical simulations. These simulations indicate that the composite's optical properties can be adjusted by manipulating the distribution of TiN particles embedded within the AlN matrix, enabling the prediction of an optimal configuration. In this paper, composite I will be regarded as the one prepared with dry powders of TiN, while composite II is the one prepared with the ready-made TiN suspensions. Figure 2i,iii show the top view of the very first layer of composites I and II, respectively, and (ii) and (iv) are the cross-sectional view of the synthesized layers of composites I and II, respectively. During SEM analysis, the surface coverage was estimated to be approximately 20% for composite I and 23% for composite II. Additionally, it was observed that the average thickness of the composites varied. This is possibly due to the presence of a large amount of agglomerations of NPs, both the dry powder and ready-made dispersion, leading to chunks of a few nanometres accumulating, one on top of the other.

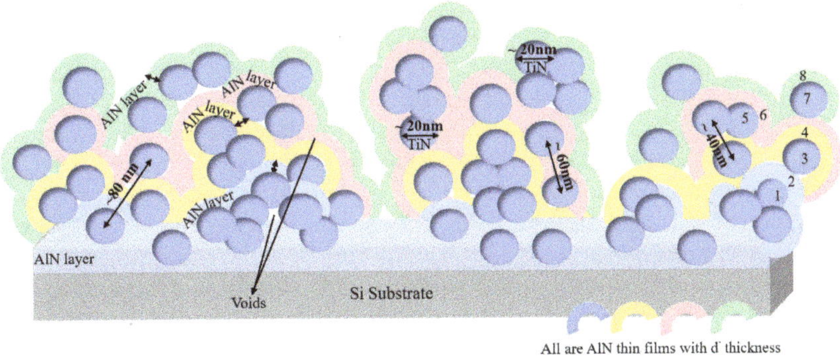

Figure 1. Schematic of the composite showing the cross-sectional view.

In addition, manual control over the TiN NPs' distribution throughout the composite leads to an inhomogeneous distribution in some places, making agglomerates of NPs, as represented in Figure 1. Overall, it was observed that the average thickness was smaller for composite I. This was verified by measuring the thickness at various positions on the sample. A key reason for this behaviour could be the addition of dispersant in the ready-made TiN suspensions. This is clearly seen in Figure 2, where TiN NPs cover more area in the case of composite II than in composite I (colour threshold for NP area calculation is shown in red). Thickness differences between the two composites were also confirmed by SE. This is explained in detail later, in Section 4.2. It is also evident from Figure 2ii,iv that the two types of composite possess a high degree of roughness due to the agglomeration of the NPs.

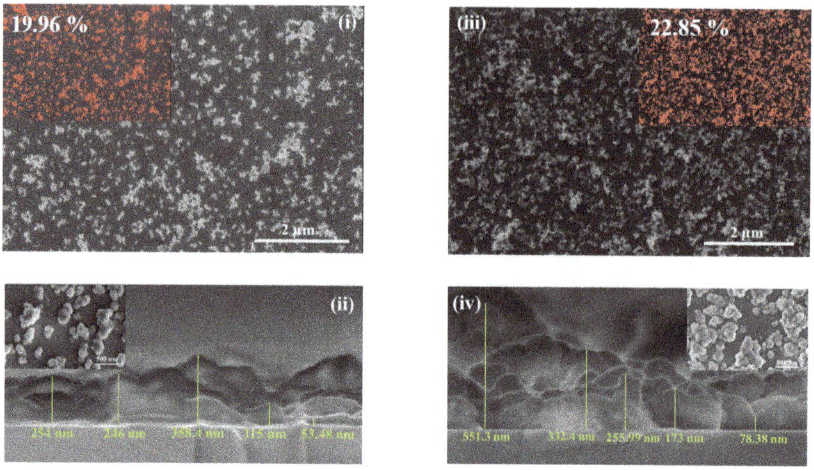

Figure 2. SEM images showing the top and cross-sectional views of the prepared samples. (**i**) Top view of the 1st deposited layer of composite I (shown in red is the area covered by the TiN nano powder). (**ii**) Cross-sectional view of composite I (total of 8 deposited layers). (**iii**) Top view of the 1st deposited layer of composite II (shown in red is the area covered by the TiN nanoparticles dispersion). (**iv**) Cross-sectional view composite II (total of 8 deposited layers).

2.2. XPS Characterization

XPS analysis was performed using a Kratos Axis Ultra DLD instrument with an Al (1486.6 eV) X-ray source and a 4 keV argon ion gun for etching. This setup enables the determination of the in-depth elemental composition and distribution of the primary constituents. XPS analysis confirms the presence of TiN, AlN, and Al_2O_3 in the core of the composite. The formation of Al_2O_3 was attributed to oxidation during the composite preparation process. This occurred due to air exposure after each AlN thin-film deposition and the post-deposition oxygen plasma treatment of AlN, intended to enhance the adhesion of TiN nanoparticles. The upper surface, which is strongly oxidised due to the direct contact with atmospheric oxygen, was a mixture of AlN and Al_2O_3. The surface of AlN films is prone to oxidation, and the emergence of different forms of oxides and hydroxides on its surface has been previously reported (Liao et al. [10], Méndez et al. [11], Jose et al. [12], Zhang [13]). Al_2O_3 and AlON are optically transparent materials over the light spectrum, near-ultraviolet, visible, and midwave-infrared regions; therefore, their presence would not have any negative impact on the optical properties of the composites. Notably, the Si signal and the oxygen associated with Si, which become prominent at longer etching times, were excluded from the percentage calculations to better highlight the trends of the elements of interest.

The XPS analysis results suggested that the composite could be considered as consisting of two distinct layers, with the upper one, referred to as layer 1, being composed of AlN and Al_2O_3, while the one below, designated as layer 2, comprised a blend of AlN, Al_2O_3, and TiN. Cross-sectional SEM images confirmed that the thickness varied between samples of composites I and II, leading us to approximate an effective thickness for each of the two layers. Table 1 gives the chemical composition at different times of etching starting from the top surface of the composite until the Si substrate is reached. Information from 0 s to 60 s (shown in red) about etching timing is completely ignored, due to the presence of carbon impurities, as shown in the C1s% column. Information from 120 s to 900 s (shown in blue) is considered as layer 1, where Ti is absent; thus, only AlN and Al_2O_3 are present. The presence of AlN and Al_2O_3 was investigated through high-resolution scans acquired for

Al2p, N1s, O1s, Ti2p, and orbitals. For layer 2, information from 1200 s to 4200 s (shown in purple), showing the presence of Ti along with Al, N, Si, and O, is considered as the main layer of the composite. The signals from 4800 s to 6000 s (shown in yellow) correspond to the oxidation layer of the silicon wafer, and the remaining data from 6600 s to 9000 s correspond to the silicon wafer substrate. We estimated the thickness of layer 1 t_1 and the thickness of layer 2 t_2 as

$$t_1 = \left[\frac{1050\ \text{s} - 120\ \text{s}}{4200\ \text{s} - 120\ \text{s}}\right] \times L \simeq 68\ \text{nm} \tag{1}$$

$$t_2 = \left[\frac{4200\ \text{s} - 1050\ \text{s}}{4200\ \text{s} - 120\ \text{s}}\right] \times L \simeq 232\ \text{nm} \tag{2}$$

Here, $L = 300$ nm represents the average thickness of the composite, while 4200 s corresponds to the etching time just before the Si signal reaches its peak. The etching begins penetrating layer 1 at 120 s. The etching time 1050 s is considered the transition between layers 1 and 2, as it lies halfway between 900 s and 1200 s. These thicknesses would provide a useful reference for subsequent FEM simulation and SEM calculations. The volume fractions of corresponding phases in layer 1 and layer 2 were estimated as outlined in the equations below. For layer 1, the average percentage composition is 44.76% for Al2p, 28.56% for N1s, and 26.67% for O1s. Assuming that the 28 atoms of N are bonded to 28 atoms of Al to form 28 molecules of AlN, this leaves 16 out of 44 atoms of Al available to bond with 26 atoms of O. For Al_2O_3, 8 atoms of Al combine with 24 atoms of O. While there are 26 oxygen atoms in total, the remaining 2 are disregarded for simplicity, as the calculations are only approximations.

$$v_{\text{AlN}} = \frac{\text{atomic mass of AlN}}{\text{density of AlN}} = 13.4\ \text{cm}^3 \tag{3}$$

$$v_{\text{Al}_2\text{O}_3} = \frac{\text{atomic mass of Al}_2\text{O}_3}{\text{density of Al}_2\text{O}_3} = 25.7\ \text{cm}^3 \tag{4}$$

where v_{AlN} represents the molar volume of AlN, and $v_{\text{Al}_2\text{O}_3}$ represents the molar volume of Al_2O_3. The total volume of AlN is determined as $V_{\text{AlN}} = 28 \times 13.4\ \text{cm}^3 = 375.2\ \text{cm}^3$, while the total volume of Al_2O_3 is calculated as $V_{\text{Al}_2\text{O}_3} = 8 \times 25.7\ \text{cm}^3 = 206.1\ \text{cm}^3$. Thus, the volume fractions in layer 1 are deduced as follows:

$$vf_{\text{AlN}} = \frac{V_{\text{AlN}}}{V_{\text{AlN}} + V_{\text{Al}_2\text{O}_3}} = 0.64 \tag{5}$$

and

$$vf_{\text{Al}_2\text{O}_3} = \frac{V_{\text{Al}_2\text{O}_3}}{V_{\text{AlN}} + V_{\text{Al}_2\text{O}_3}} = 0.35 \tag{6}$$

With same reasoning for layer 2, the average percentage composition is 35% for Al2p, 17.44% for N1s, 38.54% for O1s, and 8.10% for Ti2p. Assuming that 8 atoms of N bond to 8 atoms of Ti to form 8 molecules of TiN, this leaves 9 out of 17 atoms of N to bond with 9 atoms of Al, resulting in 9 molecules of AlN. The 26 atoms of Al left over bond to 13 atoms of O, forming 13 molecules of Al_2O_3. As the molar volume of TiN is

$$v_{\text{TiN}} = \frac{\text{atomic mass of TiN}}{\text{density of TiN}} = 11.4\ \text{cm}^3 \tag{7}$$

we deduce that $V_{\text{AlN}} = 9 \times 13.4\ \text{cm}^3 = 120.6\ \text{cm}^3$; that for Al_2O_3 is $V_{\text{Al}_2\text{O}_3} = 13 \times 25.7\ \text{cm}^3 = 334.1\ \text{cm}^3$; and that for TiN is $V_{\text{TiN}} = 8 \times 11.4\ \text{cm}^3 = 91.2\ \text{cm}^3$. Thus, the volume fractions of the phases in layer 2 are

$$vf_{\mathrm{AlN}} = \frac{V_{\mathrm{AlN}}}{V_{\mathrm{AlN}} + V_{\mathrm{Al_2O_3}} + V_{\mathrm{TiN}}} = 0.22 \tag{8}$$

,

$$vf_{\mathrm{Al_2O_3}} = \frac{V_{\mathrm{Al_2O_3}}}{V_{\mathrm{AlN}} + V_{\mathrm{Al_2O_3}} + V_{\mathrm{TiN}}} = 0.61 \tag{9}$$

and

$$vf_{\mathrm{TiN}} = \frac{V_{\mathrm{TiN}}}{V_{\mathrm{AlN}} + V_{\mathrm{Al_2O_3}} + V_{\mathrm{TiN}}} = 0.16 \tag{10}$$

Table 1. XPS analysis for composite I mentioning the elemental composition from 0 s to 9000 s of etching.

Etching Time(s)	Al (at.%)	C (at.%)	N (at.%)	O (at.%)	Si (at.%)	Ti (at.%)	
0	30.89	24.06	19.32	25.72	0	0	
60	43.74	1.87	31.77	22.60	0	0	Rows ignored-Presence of Carbon.
120	45.63	0	33.04	21.31	0	0	
180	45.91	0	33.25	20.82	0	0	
240	45.68	0	33.24	21.39	0	0	
300	45.68	0	32.74	21.39	0	0	Layer 1: 64.5% Al_2O_3, 35.5% AlN
600	44.12	0	25.66	30.21	0	0	
900	41.36	0	13.42	45.21	0	0	
1200	36.11	0	11.05	44.01	6.80	2.00	
1800	23.85	0	10.25	30.32	31.13	4.42	
2400	14.99	0	7.10	18.65	55.61	3.63	
3000	8.27	0	3.78	10.30	75.13	2.49	Layer 2: 61% Al_2O_3, 22.4% AlN, 16.5% TiN
3600	4.47	0	2.94	5.36	85.89	1.23	
4200	2.00	0	1.28	2.55	93.47	0.67	
4800	1.29	0	0.30	0.83	97.56	0	
5400	0.77	0	0	0.49	98.73	0	Subtracted: Si and SiO_2 content
6000	0	0	0	0.19	99.80	0	
6600	0	0	0	0	100	0	
7200	0	0	0	0	100	0	Rows ignored-Silicon substrate.
9000	0	0	0	0	100	0	

3. Numerical Modelling

Typical solar absorbing materials used for solar–thermal conversion are often conceived as a composite dielectric of metal cermets. Metal nitrides have been investigated for their potential advantages, such as weather resistance, high melting points, and enhanced optical and electrical properties (Khanna et al. [6], Rohmer et al. [14], Chang et al. [15], Nelson and Wagner [16]). In the past decade, the multidisciplinary element in materials science research has raised the degree of complexity in the systems under study (Dobrzański [17], Siegenfeld and Bar-Yam [18]). The use of powerful numerical modelling tools has provided exceptionally accurate predictions of the physical processes and suggested numerous advanced possibilities for optimized experimental procedures (Körkel et al. [19], Tan et al. [20]).

To simulate the interaction between light and the composite dielectric material, and subsequently the wave optics interaction with nanoparticles, we solve Maxwell's Equation (11) in the frequency domain for wave propagation through the composite film in a steady state.

$$\nabla \times \nabla \times \mathbf{E} - k_0^2 \tilde{n}^2 \mathbf{E} = 0 \tag{11}$$

Here, $\mathbf{E}$ represents the electric field vector and k_0 denotes the wave number in free space. It is also essential to qualitatively study the optical properties of the individual components of the composite. The dielectric loss is expressed within a complex wave vector as

$$\tilde{k} = k' - jk'' = \tilde{n}k_0 \quad (12)$$

Here, $\tilde{n}$ is the complex refractive index of the medium. Its real part n and imaginary part κ are both intrinsic values of a material that depend on the wavelength of the incident light and are related to the complex relative dielectric function, $\tilde{\varepsilon}$ (Batsanov et al. [21]), as

$$\tilde{n} = n - j\kappa = \sqrt{\tilde{\varepsilon_r}(\omega)} \quad (13)$$

where n is defined by Snell's law (Bahaa E. A. Saleh [22]) as the ratio between the speed of light inside the medium, v, and the speed of light in vacuum, c =299,792,458 m/s:

$$n = \frac{c}{v} \quad (14)$$

Meanwhile, κ is the wave extinction. It is related to absorption and represents dielectric loss, due to the delayed response of molecular dipoles, Joule losses from conduction, and wave scattering extinction. The attenuation of light intensity I, as the light travels through the medium, is described by the Beer–Lambert law (Mayerhöfer et al. [23]):

$$I(d) = I_0 e^{-ad} \quad (15)$$

where (a) is the attenuation coefficient and (d) is the distance travelled by the wave inside the material. The value of (a) is related to the extinction coefficient κ of the material according to

$$a = \frac{4\pi\kappa}{\lambda} \quad (16)$$

It follows that

$$ln(\frac{I}{I_0}) = -ad = -4\pi\frac{\kappa d}{\lambda} \quad (17)$$

If the film is such that $\lambda \gg \kappa d$, then it becomes transparent (Zhang et al. [24], Pérez [25], Chen and Boström [26]). Reflection occurs at the interface between two media with differing refractive indices. As light propagates through a medium, some of it is absorbed. Consequently, transmittance represents the fraction of the light beam that successfully passes through the medium, such that

$$R + T + A = 1 \quad (18)$$

In order to design a good solar light absorber with spectral selectivity, we choose materials that lead to an effective refractive index, close to that of air in the visible part of the spectrum to minimize reflection, while being higher in the infrared region to maximise reflection and reduce blackbody radiation loss. Figure 3 presents the refractive index curves of the material phases. The AlN and Al_2O_3 curves exhibit no dispersion across the visible to infrared spectrum, meaning that refractive index remains constant, and there is zero extinction coefficient. This transparency makes these materials highly efficient for light transmission. In contrast, TiN is conductive, exhibits high loss, and is dispersive, with a refractive index near unity in the visible spectrum but increasing significantly in the infrared region. When designing the composite dielectric TiN/AlN-Al_2O_3, care is taken to prevent TiN nanoparticles from reaching the percolation threshold by avoiding agglomeration that could cause excessive reflection in the visible region and considering smaller particles to enhance scattering extinction (Striebel et al. [27]).

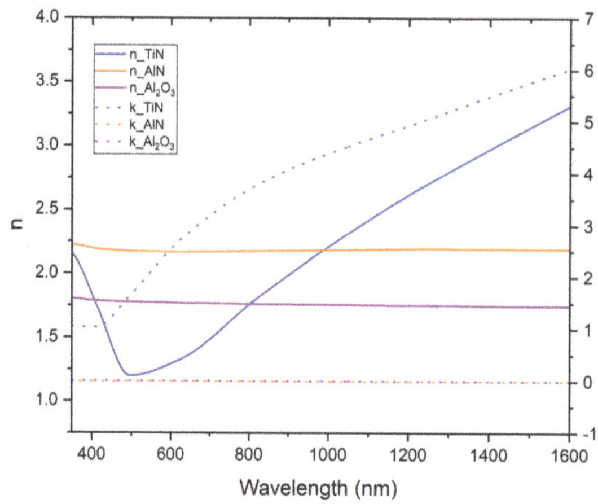

Figure 3. Optical properties n and κ for material phases TiN, AlN, and Al_2O_3.

3.1. Numerical Acquisition of $\tilde{n}(\lambda)$

The homogenization theory was introduced by Maxwell Garnett [28]. Since then, many empirical formulations have been derived using the Maxwell Garnett and Bruggeman approximations to calculate the effective dielectric constant for composites with various shapes and types of inclusions (Doyle [29], Markel [30]). Recently, however, finite-element-based methods have become the most preferred due to their simplicity (Nayak et al. [31]). Here, for the estimation of the effective optical parameters of the composite, we solve the problem of wave propagation at normal incidence, through an infinitely long dielectric slab, as sketched in Figure 4, in the frequency domain using the finite element method and calculate the scattering matrix S_{ij}. Then, the values of n and κ are derived using the algorithm described below. To achieve this, we assume a symmetric medium in which $[S_{ij}] = [S_{11}, S_{21}]$.

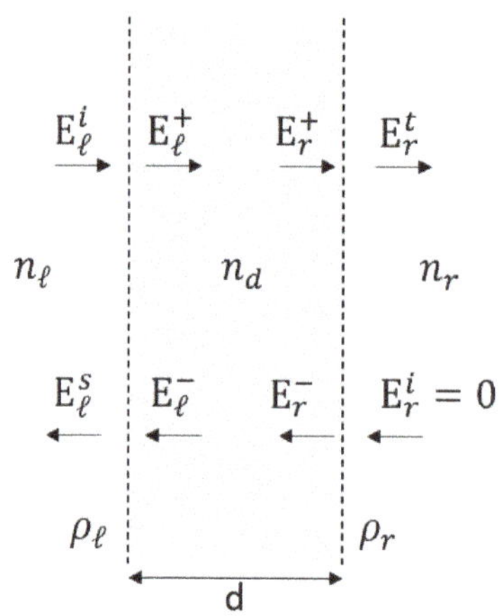

Figure 4. Illustration depicting the components of the electric field in a wave travelling from left side to right side through an infinite dielectric film with a thickness of d.

Figure 4 shows the field components, propagating from left side to right side through the dielectric film. Here, the subscripts ℓ and r denote left and right, respectively, while the superscripts $i, s,$ and t denote incident, scattered, and transmitted components, respectively.

Inside the dielectric, forward-propagating components are denoted by the superscript +, while backward-propagating components are denoted by the superscript −. The elementary reflection coefficients at the left and right sides are denoted by ρ_ℓ and ρ_r, respectively. Assuming the presence of air on both sides of the dielectric, i.e., $n_\ell = n_r = n_0$, then

$$\rho_\ell = -\rho_r = \frac{n_0 - \tilde{n}_d}{n_0 + \tilde{n}_d} \tag{19}$$

The relationships between the field components across the interfaces are given in Orfanidis [32] and are expressed in a straightforward way using the matching matrices

$$\begin{bmatrix} E_\ell^i \\ E_\ell^s \end{bmatrix} = \frac{1}{\rho_\ell + 1} \begin{bmatrix} 1 & \rho_\ell \\ \rho_\ell & 1 \end{bmatrix} \begin{bmatrix} E_\ell^+ \\ E_\ell^- \end{bmatrix} \tag{20}$$

$$\begin{bmatrix} E_r^+ \\ E_r^- \end{bmatrix} = \frac{1}{\rho_r + 1} \begin{bmatrix} 1 & \rho_r \\ \rho_r & 1 \end{bmatrix} \begin{bmatrix} E_r^t \\ 0 \end{bmatrix} \tag{21}$$

and the propagation matrix

$$\begin{bmatrix} E_\ell^+ \\ E_\ell^- \end{bmatrix} = \begin{bmatrix} e^{j\tilde{k}d} & 0 \\ 0 & e^{-j\tilde{k}d} \end{bmatrix} \begin{bmatrix} E_r^+ \\ E_r^- \end{bmatrix} \tag{22}$$

The relationships

$$E_\ell^i = \frac{1}{(\rho_\ell + 1)(\rho_\ell + 1)} \left(e^{j\tilde{k}d} + \rho_\ell \rho_r e^{-j\tilde{k}d} \right) E_r^t \tag{23}$$

and

$$E_\ell^s = \frac{1}{(\rho_\ell + 1)(\rho_\ell + 1)} \left(\rho_\ell e^{j\tilde{k}d} + \rho_r e^{-j\tilde{k}d} \right) E_r^t \tag{24}$$

lead to the expressions

$$S_{11} = \frac{E_\ell^r}{E_\ell^i} = \frac{\rho_\ell + \rho_r e^{-j2\tilde{k}d}}{1 + \rho_r \rho_\ell e^{-j2\tilde{k}d}} \tag{25}$$

and

$$S_{21} = \frac{E_r^t}{E_\ell^i} = \frac{(1 + \rho_\ell)(1 + \rho_r) e^{-j\tilde{k}d}}{1 + \rho_r \rho_\ell e^{-j2\tilde{k}d}} \tag{26}$$

for the scattering parameters. In the simulations at nanoscale, we consider the same medium on the left and right sides; therefore, we set

$$e^{-j\tilde{k}d} = \Gamma = |\Gamma| e^{+j\phi} \qquad \rho_\ell = -\rho_r = \rho \tag{27}$$

and we obtain the values

$$\zeta + \sqrt{\zeta^2 - 1} = \rho \qquad \zeta = \frac{S_{11}^2 - S_{21}^2 + 1}{2S_{11}} \tag{28}$$

and

$$\Gamma = \frac{\rho - S_{11} - S_{21}}{\rho(S_{11} + S_{21}) - 1} \tag{29}$$

The values of S_{11} and S_{21} in Equation (29) are determined from simulation on the unit cell, representing the dielectric composition at the nanoscale, as described in Section 4.1 below.

Finally, the complex refractive index of the slab (Nicolson and Ross [33], Smith et al. [34]) may be determined as

$$\tilde{n_d} = \frac{2\pi m - \phi}{k_0 d + j} \frac{\ln|\Gamma|}{k_0 d} \tag{30}$$

Here, the coefficient m denotes the branch order of the inverse tangent function.

3.2. Acquisition of $\tilde{n}(\lambda)$ from SE Measurements

SE is primarily a technique for measuring the optical properties of thin coating materials with polarized light. The measurements are based on the analysis of the polarization of the light beam reflected from the sample surface. Figure 5 shows a schematic representation of the SE measurement. A linearly polarized light beam illuminates the sample with a specified incidence angle. Upon reflection, the beam transforms into elliptically polarized light (Lissberger [35]), with the polarization change being entirely dependent on the wavelength, the angle of incidence, the thickness of the layers in the sample, and, most importantly, their optical model. The changes in polarization are expressed through ellipsometric parameters, namely the phase difference $\triangle$, which represents the difference between the s-wave and p-wave components of the reflected wave, and tan Ψ, which relates to the amplitude change. The angle of incidence for the interaction between the beam and the sample should be close to Brewster's angle, where this method is most effective (Bakker et al. [36]). A model is chosen to represent the sample's characteristics, and the parameters $\triangle$ and Ψ are compared to the measured values. This model applies mathematical dispersion formulas to determine material properties, such as thickness, surface roughness, and optical parameters, by adjusting specific fitting parameters. The constructed model describes the individual layers in terms of their thicknesses and optical parameter values (Yusoh et al. [37]).

$$\frac{R_p}{R_s} = \frac{E_p^r/E_p^i}{E_s^r/E_s^i} = \tan\Psi . e^{j\triangle} \tag{31}$$

where R_p and R_s denote the reflection coefficients of the p-polarized and s-polarized light, respectively. $E_p^{i,r}$ denotes the intensities of the incident and reflected electric field of the p-wave and $E_s^{i,r}$ denotes the intensities of the incident and reflected electric field of the s-wave.

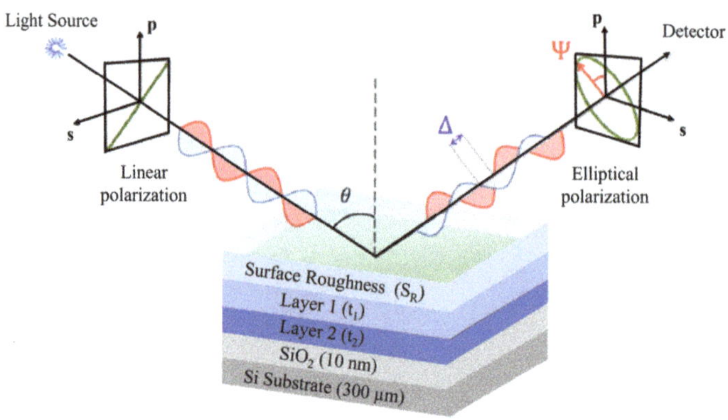

Figure 5. Schematic of the SE measurement, where the composite under study is represented by layers 1 and 2 with rough surface.

4. Results and Discussion

4.1. Numerical Simulation

Considering the volume fractions shown in the XPS table (Table 1), a unit cell model was created for layers 1 and 2 in order to calculate the effective parameters n and κ as shown in Figure 6. The left side shows details of the whole computational domain of the nanoscale model, with the Al_2O_3 phase embedded in an AlN matrix forming a representative unit cell for layer 1. On the right side of the figure, it shows TiN and AlN embedded in an Al_2O_3 matrix forming a representative unit cell for layer 2.

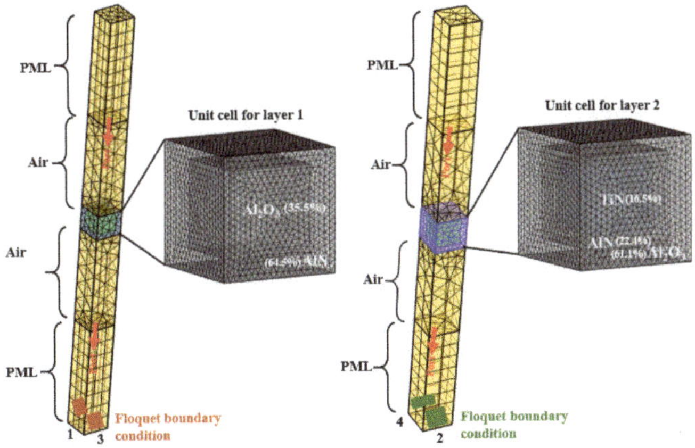

Figure 6. Details of nanoscale simulation showing a representative unit cell model for layer 1 and layer 2 used for the retrieval of n and κ.

Equation (11) is solved within the nanoscale model with the following boundary conditions:

- Perfectly Matched Layers (PMLs) were applied to the top and bottom domains of the computational domain to simulate an infinite space. Both the PMLs and air regions were assigned a thickness equal to $(\lambda/2)$ each.
- The boundary surfaces between the PML domain and the air spacing serve as ports for initiating the incident wave illumination and for post-process calculation of the scattering coefficients S_{ij}.
- Floquet periodic boundary conditions are implemented on the transverse sides of the computational domain to emulate an infinitely extensive film.

The TiN phase has a width of 20 nm and the unit cell has a width of 26 nm. These sizes are determined to comply with the volume fraction of the phases. The maximum mesh element size allowed here in the nanoscale model simulation is $\lambda/15$. The calculation takes 15 min to complete the frequency sweep over the VIS-IR range. The S-parameters are post-processed for the representative unit cell, and then the optical properties are retrieved, as explained in Section 3.1. The results from these numerical models are compared to those obtained experimentally from SE in the following sections.

4.2. Optical Measurements by Spectroscopic Ellipsometer

The numerical model for retrieving optical properties is validated by performing measurements on the synthesised samples. Due to the wide susceptibility in ellipsometry, employing a dependable and efficient numerical approach to forecast outcomes before conducting experiments is highly advantageous. This is especially useful for analysing systems with intricate nanostructures. Here, we introduce a straightforward method, offering thorough SE analysis for complex multi-layer systems. This approach allows flexibility in design regarding thickness, the number of layers, and phase composition. The SE analysis was conducted over the spectral range (370 nm to 1600 nm) allowed by the apparatus bulb. A standard model was developed to analyse composites I and II, reflecting the characteristics of the real samples. The model was initiated with a 300 μm thick Si wafer and a thin 10 nm oxidised layer of SiO_2 on top. For layer 2, a Cauchy model was created for the conductive TiN embedded within a insulating matrix of mixed AlN and Al_2O_3; thus, the Cauchy absorbent dispersion law was applied. This approach is typically used to describe the optical properties of materials that are weakly absorbent. Since the TiN content in the composite is only 20%, it qualifies as weakly absorbent. The Cauchy absorbent model is defined as

$$n(\lambda) = A + \frac{10^4}{\lambda^2}B + \frac{10^9}{\lambda^4}C \tag{32}$$

$$k(\lambda) = 10^{-5} \cdot D + \frac{10^4}{\lambda^2}E + \frac{10^9}{\lambda^4}F \tag{33}$$

Six fitting parameters are employed in this model, where the first three—A (dimensionless), B (nm^2), and C (nm^4)—have to fulfil the condition

$$0 <\mid C \mid<\mid B \mid< 1 < A \tag{34}$$

The extinction coefficient is described by the other three fit parameters: D (dimensionless), E (nm^2), and F (nm^4). Layer 1 was modelled using the 'Cauchy Transparent' model. This layer comprises only a transparent mixture of AlN and Al_2O_3. This implies that $\kappa(\lambda) = 0$.

$$n(\lambda) = A + \frac{10^4}{\lambda^2}B + \frac{10^9}{\lambda^4}C \tag{35}$$

$$\kappa(\lambda) = 0 \tag{36}$$

This model is most effective for materials with no optical absorption in the visible spectral range, typically exhibiting normal dispersion, where n decreases as the wavelength increases. Surface roughness was a significant factor in our case, as both composites demonstrated substantial roughness. Consequently, an additional layer representing the real surface roughness, modelled with a gradient, was added on top of the model. To construct the Cauchy model for both layers, the thicknesses t_1 and t_2 were initially fixed at 68 nm and 232 nm, respectively (as determined in Section 2.2). This approach allowed us to calculate the values for A, B, C, D, E, and F for both layers. Once the values of these six parameters were established, a standard model was generated. During the analysis of four samples for each composite (I and II), the thicknesses t_1 and t_2 and the surface roughness were adjusted for fitting in each iteration.

Table 2 presents the fitting parameters of the Cauchy model used to retrieve the optical functions from SE measurements. The thicknesses t_1 and t_2 and surface roughness S_R obtained for different samples are listed in the top lines. The bottom lines list the Cauchy parameter values for corresponding layers 1 and 2 of each sample. The consistent parameter values across all samples demonstrate excellent reproducibility, validating the reliability of the synthesis technique employed. The layer thicknesses align closely with averages obtained from SEM images and those derived from XPS data, reinforcing the observation that the average coating thickness, L, is greater for composite II. These results further support the predictions of the numerical model for such a complex nanostructure.

Table 2. A detailed table showcasing the thickness (t_1, t_2) and surface roughness (S_R) for three individual samples of composite I and composite II. Additionally, the table includes the Cauchy fitting parameters for each sample.

Spectroscopic Ellipsometer Model	**Composite I**						**Composite II**					
	Sample i		**Sample ii**		**Sample iii**		**Sample i**		**Sample ii**		**Sample iii**	
Surface Roughness (S_R)	68.5 nm		80 nm		64.5 nm		115.5 nm		79.2 nm		71.5 nm	
Layer 1 Cauchy Transparent (t_1)	48.3 nm		55 nm		55 nm		44 nm		84.8 nm		91.4 nm	
Layer 2 Cauchy Absorbent (t_2)	192.5 nm		180 nm		198 nm		281.8 nm		294.6 nm		296.4 nm	
SiO_2	10 nm											
Si Substrate	300 µm											
Fitting parameter	layer1	layer2	layer1	layer2	layer1	layer2	layer1	layer2	layer1	layer2	layer1	layer2
A (1)	1.809	2.973	1.809	2.973	1.809	2.973	1.809	2.973	1.809	2.973	1.809	296.43
B (nm^2)	−0.219	−0.732	−0.219	−0.732	−0.219	−0.732	−0.219	−0.732	−0.219	−0.732	−0.219	−0.732
C (nm^4)	0.025	0.097	0.025	0.097	0.025	0.097	0.025	0.097	0.025	0.097	0.025	0.097
D (1)	0.003	0.0003	0.003	0.288	0.003	0.376	0.003	0.686	0.003	0.704	0.003	0.667
E (nm^2)	0	0.092	0	−0.02	0	−0.167	0	−0.314	0	−0.316	0	−0.4
F (nm^4)	0	−0.002	0	0.006	0	0.027	0	0.057	0	0.064	0	0.095

4.3. *Analysis of the Retrieved Refractive Index for Composites I and II*

The above discussion provides a foundation for the methodology used to determine n and κ for the complex composite structure. To enhance understanding and improve accuracy in assessing the performance of the two composites regarding their light absorptance and reflectance, obtaining refractive indexes for both layer 1 and layer 2 of composites I and II is essential. This section presents a clear comparison, emphasizing differences and offering logical explanations behind the experimentally and numerically obtained optical properties. The values of n and κ from numerical simulations were derived at an incident angle of 50°.

In Section 4.1, the volume fractions used for numerical calculations were derived from XPS data shown in Table 1. The surface roughness was not incorporated within the initial model. The wave optics module in COMSOL Multiphysics software does not include a feature to directly account for roughness, yet our samples exhibit a significant degree of roughness, making it necessary to consider this factor when calculating n and κ. To address this, we included an air fraction of 25% to the unit cell of layer 1 as an alternative way to model the roughness.

Surface roughness values, approximately 60 nm as shown in Table 2, are smaller than the wavelengths in the studied spectrum (370 nm to 1600 nm), justifying the homogenization approach. This assumes that the voids contributing to roughness are part of a smooth layer 1 comprising AlN and Al_2O_3 phases. Figure 7 compares the refractive index and extinction coefficient values of layer 1 for three samples with subscripts (_i, _ii, _iii) of both composites and the numerical model predictions. The solid curves represent the refractive index n, while the dashed curves depict the extinction coefficient κ. Green curves illustrate the numerical predictions for the upper layer 1, black curves in Figure 7i correspond to composite I, and red curves in Figure 7ii correspond to composite II.

The nanoscale simulations assume a uniform mixture of Al_2O_3, AlN, and air in layer 1, leading to a quasi-constant n and a negligible κ over the spectrum, as seen in Figure 7. This prediction aligns consistently across the spectral range. The calculated n and κ values for the AlN, Al_2O_3, and air mix are well justified and align with the XPS analysis results presented in Table 1 during the initial etching phase until the composite core is reached. As a combination of three dielectric materials, the experimental κ values for layer 1 in all samples of both composites remain nearly zero, rendering layer 1 completely transparent. Notably, the n and κ values for layer 1 across all samples are identical, resulting in overlapping curves.

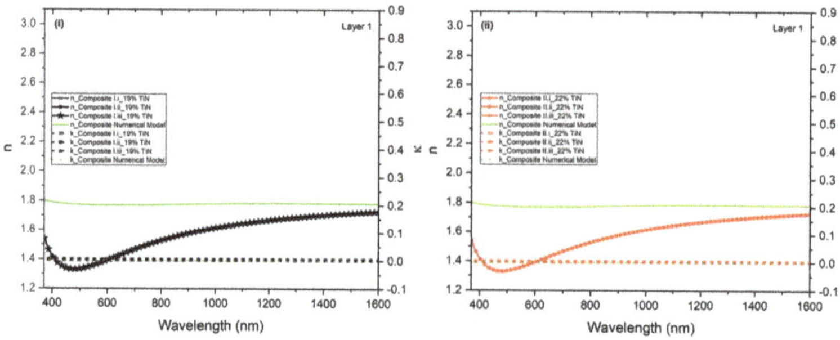

Figure 7. Comparison of optical functions n and κ determined experimentally and through nanoscale simulations for layer 1: **(i)** results corresponding to samples of composite I and **(ii)** results corresponding to samples of composite II.

Figure 8 illustrates the comparison of optical functions for each sample of both composites I and II, alongside numerical predictions for layer 2. On the left (i), the n values for samples (_i) and (_ii) closely align, while sample (iii) shows a slightly lower n in the visible region of the spectrum. The κ values for the samples of composite I range from 0.1 to 0.35 and show good agreement across the spectrum. However, the κ values obtained from the nanoscale model, represented by the dashed green curve, present high values in the visible range. This discrepancy arises because the numerical model assumes an idealized scenario with a perfect, periodic distribution of TiN inclusions within the AlN and Al_2O_3 matrix, which explains the resulting destructive interference.

All curves representing n converge in the infrared (IR) region, following a similar trend with moderate differences. Variations among the samples are attributed to the imperfect distribution of TiN NPs within the matrix. Since the TiN NPs were laid randomly in a wet solution, some inhomogeneity occurred, leading to the formation of agglomerates and clusters in specific areas, which altered light interaction behaviour. The higher κ values observed in Figure 8ii are due to a greater concentration of TiN particles in composite II. This is supported by SEM images in Figure 2, which show a TiN inclusion distribution of 22% in composite II compared to 19% in composite I.

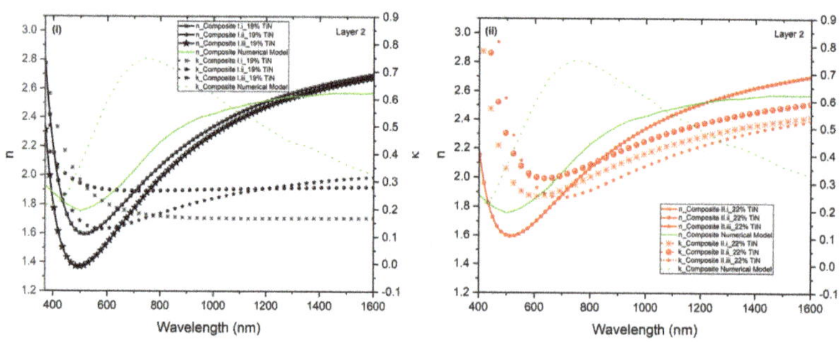

Figure 8. Comparison of optical functions (n and κ values) determined experimentally and through nanoscale simulations for layer 2: **(i)** results corresponding to samples of composite I and **(ii)** results corresponding to samples of composite II.

The electrically conducting nature of TiN nanoparticles in the layer 2 of the composite makes its κ values significantly elevated in both composites compared to the ones of layer 1. Figure 9i,ii provide a comparison of functions n and κ for composites I and II, covering, respectively, layer 1 and layer 2 alongside the numerical predictions. Figure 9i highlights the n and κ values for layer 1, showing that the refractive index values for both composites are lower than the numerical predictions in the visible range. This discrepancy is likely due to inaccuracies in the fitting model in the SE software (completeEASE M2000), while the dip seen at 500 nm is due to the TiN refractive index, as illustrated in Figure 3. Figure 9i compares the n and κ values of layer 2 for composites I and II with the nanoscale predictions. Although the distribution of TiN nanoparticles was manually controlled, it can be observed that the constant optical values for the three samples of each composite and the numerical predictions were consistent and within a comparable range.

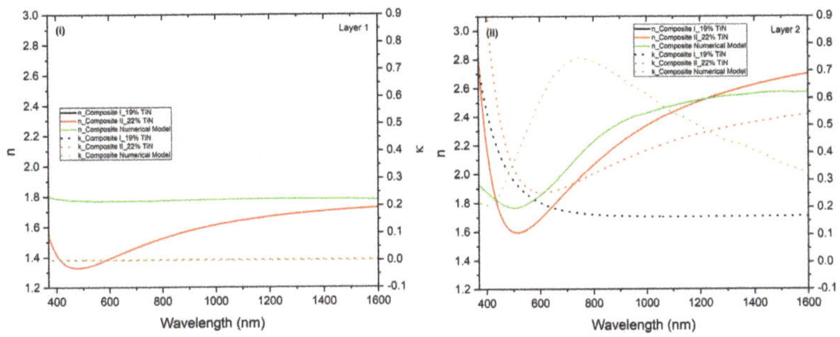

Figure 9. (**i**) Optical functions of layer 1 comparing composite I and II with numerical predictions. (**ii**) Optical functions of layer 2 comparing composite I and II with numerical predictions.

4.4. Analysis of the Reflectance Properties of Composites I and II

To validate the numerical characterization approach of the nanostructured composite, measurements of the reflectance of composites I and II were performed using a UV–visible spectrophotometer based on parameters such as the size of the TiN nanoparticles and the volume fraction of TiN within the AlN matrix. A spectrally selective filter was designed to achieve minimum reflection in the visible region and maximum reflectance in the IR region of the spectrum. The effective optical parameters of the composite, including the refractive index (n) and extinction coefficient (κ), were retrieved for both composites I and II.

The macromodel, simulating the homogeneous layers of the composite as shown in Figure 10, was refined based on the outcomes of the XPS characterization. These refinements specifically addressed the presence of a certain fraction of alumina (Al_2O_3) and the thicknesses of the layers (see XPS results in Table 1). The effective optical parameters derived from the nanoscale model were incorporated into the homogenized macromodel to evaluate the reflectance across the visible light spectrum. Since the size of the TiN NPs is extremely small compared to the wavelength, the shape effect of the NPs is negligible; therefore, the NPs were modelled as cubes in the unit cells of the nanoscale model. The side of the TiN cube was set to 20 nm, while the unit cell width was determined as 26 nm based on the volume fraction of the phases.

The composite was divided into two layers, layer 1 and layer 2, for the macromodel simulations. The n and κ values retrieved from the unit cell prepared with an AlN matrix containing a volume fraction ofAl_2O_3 and air, with the corresponding thickness deduced from XPS data, were assigned to layer 1. Similarly, the n and κ values from the unit cell prepared with an Al_2O_3 matrix containing volume fractions of AlN and TiN, with the corresponding thickness deduced from XPS data, were assigned to layer 2.

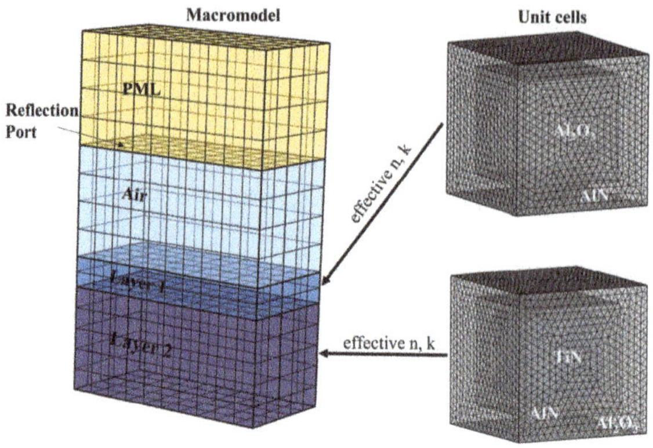

Figure 10. Details of the constructed macromodel, including the optical functions (n and k values) for each layer, are retrieved from different unit cells of the nanoscale model.

The measurements of reflectance in the VIS-IR range were performed using a Perkin Elmer Lambda 750 spectrophotometer. Figure 11 compares the measured and computed reflectance for both composites. The curves shown in black colour represent measurements performed at three different locations on composite I wafer, while the curves in red colour correspond to three different locations on the composite II wafer. Both groups of curves are closely aligned, with a reflectance difference of approximately 10% across the entire spectral range. The variation between black and red group of curves is attributed to differences in the TiN content within composites.

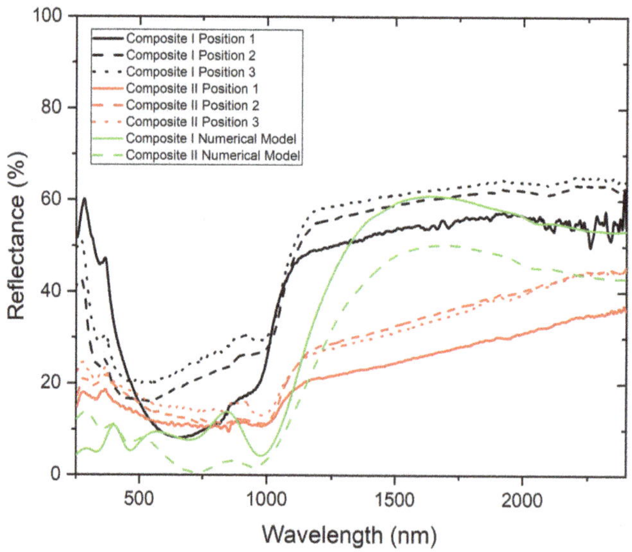

Figure 11. Reflectance properties of composites I and II: experimental measurements are represented in black and red, while the predicted reflectance from macromodel simulations is depicted in green.

The black curves show a higher reflectance in the IR region. This is primarily due to two factors: first, composite II contains more TiN and larger agglomerates compared to composite I; second, composite I has a lower AlN content (and therefore a higher refractive index, n) than composite II. The dotted green curve in the figure corresponds to predicted reflectance from the macromodel simulation for composite II, while the solid green curve corresponds to predicted reflectance from the macromodel simulation for composite I.

The predicted reflectance curve in solid green matches the measured reflectance for composite I because composite I is more homogeneous and exhibits smaller agglomerates of the TiN NPs. In contrast, the red reflectance curves differ significantly from the numerical model for composite II. This discrepancy arises because the numerical model assumes a uniformly separated distribution of TiN nanoparticles within layer 2, which was not achievable by the synthesis technique. Additionally, the nanoscale model assumes smooth surfaces, whereas the actual composites have very rough surfaces. An important observation is that the black curves exhibit a distinct filter characteristics at an infrared limit of 1 µm.

5. Conclusions

In this study, we investigated two methods for retrieving (n) and (κ): one through spectroscopic ellipsometry (SE) experimental measurements and the other via numerical simulations. These approaches provided a clear framework for making structural modifications, if required. Initially, calculations based on XPS data offered rough estimates for parameters such as (t_1), (t_2), (L), and (S_R), which facilitated simplifying the complex structure by dividing it into two distinct layers. Using the elemental composition data from XPS measurements, we also estimated the volume fraction of each component in the composite. These preliminary calculations were instrumental in developing models for both experimental and numerical analysis.

We successfully created a model for measuring (n), (κ), (t_1), (t_2), and (S_R) via SE for the complex structure of the composite. Additionally, two unit cell models were developed numerically to obtain the optical parameters. Comparative analysis revealed that all samples of composites I and II were reproducible, with their optical parameter values falling within close range of one another. Furthermore, the optical parameters obtained experimentally and numerically for both composites showed a high degree of agreement. However, minor discrepancies were observed, primarily due to differences in the experimental and numerical setup of the models.

Author Contributions: M.E.H. and N.K., methodology, validation, investigation, and writing; E.B., review, editing, supervision, and administration. All authors have read and agreed to the published version of the manuscript.

Funding: This research was funded by FNR, Luxembourg, and EPSRC, United Kingdom, under grant INTER FNR–RCUK/1611584556.

Data Availability Statement: Data are unavailable publicly but can be shared upon request.

Acknowledgments: The authors would like to thank S. Krishnamoorthy, M. Michel, O. Hassan, and R. Sevilla for their helpful discussions and guidance. We also express gratitude to LIST engineers N. Adjeroud (PE-ALD), A. El Moul (SEM), and C. Vergne (XPS) for their technical support.

Conflicts of Interest: The authors declare no conflicts of interest. The funders had no role in the design of the study; in the collection, analyses, or interpretation of data; in the writing of the manuscript; or in the decision to publish the results.

References

1. Knöner, G.; Parkin, S.; Nieminen, T.A.; Heckenberg, N.R.; Rubinsztein-Dunlop, H. Measurement of the Index of Refraction of Single Microparticles. *Phys. Rev. Lett.* **2006**, *97*, 157402. [CrossRef]
2. Brown, R. Absorption and Scattering of Light by Small Particles. *Opt. Acta Int. J. Opt.* **1984**, *31*, 3. [CrossRef]
3. Bohren, C.F.; Huffman, D.R. *Absorption and Scattering of Light by Small Particles*; Wiley: Hoboken, NJ, USA, 1998. [CrossRef]

4. Righini, M.; Ghenuche, P.; Cherukulappurath, S.; Myroshnychenko, V.; de Abajo, F.J.G.; Quidant, R. Nano-optical Trapping of Rayleigh Particles and Escherichia coli Bacteria with Resonant Optical Antennas. *Nano Lett.* **2009**, *9*, 3387–3391. [CrossRef]
5. Bendix, P.M.; Oddershede, L.B. Expanding the Optical Trapping Range of Lipid Vesicles to the Nanoscale. *Nano Lett.* **2011**, *11*, 5431–5437. [CrossRef]
6. Khanna, N.; Hachemi, M.E.; Sevilla, R.; Hassan, O.; Morgan, K.; Barborini, E.; Belouettar, S. Multi-physical modelling, design optimisation and manufacturing of a composite dielectric solar absorber. *Compos. Part C Open Access* **2022**, *8*, 100282. [CrossRef]
7. Bilokur, M.; Gentle, A.R.; Arnold, M.D.; Cortie, M.B.; Smith, G.B. High temperature optically stable spectrally-selective Ti1xAlxN-based multilayer coating for concentrated solar thermal applications. *Sol. Energy Mater. Sol. Cells* **2019**, *200*, 109964. [CrossRef]
8. Du, M.; Liu, X.; Hao, L.; Wang, X.; Mi, J.; Jiang, L.; Yu, Q. Microstructure and thermal stability of Al/Ti0.5Al0.5N/Ti0.25Al0.75N/AlN solar selective coating. *Sol. Energy Mater. Sol. Cells* **2013**, *111*, 49–56. [CrossRef]
9. Khanna, N. Metamaterial Design and Elaborative Approach for Efficient Selective Solar Absorber. Ph.D. Thesis , Unilu-University of Luxembourg, Belval, Luxembourg, 2022.
10. Liao, H.M.; Sodhi, R.N.S.; Coyle, T.W. Surface composition of AlN powders studied by x-ray photoelectron spectroscopy and bremsstrahlung-excited Auger electron spectroscopy. *J. Vac. Sci. Technol. A Vac. Surfaces Film.* **1993**, *11*, 2681–2686. [CrossRef]
11. Méndez, M.G.; Rodríguez, S.M.; Shaji, S.; Krishnan, B.; Pérez, P.B. Structural Properties of AlN Films with Oxygen Content Deposited by Reactive Magnetron Sputtering: XRD and XPS Characterisation. *Surf. Rev. Lett.* **2011**, *18*, 23–31. [CrossRef]
12. Jose, F.; Ramaseshan, R.; Dash, S.; Bera, S.; Tyagi, A.K.; Raj, B. Response of magnetron sputtered AlN films to controlled atmosphere annealing. *J. Phys. D Appl. Phys.* **2010**, *43*, 075304. [CrossRef]
13. Zhang, Y. Characterization of as-received hydrophobic treated AlN powder using XPS. *J. Mater. Sci. Lett.* **2002**, *21*, 1603–1605. [CrossRef]
14. Rohmer, M.M.; Bénard, M.; Poblet, J.M. Structure, Reactivity, and Growth Pathways of Metallocarbohedrenes M8C12 and Transition Metal/Carbon Clusters and Nanocrystals: A Challenge to Computational Chemistry. *Chem. Rev.* **2000**, *100*, 495–542. [CrossRef]
15. Chang, Y.H.; Chiu, C.W.; Chen, Y.C.; Wu, C.C.; Tsai, C.P.; Wang, J.L.; Chiu, H.T. Syntheses of nano-sized cubic phase early transition metal carbides from metal chlorides and n-butyllithium. *J. Mater. Chem.* **2002**, *12*, 2189–2191. [CrossRef]
16. Nelson, J.A.; Wagner, M.J. High Surface Area Mo2C and WC Prepared by Alkalide Reduction. *Chem. Mater.* **2002**, *14*, 1639–1642. [CrossRef]
17. Dobrzański, L.A. Significance of materials science for the future development of societies. *J. Mater. Process. Technol.* **2006**, *175*, 133–148. [CrossRef]
18. Siegenfeld, A.F.; Bar-Yam, Y. An Introduction to Complex Systems Science and Its Applications. *Complexity* **2020**, *2020*, 1–16. [CrossRef]
19. Körkel, S.; Kostina, E.; Bock, H.G.; Schlöder, J.P. Numerical methods for optimal control problems in design of robust optimal experiments for nonlinear dynamic processes. *Optim. Methods Softw.* **2004**, *19*, 327–338. [CrossRef]
20. Tan, G.; Lehmann, A.; Teo, Y.M.; Cai, W. *Methods and Applications for Modeling and Simulation of Complex Systems*; Springer Nature: Singapore, 2019.
21. Batsanov, S.S.; Ruchkin, E.D.; Poroshina, I.A. *Refractive Indices of Solids*; Springer: Berlin/Heidelberg, Germany, 2016. Available online: https://www.ebook.de/de/product/26683914/stepan_s_batsanov_evgeny_d_ruchkin_inga_a_poroshina_refractive_indices_of_solids.html (accessed on 19 January 2025).
22. Bahaa, E.A.; Saleh, M.C.T. *Fundamentals of Photonics*; Wiley: Hoboken, NJ, USA, 2019.
23. Mayerhöfer, T.G.; Pahlow, S.; Popp, J. The Bouguer-Beer-Lambert Law: Shining Light on the Obscure. *ChemPhysChem* **2020**, *21*, 2029–2046. [CrossRef]
24. Zhang, K.; Hao, L.; Du, M.; Mi, J.; Wang, J.; Meng, J. A review on thermal stability and high temperature induced ageing mechanisms of solar absorber coatings. *Renew. Sustain. Energy Rev.* **2017**, *67*, 1282–1299. [CrossRef]
25. Pérez, I.H. *Multilayer Selective Coatings for High Temperature Solar Applications*; Editions Universitaires Européennes: Londonm UK, 2017. Available online: https://www.ebook.de/de/product/29245585/irene_heras_perez_multilayer_selective_coatings_for_high_temperature_solar_applications.html (accessed on 19 January 2025).
26. Chen, Z.; Boström, T. Electrophoretically deposited carbon nanotube spectrally selective solar absorbers. *Sol. Energy Mater. Sol. Cells* **2016**, *144*, 678–683. [CrossRef]
27. Striebel, M.; Wrachtrup, J.; Gerhardt, I. Absorption and Extinction Cross Sections and Photon Streamlines in the Optical Near-field. *Sci. Rep.* **2017**, *7*, 15420. [CrossRef] [PubMed]
28. Garnett, J.C.M. Colours in metal glasses and in metallic films. *Philos. Trans. R. Soc* **1904**, *203*, 385–420. [CrossRef]
29. Doyle, W.T. Optical properties of a suspension of metal spheres. *Phys. Rev. B* **1989**, *39*, 9852–9858. [CrossRef]
30. Markel, V.A. ntroduction to the Maxwell Garnett approximation tutorial. *J. Opt. Soc. Am. A* **2016**, *33*, 1244. [CrossRef]

31. Nayak, J.K.; Chaudhuri, P.R.; Ratha, S.; Sahoo, M.R. A comprehensive review on effective medium theories to find effective dielectric constant of composites. *J. Electromagn. Waves Appl.* **2022**, *37*, 1–41. [CrossRef]
32. Orfanidis, S.J. *Electromagnetic Waves and Antennas*; 2016 Available online: https://eceweb1.rutgers.edu/~orfanidi/ewa/ (accessed on 19 January 2025).
33. Nicolson, A.M.; Ross, G.F. Measurement of the Intrinsic Properties of Materials by Time-Domain Techniques. *IEEE Trans. Instrum. Meas.* **1970**, *19*, 377–382. [CrossRef]
34. Smith, D.R.; Vier, D.C.; Koschny, T.; Soukoulis, C.M. Electromagnetic parameter retrieval from inhomogeneous metamaterials. *Phys. Rev. E* **2005**, *71*, 036617. [CrossRef]
35. Lissberger, P.H. Ellipsometry and polarised light. *Nature* **1977**, *269*, 270. [CrossRef]
36. Bakker, J.; Bryntse, G.; Arwin, H. Determination of refractive index of printed and unprinted paper using spectroscopic ellipsometry. *Thin Solid Film.* **2004**, *455-456*, 361–365. [CrossRef]
37. Yusoh, R.; Horprathum, M.; Eiamchai, P.; Chindaudom, P.; Aiempanakit, K. Determination of Optical and Physical Properties of ZrO_2 Films by Spectroscopic Ellipsometry. *Procedia Eng.* **2012**, *32*, 745–751. [CrossRef]

crystals

MDPI

Article

Interactions Between Sc_2O_3 Ceramics and Calcium–Magnesium–Alumina–Silicate (CMAS) at Elevated Temperature

Zupeng Mo [1], Zijian Mo [1], Zhiyun Yu [1], Yifan Cheng [1], Yiling Miao [1] and Tianquan Liang [1,2,3,*]

[1] School of Resources, Environment and Materials, Guangxi University, Nanning 530004, China; 2215391077@st.gxu.edu.cn (Z.M.); 2215391076@st.gxu.edu.cn (Z.M.); 2115391083@st.gxu.edu.cn (Z.Y.); 2239200122@st.gxu.edu.cn (Y.C.); 2239200631@st.gxu.edu.cn (Y.M.)

[2] State Key Laboratory of Featured Metal Materials and Life-Cycle Safety for Composite Structures, Guangxi University, Nanning 530004, China

[3] MOE Key Laboratory of New Processing Technology for Nonferrous Metals and Materials, Guangxi University, Nanning 530004, China

* Correspondence: liangtianquan@gxu.edu.cn

Abstract: The thermochemical interactions of Sc_2O_3 ceramics with CMAS at 1250 °C and 1300 °C were investigated in this paper. A continuously dense reaction layer (DRL) forms on the surface of the ceramic at the beginning of the reaction within 15 min, and temperature significantly affects the components of the DRL. The DRL is mainly composed of a diopside phase at 1250 °C, whose thickness decreases with reaction time, while it is composed of garnet and minor diopside phases at 1300 °C, and thickens in accordance with the parabolic law with exposure time. The DRL shows good effect on alleviating Mg^{2+} infiltration and some mitigating effect to Al^{3+}, and relatively inferior resistance to Ca^{2+} and Si^{4+} penetration. The concentration of Sc^{3+} in the residual CMAS increases with reaction temperature and time, and the average contents are about 0.7 at% and 3.7 at% after reactions at 1250 °C and 1300 °C, respectively. The mechanism is discussed systematically.

Keywords: Sc_2O_3; CMAS; thermochemical interaction; DRL; infiltration

1. Introduction

Thermal barrier coatings (TBCs) are widely applied to gas turbine engines, resulting from the advantages of effectively insulating and protecting hot components during service [1,2]. The 7–8 wt.% yttria-stabilized zirconia (YSZ) is one of the current commercial TBC materials for excellent mechanical properties [3,4]. However, it is difficult for YSZ to meet the operating temperature requirement with the development of the advanced gas turbine engine. Particularly, the CMAS glass formed by environmental deposits such as sand and dust in the air will melt, penetrate and damage the YSZ coating through dissolution-precipitation within a very short period, leading to phase transition, stress cracking and eventual failure of the TBC [5–7]. This significantly damages the service performance and safety of the TBC. It is urgent to propose a new strategy or develop new TBCs to be used against molten CMAS attack.

Extensive research has been conducted to reduce CMAS infiltration and corrosion, including modification of the compositions and microstructures of the coatings [8,9], the application of overlayered protective coatings [10,11] and the development of novel ceramic coating materials [12,13]. It is reported that co-doping Sc_2O_3–Y_2O_3 or co-doping Yb_2O_3–Y_2O_3 in ZrO_2 exhibit superior phase stability and can significantly reduce CMAS corrosion and infiltration compared to YSZ [14,15]. The former is attributed to the low solubility of Sc^{3+} as well as its weak affinity with Ca^{2+}, while the latter is a result of the low diffusion rate of Yb^{3+} in CMAS. The zirconates with a high concentration of RE_2O_3 (RE means the rare earth elements), such as $La_2Zr_2O_7$ [16,17], $Gd_2Zr_2O_7$ [18,19] and $Y_2Zr_2O_7$ [20], also show exceptional CMAS infiltration owing to forming a dense reaction layer (DRL) composed of an RE-apatite phase ($Ca_4RE_6(SiO_4)_6O$ or $Ca_2RE_8(SiO_6)_6O_2$) after the reaction of $RE_2Zr_2O_7$ and CMAS. The modification in microstructures of the coatings also show meaningful academic reference for the development of TBCs.

Overlayered protective coatings, such as Al_2O_3 [21,22], Y_2O_3 [23] and Pt [24,25], are also feasible methods to mitigate the attack of CMAS at high temperature. The overlayer can rapidly react with CMAS to form a continuous DRL composed of anorthite or an RE-apatite phase, which can greatly inhibit the CMAS infiltration and corrosion for weak wettability to the melt. Other materials such as RE-silicates [26,27], Ba_2REAlO_5 [28,29], $BaLn_2Ti_3O_{10}$ [30], $LaMgAl_{11}O_{19}$ [31,32], $LnPO_4$ (Ln refers to the rare earth lanthanides) [33], $La_2Ce_2O_7$ [34], $YAlO_3$ [35] and RE high-entropy ceramics [36,37] also can reduce the high-temperature penetration and corrosion of CMAS by forming an RE-apatite or anorthite DRL. In addition, the reaction of the coatings with CMAS leads to composition change and crystallization of the melts, which further inhibits CMAS infiltration [21,23]. The RE-apatite or anorthite DRL shows effective resistance to the molten CMAS attack, which is regarded as a promising method to protect against the penetration and corrosion of CMAS.

According to the present reports, zirconate doped with rich RE_2O_3 remains a valuable TBC material for its superior mechanical property and excellent resistance to infiltration and corrosion of CMAS [13,20]. Particularly, the addition of Sc_2O_3 can obviously enhance the phase stability and CMAS corrosion resistance of ZrO_2-based materials. Wang [38,39] reported that the co-doping 2 mol.% Sc_2O_3–16 mol.% CeO_2 in ZrO_2 shows better tetragonal phase (t-phase) stability and resistance to infiltration and corrosion of CMAS compared to YSZ, resulting from the low solubility of Sc^{3+} and chemical inertness of Ce^{2+} in CMAS. Fan [40] and Liao [41] revealed that co-doping Sc_2O_3–Y_2O_3 in ZrO_2 exhibits good CMAS infiltration and corrosion resistance owing to the low solubility of Sc^{3+}, the difference in chemical potential and the poor affinity of Sc^{3+} with Ca^{2+}. Sc_2O_3-based multi-R_2O_3 doping ZrO_2 still is an effective design for the new TBC ceramics for excellent comprehensive performance.

The addition of Sc_2O_3 is a feasible method for improving the resistance to melt corrosion of ZrO_2-based TBCs due to the special properties of Sc^{3+}. However, research on the resistance of Sc_2O_3 to CMAS corrosion is relatively scarce, and its reaction products with CMAS, its effect against CMAS corrosion as well as its mechanism of corrosion resistance still need to be further investigated and improved. Therefore, it is of great significance to explore the thermochemical reaction of Sc_2O_3 with CMAS for the formation kinetics of the DRL and its inhibition on the infiltration of the melt. Aiming to reveal the thermochemical interactions of the Sc_2O_3 ceramic with the melt, the reaction products and penetration behavior of the CMAS at 1250 °C and 1300 °C are systematically investigated in this work. The growing of the DRL and its effect on inhibiting the melt infiltration also are analyzed. The interaction mechanism is discussed in detail.

2. Experimental Design

2.1. Ceramic Samples and CMAS Preparation

The Sc_2O_3 powders (99.9%, Aladdin reagents Co., Ltd., Shanghai, China) were cold-pressed at 250 MPa for 10 min to prepare a disc with dimensions of around φ15 mm × 2 mm. Then, the Sc_2O_3 discs were sintered in a furnace at 1600 °C for 5 h to obtain a ceramic pellet. The chemical composition of glass CMAS in this work was

$33CaO$-$9MgO$-$13AlO_{1.5}$-$45SiO_2$ [42] (mol.%). The CaO, Si_2, Al_2O_3 and MgO powders (Aladdin Reagents Ltd.) were slurry-mixed with appropriate molar ratios thoroughly in full ethyl alcohol, through ball-milling for 6 h at 300 rph, followed by drying for 6 h, then annealed at 1300 °C for 4 h to form a transparent glass block. After crushing, grinding and 200 mesh sieve filtering, glass CMAS powder was obtained.

The Sc_2O_3 ceramics were uniformly coated by CMAS, and the content was 20 mg/cm^2. Then, the samples were heat-treated at various temperatures (1250 °C and 1300 °C) for 15 min, 4 h, 8 h and 24 h, respectively. To further understand the thermochemical behaviors and the reaction products between the Sc_2O_3 and CMAS, the mixture of the CMAS–Sc_2O_3 powders (1:1 wt.%) was heat-treated at 1250 °C and 1300 °C for different times (15 min and 4 h). Here, a furnace (KSL-1700X, MTI, Richmond, CA, USA) was used for the thermal reaction tests.

2.2. Characterizations

An X-ray diffractometer (XRD, Rigaku D/MAX2500V, Tokyo, Japan, Cu Kαradiation) with a scan rate of 6°/min was used to identify the phase constitution of the sample after the high-temperature reaction. The cross-sectional morphology of Sc_2O_3 ceramics sheets after reaction with CMAS was examined by a PhenomProx type integrated scanning electron microscope (FESEM, SU8020, HITACHI, Tokyo, Japan), and energy dispersive spectroscopy (EDS, OXFORD X-MAX 80, Birmingham, UK) was used to observe the elements' distribution. Simultaneously, the reaction behavior of Sc_2O_3 with CMAS powder (1:1, wt.%) was carried out in air by differential scanning calorimetry (DSC, STA 449F3, NETZSCH, Selb, Germany) from room temperature to 1450 °C at a rate of 10 °C/min to further understand the thermochemical behaviors between Sc_2O_3 and CMAS.

3. Results and Discussion

3.1. Phase Analysis

The phase constituents of each set of the mixed powders of Sc_2O_3–CMAS (1:1 wt.%) under high temperatures for different times are shown in Figure 1. According to Figure 1a, the diopside ($CaMgSi_2O_6$, PDF#086-0932) and garnet ($Ca_3Sc_2Si_3O_{12}$, PDF#04-008-4631) phases and minor $Sc_2Si_2O_7$ (PDF#031-1225) phase are observed in the Sc_2O_3–CMAS mixed powder after reaction for 15 min, and Sc_2O_3 (PDF#023-1402) is still detected after reaction with CMAS at elevated temperature. The phases of the mixture after exposure for 4 h is similar to that after reaction for 15 min (Figure 1b), showing that the reaction of the Sc_2O_3–CMAS mixed powder is not drastic [43]. Note that the amount of the diopside phase is more than that of the garnet according to the higher diffraction peaks of diopside around 25–35°. It is indicated that the reaction of the mixed powder mainly causes CMAS crystallization when below 1300 °C. Furthermore, Sc-apatite could not be detected in Figure 1, resulting from the low reactivity of Sc_2O_3 in the CMAS and the poor affinity between the Ca^{2+} and Sc^{3+} [41,44]. The whole reactions are summarized as:

$$Sc_2O_3 + CMAS \rightarrow Ca_3Sc_2Si_3O_{12} + CaMgSi_2O_6 + Sc_2Si_2O_7.$$

Figure 2 shows the reaction products of the reactions of the Sc_2O_3 ceramics with CMAS for different times at 1250 °C and 1300 °C. The samples are of single Sc_2O_3 phase before reaction (Figure 2a). According to Figure 2b, the products are the main diopside, minor garnet and anorthite ($CaAl_2Si_2O_8$, PDF#073-0264) phases on the surfaces of the ceramics after interactions at 1250 °C, while they are the diopside and garnet at 1300 °C (Figure 2c). Moreover, the reaction products of the ceramics with the melt are different to those of mixed powder, which is attributed to the different thermochemical reaction.

For the Sc_2O_3 ceramics, the anorthite phase is formed owing to the composition change to the melts [44]. Additionally, it has been reported that $Sc_2Si_2O_7$ will react with CMAS to form $Ca_3Sc_2Si_3O_{12}$ [45]. The formation of the DRL and the reduction in the grain boundary owing to the grain coarsening limit the diffusion of Sc^{3+}, leading to a decrease in the diffusion flux of Sc^{3+}. Combined with the low solubility of the Sc^{3+} in CMAS [40,41], the small amount of $Sc_2Si_2O_7$ formed in CMAS subsequently disappeared by continuing to react with CMAS to form garnet. It can be seen that the amount of diopside phase decreases, when the garnet phase increases with higher temperature. This is related to the gradual dissolution of the diopside phase [46]. Furthermore, Sc_2O_3 also exists after reaction with CMAS. It is a result of the low reactivity of Sc^{3+} in CMAS [40]. The reactions are summarized as:

$$Sc_2O_3 + CMAS \rightarrow Ca_3Sc_2Si_3O_{12} + CaMgSi_2O_6 + Sc_2Si_2O_7 + CaAl_2Si_2O_8.$$

$$Sc_2Si_2O_7 + CMAS \rightarrow Ca_3Sc_2Si_3O_{12}.$$

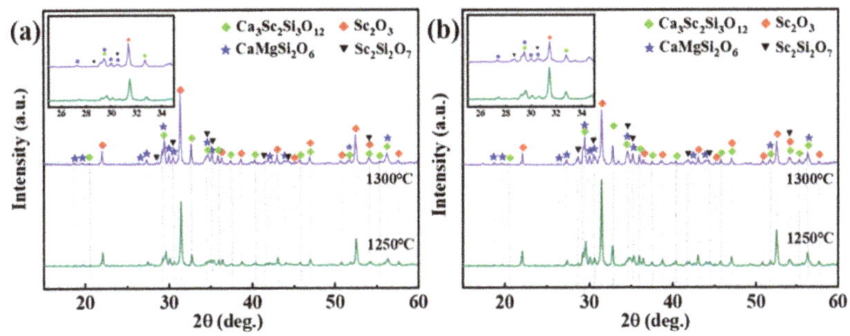

Figure 1. XRD patterns of mixed powders with Sc_2O_3–CMAS (1:1 wt.%) after reaction at 1250 °C and 1300 °C for (**a**) 15 min and (**b**) 4 h.

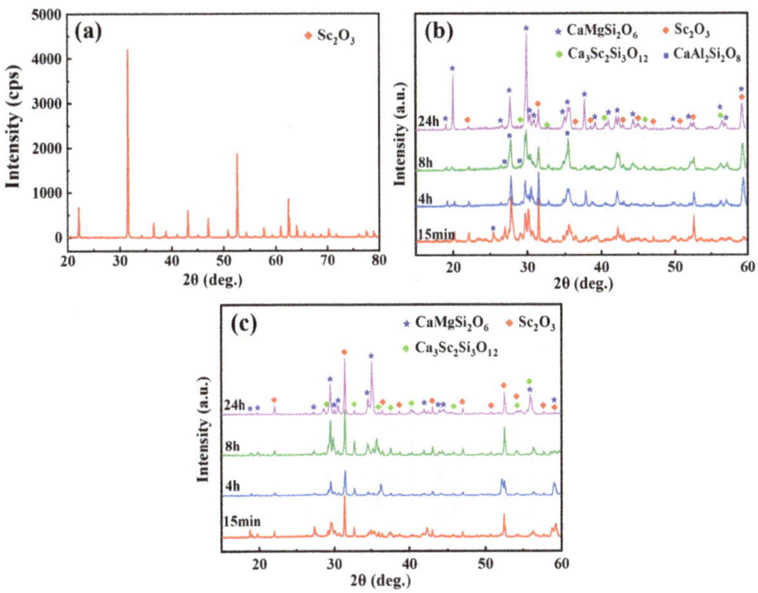

Figure 2. XRD patterns of the Sc_2O_3 ceramics after reaction with the melt for different times. (**a**) Sc_2O_3 ceramic, (**b**) 1250 °C and (**c**) 1300 °C.

3.2. Interactions of CMAS/Sc_2O_3

The SEM graphs of the Sc_2O_3 ceramics after reaction with the melt at 1250 °C are presented in Figure 3 (the symbol "+" is used to highlight the zones, and the number is the name for the zones). It is clear that a distinct continuous DRL forms at the boundary of the Sc_2O_3 ceramic and CMAS. The DRL is mainly composed of the diopside phase (corresponding to regions #4, 8, 12 and 15 in Figure 3 and Table 1). The rod-like or block-like crystallization phases in CMAS mainly are diopside and a small amount of anorthite, which vertically grow into the CMAS. The diopside phase shows coarsening during the initial reaction stage, while it shows a short/small lumpy phase after long exposure. The diopside phase undergoes dissolution during reaction [46]. It is worth noting that no anorthite phase can be detected after reaction for 24 h, while it can be observed within 8 h (Figure 3a–c), indicating that the diopside and anorthite show poor thermostability in the molten CMAS.

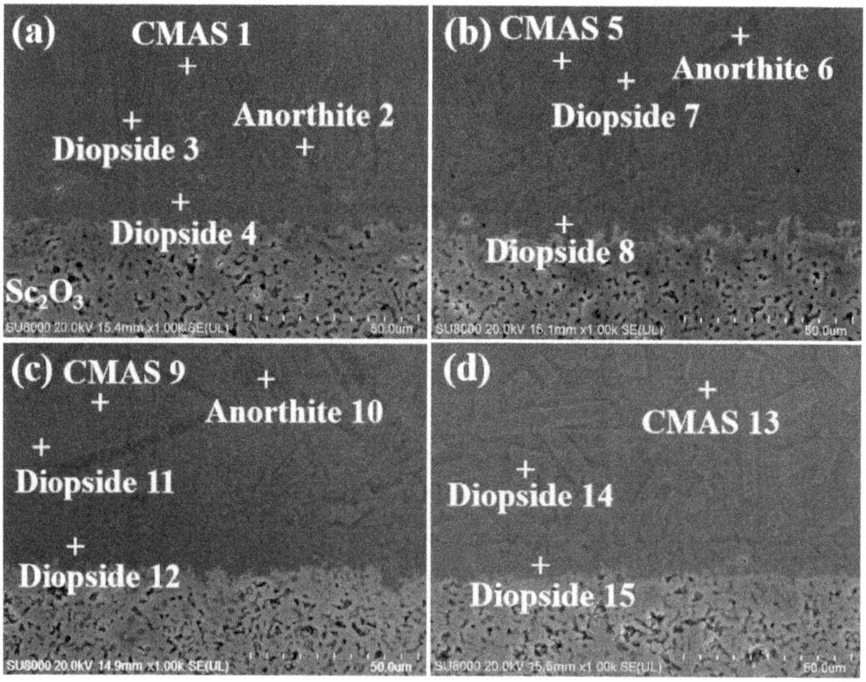

Figure 3. SEM graphs of the Sc_2O_3 ceramics after reaction with the melt at 1250 °C for (**a**) 15 min, (**b**) 4 h, (**c**) 8 h and (**d**) 24 h.

Table 1. Chemical compositions of the marked areas in Figure 3 (at%).

Zone #	Ca	Mg	Al	Si	Sc	Phase
1	33.4	3.1	16.5	46.7	0.3	CMAS
2	20.5	1.4	34.3	43.5	0.3	Anorthite
3	25.9	19.7	5.0	46.6	2.8	Diopside
4	25.8	19.6	6.4	43.9	4.3	Diopside
5	32.2	1.7	17.3	48.3	0.5	CMAS
6	19.6	1.4	33.5	45.2	0.3	Anorthite
7	25.2	18.3	6.0	45.4	5.1	Diopside
8	25.5	17.5	7.2	44.9	4.9	Diopside
9	32.0	1.7	18.2	47.5	0.6	CMAS

Table 1. *Cont.*

Zone #	Ca	Mg	Al	Si	Sc	Phase
10	19.1	1.5	32.1	47.0	0.3	Anorthite
11	25.3	17.3	7.4	44.4	5.6	Diopside
12	25.6	17.7	7.1	44.1	5.5	Diopside
13	25.7	2.3	19.2	51.8	1.0	CMAS
14	24.2	19.7	5.3	46.6	4.2	Diopside
15	24.1	20.8	4.3	47.7	3.1	Diopside

The reaction of the Sc_2O_3 ceramic with CMAS results in the crystallization of residual CMAS, resulting from the change in chemical composition [44]. According to Table 1, it can be seen that Ca^{2+} is one of the key ions from the CMAS. The concentration of Ca^{2+} in CMAS in region #1 (Figure 3a) is about 33.4 at% after reaction at 1250 °C for 15 min (Table 1), while it is 32.2, 32.0 and 25.7 (at%) for residual CMAS after reaction for 4 h, 8 h and 24 h (regions #5, 9 and 13 in Table 1), respectively. It decreases with reaction time, attributing to the formation of diopside and garnet phases. Furthermore, the content of Sc^{3+} in the remaining melt increases with time, resulting from the dissolution of the Sc_2O_3 ceramic when exposed to the molten CMAS, and the average content of Sc^{3+} is about 0.7 at% (except for that after 15 min). The contents of Si^{4+}, Mg^{2+} and Al^{3+} also vary with exposure time. The change in composition leads to the crystallization of CMAS, which increases the melt viscosity and mitigates the infiltration [23].

Figure 4 shows the elemental mappings of the Sc_2O_3 ceramics after reaction with the melt at 1250 °C (the dotted lines refer to the thickness of the reaction layer). The DRL exhibits an excellent effect on mitigating Mg^{2+} infiltration, and some blocking effect on Al^{3+} penetration within 8 h. However, the resistance to Ca^{2+} and Si^{4+} infiltration is relatively poor compared to Mg^{2+} and Al^{3+}. According to the Ca mapping, the infiltration depth of Ca^{2+} increases with reaction time, resulting in forming a Ca-rich zone beneath the DRL. The Ca^{2+} infiltration depths are 13.1 μm, 15.6 μm, 16.8 μm and 17.5 μm after reaction for 15 min, 4 h, 8 h and 24 h, respectively. As for the Al mapping, there is a localized enrichment of Al, showing the formation of the anorthite phase [21]. It is noted that the Sc content in CMAS is about 1.0 at% after 24 h (Table 1), suggesting that the amount of Sc^{3+} in the residual CMAS is very low at 1250 °C [40].

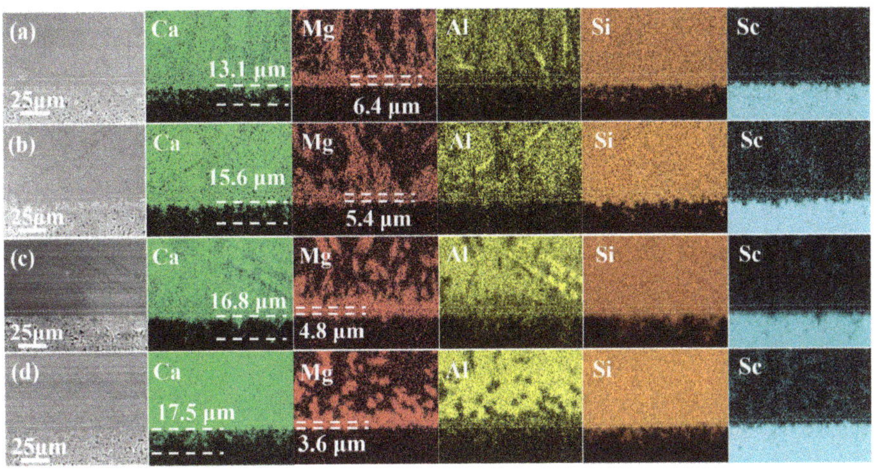

Figure 4. Elemental mappings of Sc_2O_3 ceramics after reaction with the melt at 1250 °C for (**a**) 15 min, (**b**) 4 h, (**c**) 8 h and (**d**) 24 h.

Figure 5 shows the SEM graphs of the Sc_2O_3 ceramics reacting with CMAS at 1300 °C. A DRL forms on the surface of the ceramic after 15 min reaction, which is mainly composed of the garnet ($Ca_3Sc_2Si_3O_{12}$) and minor diopside phases according to the compositions of regions #3, 6, 9 and 13 in Figure 5 and Table 2. The garnet phase in the DRL and diopside phase in the residual CMAS increase and coarsen with prolonged exposure time [47]. No anorthite phase can be detected, consistent with the result in Figure 2c. According to Table 2, the concentration of Ca^{2+} in the melt as region #1 (Figure 5a) is about 30.2 at% after 15 min, and they are 30.1, 29.7 and 26.1 for the residual CMAS after reaction for 4 h, 8 h and 24 h (regions #4, 7 and 10), respectively. It decreases with time, resulting from the formation of more garnet and diopside phases. The content of Sc^{3+} in the remaining CMAS increases with time. The average content of Sc^{3+} is about 3.7 at% (except for that after 15 min), which is higher than that at 1250 °C. It is attributed to a more outward diffusion of Sc^{3+} from the substrate, as we know that the infiltration of the melt can greatly be improved due to the lower viscosity at higher temperature [23]. The penetrated melt reacts with the substrate Sc_2O_3 ceramic, resulting in generation and diffusion of Sc^{3+}. This then leads to a concentration increase in Sc^{3+} in the CMAS even despite the formation of the various crystallization phases.

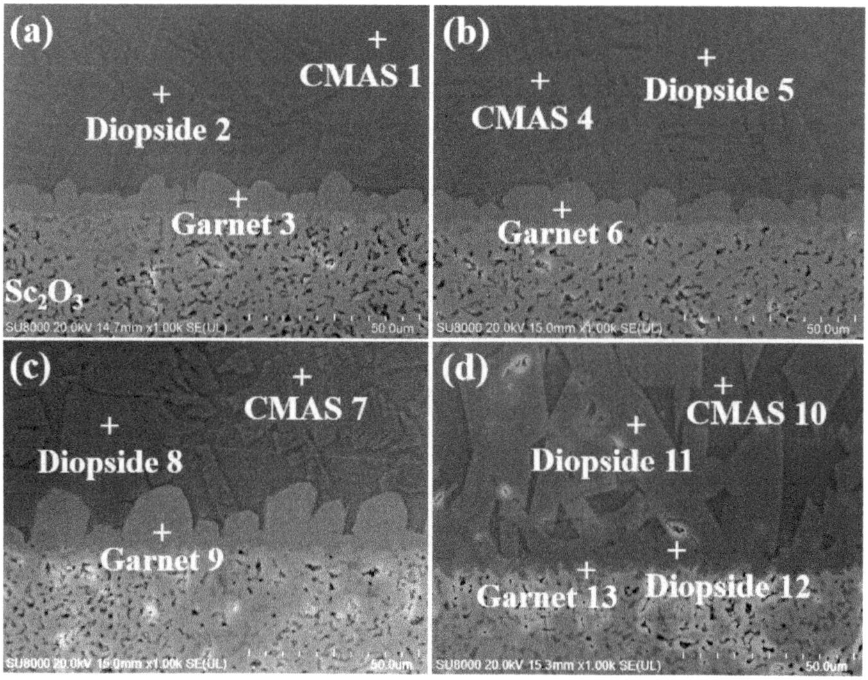

Figure 5. SEM graphs of the Sc_2O_3 ceramics after reaction with the melt at 1300 °C for (**a**) 15 min, (**b**) 4 h, (**c**) 8 h and (**d**) 24 h.

The elemental mappings of the Sc_2O_3 ceramics after reaction with the melt at 1300 °C are presented in Figure 6. The DRL shows a good effect on alleviating the Mg^{2+} infiltration and some mitigating effect on Al^{3+}. An infiltration zone with some Ca forms beneath the DRL according to the Ca mapping. The penetration depth increases with reaction time, whose depths are 13.6 μm, 15.0 μm, 28.9 μm and 68.5 μm after reaction for 15 min, 4 h, 8 h and 24 h, respectively. The infiltration of Si is similar to that of Ca. More melt crystallization for more diopside phases is detected in the CMAS from the elemental mapping of Mg. It is a result of the composition change of the melt such as the decrease in Ca as well as the increase in Sc [21] (zones #1, 4, 7 and 10 in Table 2).

Table 2. Chemical compositions of the marked areas in Figure 5 (at%).

Zone #	Ca	Mg	Al	Si	Sc	Phase
1	30.2	3.3	15.7	48.2	2.6	CMAS
2	24.2	16.2	10.3	43.5	5.8	Diopside
3	34.4	1.5	2.2	38.1	23.8	Garnet
4	30.1	2.7	15.2	48.5	3.5	CMAS
5	24.3	16.2	10.6	42.8	6.1	Diopside
6	35.2	1.6	2.5	37.2	23.5	Garnet
7	29.7	3.6	15.0	48.1	3.6	CMAS
8	24.5	15.8	11.2	42.4	6.1	Diopside
9	34.7	1.8	2.6	37.4	23.5	Garnet
10	26.1	1.7	17.4	49.8	4.0	CMAS
11	23.8	10.2	13.2	38.9	13.9	Diopside
12	22.9	11.0	12.0	40.0	14.1	Diopside
13	31.0	4.0	4.9	38.1	22.0	Garnet

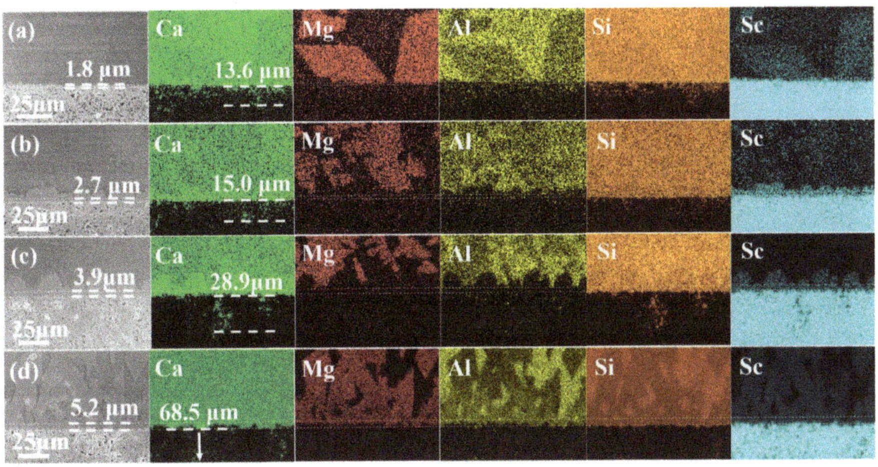

Figure 6. Elemental mappings of the Sc_2O_3 ceramics after reaction with the melt at 1300 °C for (**a**) 15 min, (**b**) 4 h, (**c**) 8 h and (**d**) 24 h.

A DRL forms at the Sc_2O_3/CMAS interface after exposure to the molten CMAS for 15 min, and the changes of thickness of the DRL with reaction temperature and time are shown in Figure 7a. The growing of the thickness follows the parabolic law, and the corresponding equations are well fitted with $y = Ax^2 + Bx + C$. It is obvious that the DRL decreases with time at 1250 °C, resulting from the dissolution of the diopside phase, while the DRL thickens with time at 1300 °C. This is a result of the lower viscosity of CMAS [23] and the increasing concentration of the Sc^{3+} in the residual CMAS (Table 2) [48]. The growing DRL becomes steady after reaction for 24 h, indicating that there is a dynamic balancing state of thickening and dissolution for the DRL during exposure to the melt.

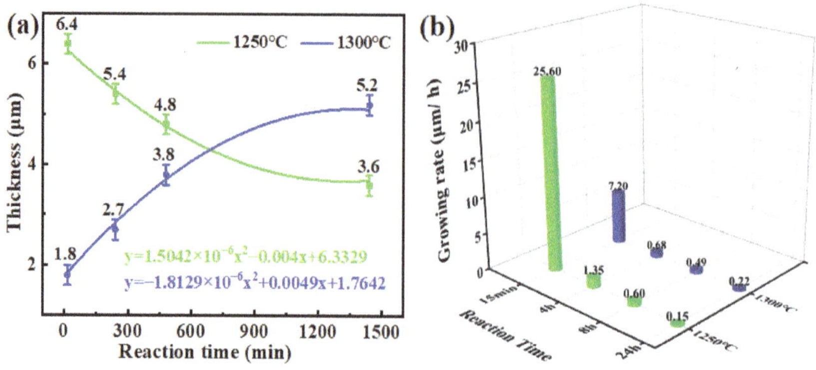

Figure 7. The thicknesses (**a**) and the growing rates (**b**) of the DRLs with temperature and time.

Figure 7b exhibits the growing speeds of the DRLs at 1250 °C and 1300 °C. The thickening rates of the DRL are about 25.6 and 7.20 μm/h for the Sc_2O_3 ceramics within the first 15 min of the reaction at two temperatures (1250 °C, 1300 °C), respectively, and the growing speeds are only around 0.15 and 0.22 μm/h after reaction for 24 h. It is obvious that the growing speed of the DRL decreases with time at high temperature. As for 1250 °C, it is attributed to the dissolution of the diopside phase [46]. At 1300 °C, it is related to the increase in DRL thickness and garnet grain coarsening, resulting in an increase in the diffusion path and fewer diffusion channels. Furthermore, the growing rate of the DRL at 1300 °C is slower compared to that at 1250 °C during the initial 8 h. It may be related to the growing behavior of the DRL and thermochemical stability of the diopside and garnet phases in the molten CMAS. While the growing speed of the DRL at 1300 °C is relatively higher after more than 8 h, it may be due to the decreasing viscosity and the severe infiltration of the melt.

3.3. Mechanism Analysis

To further understand the reactions of the Sc_2O_3 ceramic with the CMAS at high temperature, the DSC tests were conducted on the CMAS and the mixed powders with Sc_2O_3–CMAS (1:1 wt.%) during heating from room temperature to 1450 °C, and the corresponding patterns are exhibited in Figure 8. A small peak exists at around 800 °C, representing the glass transition (GT) of CMAS [45]. Moreover, the CMAS begins to melt at about 1207.6 °C, and the endothermic peak P1 of the CMAS curve is around 1229.5 °C, close to that in Ref. [49]. The peak P1 of the mixed powder is around 1225.3 °C, indicating that Sc_2O_3 shows a small effect for the melting of CMAS. The reaction of Sc_2O_3 with CMAS causes peak P2, forming garnet, diopside and $Sc_2Si_2O_7$ phases. The $Sc_2Si_2O_7$ will react with CMAS to form $Ca_3Sc_2Si_3O_{12}$ [45], suggesting that $Sc_2Si_2O_7$ is a transition phase during thermochemical reactions. Peak P3 (1350.7 °C) indicates the melting of diopside [50]. Additionally, peak P4 may be related to the melting behavior of garnet. The whole thermochemical reaction processes are summarized as:

GT: CMAS undergoes glass transition;

P1: Sc_2O_3 and CMAS powders melt;

P2:

$$Sc_2O_3 + CMAS \rightarrow Ca_3Sc_2Si_3O_{12} + CaMgSi_2O_6 + Sc_2Si_2O_7.$$

$$Sc_2Si_2O_7 + CMAS \rightarrow Ca_3Sc_2Si_3O_{12};$$

P3: The melting of $CaMgSi_2O_6$;

P4: The melting of $Ca_3Sc_2Si_3O_{12}$.

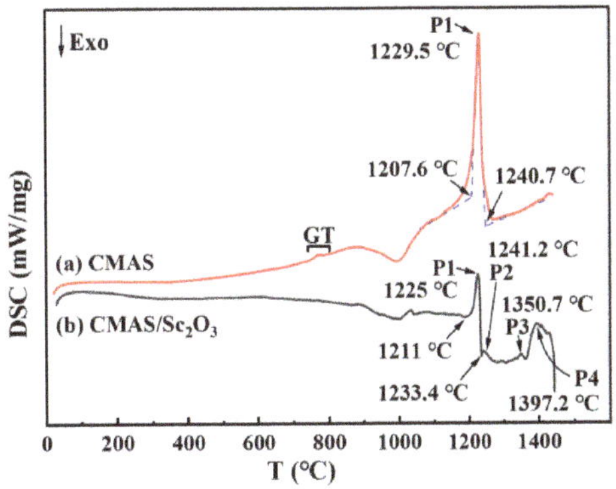

Figure 8. DSC curves of (**a**) CMAS and (**b**) mixture of CMAS–Sc_2O_3 (1:1 wt.%).

Figure 9 illustrates the thermochemical reaction mechanisms of the Sc_2O_3 ceramic and the CMAS at high temperature. At the beginning of the reaction at 1250 °C, the Sc_2O_3 ceramic reacts with the melt to form a DRL within the initial 15 min, and it is mainly the diopside phase. Then, the anorthite phase precipitates in the residual CMAS owing to the composition change in the CMAS (Figure 9b). The DRL becomes thin with reaction time, attributed to the dissolution of the diopside [46], and the anorthite phase disappears after 24 h. As for 1300 °C, the DRL is composed of the dominating garnet and minor diopside phases. The DRL thickens with reaction time, resulting from the good stability of the garnet phase and the increasing Sc^{3+} infiltration from the substrate [48]. The DRL shows good effect on inhibiting Mg^{2+} penetration and some blocking effect on Al^{3+}, and relatively inferior resistance to Ca^{2+} and Si^{4+} penetration.

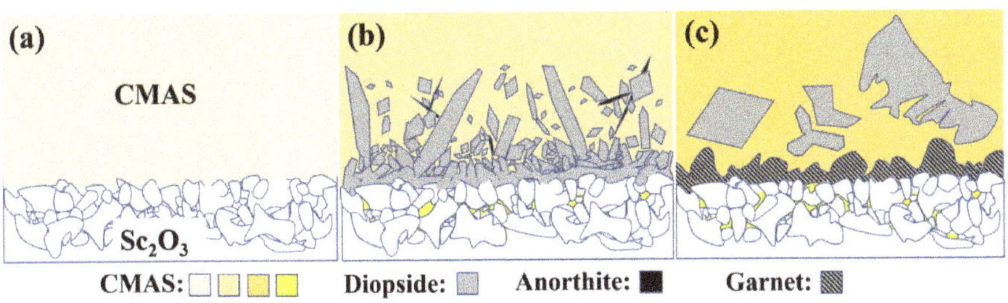

Figure 9. Schematics of the thermochemical interactions of the CMAS and Sc_2O_3 ceramic at high temperature (**a**) before interaction, (**b**) 1250 °C and (**c**) 1300 °C.

4. Conclusions

The thermochemical interactions of the Sc_2O_3 ceramic with CMAS at 1250 °C and 1300 °C were investigated in this work. Different reaction products and the CMAS crystallization phases form at various temperatures, resulting in distinct phase components for the dense reaction layer (DRL). The DRL is mainly composed of the diopside phase ($CaMgSi_2O_6$) at 1250 °C, decreasing with reaction time, while it is of the dominant garnet phase ($Ca_3Sc_2Si_3O_{12}$) at 1300 °C, increasing with exposure time. It is attributed to the difference in stability between the garnet and diopside phases in the molten CMAS. The thickening rate of the DRL decreases and following the parabolic law with duration time. They are about 25.6 and 7.20 (μm/h) after reaction for the initial 15 min, and only 0.15 and 0.22 (μm/h) after 24 h reaction at 1250 °C and 1300 °C, respectively. The DRL shows good effect on inhibiting Mg^{2+} and some blocking effect on Al^{3+} penetration, and relatively inferior resistance to Ca^{2+} and Si^{4+} infiltration. It is indicated that a DRL composed of diopside and/or garnet phases is a potential strategy for protective coating against molten CMAS attack.

Author Contributions: Conceptualization, T.L.; formal analysis, Z.M. (Zupeng Mo) and Z.Y.; investigation, Z.M. (Zupeng Mo), Z.M. (Zijian Mo), Z.Y., Y.C. and Y.M.; data curation, Z.M. (Zupeng Mo) and Z.M. (Zijian Mo); writing—initial draft preparation, Z.M. (Zupeng Mo); writing—review and editing, Z.M. (Zupeng Mo) and T.L.; supervision, T.L. All authors have read and agreed to the published version of the manuscript.

Funding: The research was funded by the National Natural Science Foundation of China, grant number 52061004.

Data Availability Statement: Data are contained within the article.

Conflicts of Interest: The authors declare no conflicts of interest.

References

1. Fang, H.J.; Wang, W.Z.; Deng, S.J.; Yang, T.; Zhu, H.; Huang, J.B.; Ye, D.D.; Guo, X.P. Interaction between Yb_2O_3-Y_2O_3 co-stabilized ZrO_2 ceramic powder and molten silicate deposition, and its implication on thermal barrier coating application. *Mater. Charact.* **2021**, *180*, 111418. [CrossRef]
2. Darolia, R. Thermal barrier coatings technology: Critical review, progress update, remaining challenges and prospects. *Int. Mater.* **2013**, *58*, 315–348. [CrossRef]
3. Dai, J.W.; Huang, B.; He, L.M.; Mu, R.D.; Tian, H.; Xu, Z.H. Thermal cycling behavior and failure mechanism of Yb_2O_3-doped yttria-stabilized zirconia thermal barrier coatings. *Mater. Today Commun.* **2023**, *34*, 105–409. [CrossRef]
4. Rai, A.K.; Schmitt, M.P.; Bhattacharya, R.S.; Zhu, D.M.; Wolfe, D.E. Thermal conductivity and stability of multilayered thermal barrier coatings under high temperature annealing conditions. *J. Eur. Ceram. Soc.* **2015**, *35*, 1605–1612. [CrossRef]
5. Jackson, R.W.; Zaleski, E.M.; Poerschke, D.L.; Hazel, B.T.; Begley, M.R.; Levi, C.G. Interaction of molten silicates with thermal barrier coatings under temperature gradients. *Acta. Mater.* **2015**, *89*, 396–407. [CrossRef]
6. Naraparaju, R.; Chavez, J.J.G.; Schulz, U.; Ramana, C.V. Interaction and infiltration behavior of Eyjafjallajőkull, Sakurajima volcanic ashes and a synthetic CMAS containing FeO with/in EB-PVD ZrO_2-65 wt% Y_2O_3 coating at high temperature. *Acta. Mater.* **2017**, *136*, 164–180. [CrossRef]
7. Holgate, C.S.; Seward, G.G.E.; Ericks, A.R.; Poerschke, D.L.; Levi, C.G. Dissolution and diffusion kinetics of yttria-stabilized zirconia into molten silicates. *J. Eur. Ceram. Soc.* **2021**, *41*, 1984–1994. [CrossRef]
8. Zhang, B.P.; Song, W.J.; Wei, L.L.; Xiu, Y.X.; Xu, H.B.; Dingwell, D.B.; Guo, H.B. Novel thermal barrier coatings repel and resist molten silicate deposits. *Scr. Mater.* **2019**, *163*, 71–76. [CrossRef]
9. He, Y.X.; Xiao, G.Z.; Wang, C.; Lu, X.F.; Li, L.Y.; Liu, S.Y.; Wu, Y.S.; Wang, Z.J. Improved thermal properties and CMAS corrosion resistance of rare-earth monosilicates by adjusting the configuration entropy with RE-doping. *Corros. Sci.* **2024**, *226*, 111664. [CrossRef]
10. Rai, A.K.; Bhattacharya, R.S.; Wolfe, D.E.; Eden, T.J. CMAS-resistant thermal barrier coatings (TBC). *Int. J. Appl. Ceram. Technol.* **2010**, *7*, 662–674. [CrossRef]
11. Fang, H.J.; Wang, W.Z.; Huang, J.B.; Ye, D.D. Investigation of CMAS resistance of sacrificial plasma-sprayed mullite-YSZ protective layer on 8YSZ thermal barrier coating. *Corros. Sci.* **2020**, *173*, 108764. [CrossRef]
12. Wu, D.; Zhang, H.; Shan, X.; Yang, F.; Guo, F.W.; Xiao, P.; Gong, S.K. A novel CMAS-resistant material based on thermodynamic equilibrium design: Apatite-type $Gd_{10}(SiO_4)_6O_3$. *J. Am. Ceram. Soc.* **2020**, *103*, 3401–3415. [CrossRef]
13. Wang, Y.H.; Ma, Z.; Liu, L.; Liu, Y.B. Reaction products of $Sm_2Zr_2O_7$ with calcium–magnesium–aluminum–silicate (CMAS) and their evolution. *J. Adv. Ceram.* **2021**, *10*, 1389–1397. [CrossRef]
14. Fan, W.; Wang, Z.Z.; Bai, Y.; Che, J.W.; Wang, R.J.; Ma, F.; Tao, W.Z.; Liang, G.Y. Improved properties of scandia and yttria co-doped zirconia as a potential thermal barrier material for high temperature applications. *J. Eur. Ceram. Soc.* **2018**, *38*, 4502–4511. [CrossRef]
15. Fang, H.J.; Wang, W.Z.; Huang, J.B.; Li, Y.J.; Ye, D.D. Corrosion behavior and thermos-physical properties of a promising Yb_2O_3 and Y_2O_3 co-stabilized ZrO_2 ceramic for thermal barrier coatings subject to calciummagnesium-aluminum-silicate (CMAS) deposition: Experiments and first-principles calculation. *Corros. Sci.* **2021**, *128*, 111418. [CrossRef]
16. Ramachandran, C.S.; Balasubramaniana, V.; Ananthapadmanabhan, P.V. Thermal cycling behavior of plasma sprayed lanthanum zirconate based coatings under concurrent infiltration by a molten glass concoction. *Ceram. Int.* **2013**, *39*, 1413–1431.

17. Gok, M.G.; Karabas, M. Production of Re doped $La_2Zr_2O_7$ based TBCs and numerical analysis of their use on IC engine piston surface. *Ceram. Int.* **2022**, *48*, 11173–11180. [CrossRef]
18. Dolmaire, A.; Béchade, E.; Geffroy, P.M.; Goutier, S.; Vardelle, M.; Vilasi, M.; Joulia, A. Reaction mechanisms of $Gd_2Zr_2O_7$ in silicate melts derived from CAS. *J. Eur. Ceram. Soc.* **2022**, *42*, 7247–7257. [CrossRef]
19. Bahamirian, M. Nanostructured $Gd_2Zr_2O_7$: A promising thermal barrier coating with high resistance to CaO-MgO-Al_2O_3-SiO_2 corrosion. *J. Aust. Ceram. Soc.* **2023**, *59*, 165–177. [CrossRef]
20. Drexler, J.M.; Ortiz, A.L.; Padture, N.P. Composition effects of thermal barrier coating ceramics on their interaction with molten Ca–Mg–Al–silicate (CMAS) glass. *Acta. Mater.* **2012**, *60*, 5437–5447. [CrossRef]
21. Naraparaju, R.; Pubbysetty, R.P.; Mechnich, P.; Schulz, U. EB-PVD alumina (Al_2O_3) as a top coat on 7YSZ TBCs against CMAS/VA infiltration: Deposition and reaction mechanisms. *J. Eur. Ceram. Soc.* **2018**, *38*, 3333–3346. [CrossRef]
22. Guo, Y.Q.; Wei, L.L.; He, Q.; Deng, Y.P.; He, W.T.; Guo, H.B. PS–PVD alumina overlayer on thermal barrier coatings against CMAS attack. *J. Therm. Spray Technol.* **2021**, *30*, 864–872. [CrossRef]
23. Liang, T.Q.; Huang, Z.H.; Li, M.H.; Xiao, W.T.; He, H.; He, A.P.; Chen, X.Y.; Luo, N.N. Thermochemical reaction behavior between Y_2O_3 and calcium-magnesium-aluminosilicate (CMAS) at elevated temperature. *Corros Sci.* **2022**, *208*, 110631. [CrossRef]
24. Liu, H.; Cai, J.; Zhu, J.H. CMAS (CaO-MgO-Al_2O_3-SiO_2) resistance of Y_2O_3-stabilized ZrO_2 thermal barrier coatings with Pt layers. *Ceram. Int.* **2018**, *44*, 452–458. [CrossRef]
25. Zhao, H.B.; Levi, C.G.; Wadley, H.N.G. Molten silicate interactions with thermal barrier coatings. *Int. J. Appl. Ceram. Technol.* **2014**, *251*, 74–86. [CrossRef]
26. Tian, Z.L.; Zhang, J.; Zheng, L.Y.; Hu, W.P.; Ren, X.M.; Lei, Y.M. General trend on the phase stability and corrosion resistance of rare earth monosilicates to molten calcium–magnesium–aluminosilicate at 1300 °C. *Corros. Sci.* **2019**, *148*, 281–292. [CrossRef]
27. Tian, Z.L.; Ren, X.M.; Lei, Y.M.; Zheng, L.Y.; Geng, W.R.; Zhang, J. Corrosion of $RE_2Si_2O_7$ (RE = Y, Yb, and Lu) environmental barrier coating materials by molten calcium-magnesium-alumino-silicate glass at high temperatures. *J. Eur. Ceram. Soc.* **2019**, *39*, 4245–4254. [CrossRef]
28. Wei, L.L.; Guo, L.; Li, M.Z.; Guo, H.B. Calcium-magnesium-alumina-silicate (CMAS) resistant Ba_2REAlO_5 (RE = Yb, Er, Dy) ceramics for thermal barrier coatings. *J. Eur. Ceram. Soc.* **2018**, *37*, 4991–5000. [CrossRef]
29. Guo, L.; Zhang, C.L.; He, Q.; Yu, J.X.; Yan, Z.; Ye, F.X.; Dan, C.Y.; Ji, V. Microstructure evolution and hot corrosion mechanisms of Ba_2REAlO_5 (RE = Yb, Er, Dy) exposed to V_2O_5 + Na_2SO_4 molten salt. *J. Eur. Ceram. Soc.* **2018**, *38*, 3555–3563. [CrossRef]
30. Guo, L.; Li, M.Z.; Yang, C.X.; Zhang, C.L.; Xu, L.M.; Ye, F.X.; Dan, C.Y.; Ji, V. Calcium-magnesium-alumina-silicate (CMAS) resistance property of $BaLn_2Ti_3O_{10}$ (Ln=La, Nd) for thermal barrier coating applications. *Ceram. Int.* **2017**, *43*, 10521–10527. [CrossRef]
31. Zeng, J.Y.; Sun, J.B.; Zhang, H.; Yang, X.; Qiu, F.Y.; Zhou, P.F.; Niu, W.B.; Dong, S.J.; Zhou, X.; Cao, X.Q. Lanthanum magnesium hexaluminate thermal barrier coatings with pre-implanted vertical microcracks: Thermal cycling lifetime and CMAS corrosion behavior. *Ceram. Int.* **2018**, *44*, 11472–11485. [CrossRef]
32. Song, C.X.; Qi, X.; Huang, L.; Sun, J.B.; Li, L.Z.; Li, C.G.; Lu, W.H. CaO-MgO-Al_2O_3-SiO_2 (CMAS) corrosion behaviour of $LaMgAl_{11}O_{19}$/$GdPO_4$ thermal barrier coating materials. *Ceram. Int.* **2023**, *49*, 26578–26588. [CrossRef]
33. Guo, L.; Yan, Z.; Yu, Y.; Yang, J.; Li, M.Z. CMAS resistance characteristics of $LaPO_4$/YSZ thermal barrier coatings at 1250 °C–1350 °C. *Corros. Sci.* **2019**, *154*, 111–122. [CrossRef]
34. Kang, Y.X.; Bai, Y.; Fan, W.; Yuan, T.; Gao, Y.; Bao, C.G.; Li, B.Q. Thermal cycling performance of $La_2Ce_2O_7$/50 vol.% YSZ composite thermal barrier coating with CMAS corrosion. *J. Eur. Ceram. Soc.* **2018**, *38*, 2851–2862. [CrossRef]
35. Turcer, L.R.; Krause, A.R.; Garces, H.F.; Zhang, L.; Padture, N.P. Environmental barrier coating ceramics for resistance against attack by molten calcia-magnesia-aluminosilicate (CMAS) glass: Part I, $YAlO_3$ and γ-$Y_2Si_2O_7$. *J. Eur. Ceram. Soc.* **2018**, *38*, 3905–3913. [CrossRef]
36. Yan, R.X.; Liang, W.P.; Miao, Q.; Zhao, H.; Liu, R.X.; Li, J.L.; Zang, K.; Dong, M.J.; He, X.P.; Gao, X.G.; et al. Mechanical, thermal and CMAS resistance properties of high-entropy $(Gd_{0.2}Y_{0.2}Er_{0.2}Tm_{0.2}Yb_{0.2})_2Zr_2O_7$ ceramics. *Ceram. Int.* **2023**, *49*, 20729–220741. [CrossRef]
37. Deng, S.X.; He, G.; Yang, Z.C.; Wang, J.X.; Li, J.T.; Jiang, L. Calcium-magnesium-alumina-silicate (CMAS) resistant high entropy ceramic $(Y_{0.2}Gd_{0.2}Er_{0.2}Yb_{0.2}Lu_{0.2})_2Zr_2O_7$ for thermal barrier coatings. *J. Mater. Sci. Technol.* **2022**, *107*, 259–265. [CrossRef]
38. Wang, J.S.; Chen, M.D.; Li, C.Z.; Chen, L.Y.; Yu, Y.S.; Wang, Y.H.; Liu, B.; Jing, Q.S. Comparison of corrosion behaviors of Sc_2O_3-CeO_2 co-stabilized ZrO_2 and YSZ ceramics exposed to CMAS at 1250 °C. *Surf. Coat. Technol.* **2021**, *428*, 127879. [CrossRef]
39. Wang, J.S.; Lu, X.J.; Hu, M.Q.; Chen, M.D.; Sun, J.R.; Wang, Y.H.; Shu, C.X.; Zhang, H.; Liu, B.; Sun, J.B.; et al. Phase stability, thermophysical properties, thermal shock behavior and CMAS resistance of Sc_2O_3-CeO_2 co-stabilized ZrO_2 TBCs. *Surf. Coat. Technol.* **2023**, *467*, 129679. [CrossRef]
40. Fan, W.; Bai, Y.; Liu, Y.F.; Kang, Y.X.; Wang, Y.; Wang, Z.Z.; Tao, W.Z. Corrosion behavior of Sc_2O_3-Y_2O_3 co-stabilized ZrO_2 thermal barrier coatings with CMAS attack. *Ceram. Int.* **2019**, *45*, 15763–15767. [CrossRef]

41. Liao, Y.X.; Dai, Y.F.; Zhai, Y.F.; He, A.P.; He, H.; Liang, T.Q. The corrosion behavior of Sc_2O_3-Y_2O_3 co-doped ZrO_2 influenced by Sc_2O_3 content in CMAS at 1300 °C. *J. Eur. Ceram. Soc.* **2024**, *44*, 1179–1187. [CrossRef]
42. Yan, Z.; Guo, L.; Li, Z.; Yu, Y.; He, Q. Effects of laser glazing on CMAS corrosion behavior of Y_2O_3 stabilized ZrO_2 thermal barrier coatings. *Corros. Sci.* **2019**, *157*, 450–461. [CrossRef]
43. Ye, F.X.; Yuan, Y.H.; Yan, S.; Guo, L.; Yu, J.X. High-temperature corrosion mechanism of a promising scandium tantalate ceramic for next generation thermal barrier coating under molten calcium–magnesium-aluminosilicate (CMAS). *Mater. Chem. Phys.* **2020**, *256*, 123679. [CrossRef]
44. Poerschke, D.L.; Barth, T.L.; Levi, C.G. Equilibrium relationships between thermal barrier oxides and silicate melts. *Act. Mater.* **2016**, *120*, 302–314. [CrossRef]
45. Liu, J.; Zhang, L.T.; Liu, Q.M.; Cheng, L.F.; Wang, Y.G. Calcium–magnesium–aluminosilicate corrosion behaviors of rare-earth disilicates at 1400 °C. *J. Eur. Ceram. Soc.* **2013**, *33*, 3419–3428. [CrossRef]
46. Guo, L.; Xin, H.; Li, Y.Y.; Yu, Y.; Yan, Z.; Hu, C.W.; Ye, F.X. Self-crystallization characteristics of calcium-magnesium-alumina-silicate (CMAS) glass under simulated conditions for thermal barrier coating applications. *J. Eur. Ceram. Soc.* **2020**, *40*, 5683–5691. [CrossRef]
47. Zhou, B.Y.; Wu, Y.; Ke, X.J.; Zhou, Q.J.; Cui, Y.J.; Wang, C.L. Resistance of ytterbium silicate environmental barrier coatings against molten calcium-magnesium-aluminosilicate (CMAS): A comprehensive study. *Surf. Coat. Technol.* **2024**, *479*, 130540. [CrossRef]
48. Wang, Z.C.; Zhou, M.F.; Li, M.Y.H.; Robinson, P.T.; Harlov, D.E. Kinetic controls on Sc distribution in diopside and geochemical behavior of Sc in magmatic systems. *Geochim. Cosmochim. Acta* **2022**, *325*, 316–332. [CrossRef]
49. Zaleski, E.M.; Ensslen, C.; Levi, C.G. Melting and Crystallization of Silicate Systems Relevant to Thermal Barrier Coating Damage. *J. Am. Ceram. Soc.* **2015**, *98*, 1642–1649. [CrossRef]
50. Reinsch, S.; Nascimento, M.L.F.; Müller, R.; Zanotto, E.D. Crystal growth kinetics in cordierite and diopside glasses in wide temperature ranges. *J. Non-Cryst. Solids* **2008**, *354*, 5386–5394. [CrossRef]

crystals

MDPI

Article

The Typical Microstructure of Twin-Roll Cast 2139 Alloy and Its Impact on Mechanical Properties

Zhenkuan Liu [1,2,†], Yuxiao Wang [1,2,†], Qiaoning Chen [1,2], Longzhou Meng [3], Zhengcheng Yang [1,2], Hongqun Tang [1,2,*], Xiaoming Qian [3,*], Yifei Xu [3], Yong Li [4,5] and Xu Li [5]

1 Guangxi Key Laboratory of Processing for Non-Ferrous Metals and Featured Materials, Guangxi University, Nanning 530004, China; liuzhenkuan_work@163.com (Z.L.)
2 State Key Laboratory of Featured Metal Materials and Life-Cycle Safety for Composite Structures, Guangxi University, Nanning 530004, China
3 National Engineering Research Center for Equipment and Technology of Cold Strip Rolling, Yanshan University, Qinhuangdao 066004, China
4 Guangxi Advanced Aluminum Processing Innovation Center Co., Ltd., Nanning 530007, China
5 State Key Laboratory of Rolling and Automation, Northeastern University, Shenyang 110819, China
* Correspondence: hqtang@gxu.edu.cn (H.T.); qxm@ysu.neu.edu.cn (X.Q.)
† These authors contributed equally to this work.

Abstract

The typical microstructure and mechanical properties of twin-roll cast (TRC) 2139 aluminum alloy were investigated and compared with mold casting (MC) 2139 alloy. This work pioneers the application of TRC to produce 2139 Al-Cu-Mg alloy, a material that is challenging for rapid solidification. The TRC process resulted in a denser dendritic structure, with the composition of intermetallic compounds, primarily Al_2Cu and Al_2CuMg, remaining largely stable throughout the casting process. After solution treatment, the recrystallized grains in the MC sheets were uniformly distributed, while the TRC sheets exhibited a more localized and refined recrystallized microstructure, particularly within coarse second-phase regions. Following heat treatments, the TRC sheets showed a significant increase in the Ω phase after T6, with a slight growth in size and a uniform distribution, while the Ω phase in T8 showed an increased density and smaller size, which diffused evenly across the material. The TRC process uniquely refines the microstructure and enhances Ω phase precipitation, yielding a 10%+ improvement in strength and ductility over conventional casting. The mechanical properties of the TRC sheets improved significantly: tensile and yield strengths increased by over 10% after T6, compared to MC sheets, with elongation slightly higher in TRC. T8 treatment further enhanced the mechanical properties of the TRC sheets, achieving an improvement in strength with only a minor trade-off in elongation. This establishes TRC as a superior industrial route for high-performance aluminum sheets, offering a promising industrial route, delivering substantial improvements in both strength and ductility over conventional casting methods.

Keywords: 2139 Al-Cu-Mg alloy; twin-roll casting; mechanical properties

1. Introduction

Al-Cu-Mg alloy has advantages such as high temperature resistance, high specific strength, creep resistance, oxidation resistance, and good processing performance. It is expected to partially replace titanium alloys used in 300~400 °C heat-resistant aluminum alloys [1,2]. In recent years, in order to further improve the high-temperature mechanical properties of Al-Cu-Mg alloys, the microstructure and mechanical properties of the alloys have been altered by microalloying, compositing, and process improvement, and certain research progress has been made [3–5]. Among these, 2139 aluminum alloy, a new high Cu/Mg ratio alloy developed by Alcan, has good damage tolerance properties and still exhibits high fracture toughness and creep resistance with low strength damage after thermal exposure at 200 °C [6]. At the engineering application level, 2139 aluminum alloy has been utilized in weight-reduction designs for armored vehicle structures and aerospace components due to its balanced strength-toughness ratio and ballistic damage resistance. Sabry demonstrated through the friction stir welding of TiB_2/Al_2O_3-reinforced Al6061-Al6082 alloys that ceramic particle reinforcement can simultaneously improve both strength and tribological performance in dissimilar alloys, offering valuable insights for strengthening strategies applicable to 2139 alloy in high-performance aerospace and armored vehicle applications [7]. Third-generation Al-Cu-Li alloys, enhanced by the T1 (Al_2CuLi) phase and featuring high specific stiffness, are employed in critical components such as wing skins, wall panels, and stiffeners. This achieves 5–15% structural weight reduction while improving fatigue and damage tolerance [8,9].

Twin-roll casting (TRC) is a short-flow, low-cost molding method, which ensures uniform composition and structure refinement and greatly improves the mechanical properties of the strips compared to the traditional preparation process [10,11]. TRC is a combination of casting and rolling to produce strips by pouring the molten metal directly onto water-cooled rolls [12,13].

The process involves rapid solidification through casting rolls, achieving cooling rates 2–3 orders of magnitude higher than conventional direct-chill casting. This high cooling rate, combined with enhanced solid solubility, promotes the formation of a supersaturated solid solution, which is crucial for subsequent precipitation strengthening. Li Shiju et al. successfully produced AA2099 aluminum-lithium alloy plates using a TRC experimental platform and optimized the corresponding process parameters. Similarly, Sabry demonstrated through friction stir welding of SiCp-reinforced AZ31C magnesium alloy that advanced solid-state forming processes can achieve grain refinement and uniform distribution of reinforcement phases via parameter optimization [14], significantly improving mechanical properties—providing a valuable reference for microstructural regulation of 2139 alloy via the TRC process [15]. Compared with traditional casting, TRC also refines eutectic phases and introduces a high density of dislocations into the alloy matrix.

Although TRC has the advantages of a simple production process, short production cycle, high yield, and high production efficiency, it mainly produces some low-alloy aluminum alloys with a narrow crystallization temperature range in the 1xxx series, 3xxx series, and 8xxx series [16]. For alloys with a wide liquid-solid zone, it is difficult to form sheets in a rapidly solidifying environment, requiring lower casting and rolling speeds and increased cooling intensity [17]. In this study, 2139 aluminum alloy sheets with wide solidification intervals were successfully produced by a TRC process for the first time. In order to investigate the structure and properties of the TRC samples, conventional sheets were also produced by an MC process, and the effects of the rolling heat treatment process on the structure and mechanical properties of the two sheets were discussed. In addition, the mechanical properties of the TRC sheets in the T8 state were also explored. Therefore, this study aims to compare 2139 alloy sheets produced via the TRC and MC processes, focusing on exploring the feasibility of the TRC process for industrial applications in high-performance aluminum alloy sheets. The research outcomes are expected to provide a theoretical basis and technical guidance for the industrial production of 2139 alloy sheets using the TRC process.

https://doi.org/10.3390/cryst16010013

2. Materials and Methods

The chemical composition of the 2139 aluminum alloy used in this study is shown in Table 1, and the raw materials produced by the alloy include industrially pure Al ingot (99.9%), Al-50% Cu, pure Mg ingot (99.9%), Al-10% Mn, pure silver grains (99.9%), and pure Zn ingot (99.9%) intermediate alloys.

Table 1. Chemical composition of 2139 alloy (wt%).

Cu	Mg	Mn	Ag	Zn	Fe	Si	Al
0.5	0.5	0.4	0.4	0.25	<0.01	<0.01	Balance

A schematic diagram of the TRC process is shown in Figure 1. In the melting process, pure aluminum ingots were first placed in TiO2-coated stainless steel crucibles and heated to 750 °C, and then Al-50%Cu and Al-10%Mn intermediate alloys were added. The melt is held at 720 °C for 1 h to ensure that the intermediate alloy is fully dissolved, and then pure magnesium ingots, zinc ingots, and silver grains are carefully pressed into the melt in sequence, followed by degassing and slag removal. Once the alloy was fully melted and held for 30 min, the casting and rolling machine was activated. Research by Sabry on friction stir welding of dissimilar AA2024 and A356-T6 alloys indicates that process parameters significantly affect mechanical properties by regulating temperature distribution and recrystallization behavior in the weld zone. Based on this insight, the TRC rolling speed in this study was controlled at 0.9 m/min to balance cooling efficiency and microstructural uniformity [18]. When the melt temperature was steadily lowered to 690 °C, the melt was poured into the flow channels and nozzles made of heat-resistant material of MgO, which then flowed into the roll seams. The width of the nozzle was 200 mm, and 6 mm thick sheets were produced.

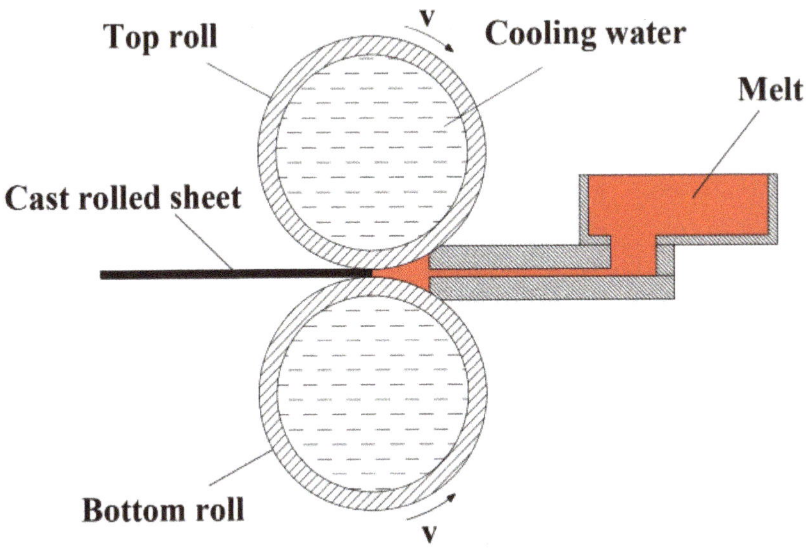

Figure 1. Schematic diagram of TRC process.

A cutting of 10 mm × 8 mm × 6 mm was taken as a specimen, which was then ground and polished in different states for microstructure observation. An OLYMPUSBX53M optical microscope (OM) (OLYMPUS, Tokyo, Japan) and a scanning electron microscope SSX-550 (SEM) equipped with an X-ray spectrometer (Shimadzu, Kyoto, Japan) were

https://doi.org/10.3390/cryst16010013

used to observe the microstructure, and electron back-scattering diffraction (EBSD) using a Zeiss Ultra-55 SEM (Carl Zeiss AG, Oberkochen, Germany) was used to analyze the grain distribution of the samples. The EBSD samples were prepared using electrolytic polishing, and the electrolytic polishing solution. The electrolytic polishing solution was a 10% perchloric acid alcohol solution (v/v), the voltage was 25 V, the time was 30 s, and the EBSD scanning step was 0.6 μm. In analyzing the EBSD results, orientation error angles of 2–15° and greater than 15° were defined as low- and high-angle boundaries, respectively. Orientation error angles smaller than 2° were eliminated to avoid errors arising from the sample surface and experimental polishing.

The cast slabs were homogenized at 510 °C for 24 h, then heated to 460 °C for 1 h. The slabs, initially 8 mm thick, were hot-rolled to 2.0 mm, with a 75% deformation. Since the recrystallization temperature for 2xxx aluminum alloys is typically 400–450 °C, 440 °C was selected to promote recovery and controlled recrystallization, relieving residual stresses and preventing excessive grain growth [19]. After heating to 440 °C for 1 h, the slabs were air-cooled and cold-rolled to 1.5 mm with 25% deformation.

After cold rolling, the samples were solution-treated at 510 °C for 0.5 h, quenched, and aged at 160 °C for 12 h (T6 state), as shown in Figure 2a. To investigate the effect of different heat treatments, a 5% pre-stretch deformation was applied before aging, followed by homogenizing heat treatment at 160 °C for 12 h to achieve the T8 state (Figure 2b).

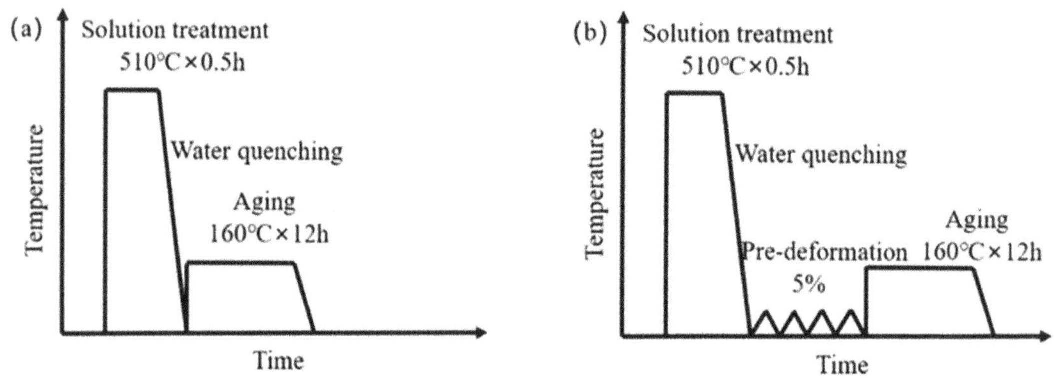

Figure 2. Heat treatment process diagram (single-stage aging): (**a**) MC/TRC sample T6; (**b**) TRC sample T8.

Tensile experiments were carried out on samples in different states, and the dimensions of the specimen are shown in Figure 3. The normal temperature tensile experiments were carried out using a 100 kN tensile machine, model INSTRON-4206 (Instron, Norwood, United States). The tensile rate of 1 mm/min was adopted in accordance with ASTM E8/E8M standard [20] for metallic materials, ensuring consistent strain rate conditions for accurate comparison of mechanical properties. Additionally, three standard tensile specimens were taken from each group, and the average value of each group was used as the final result. The same shape and size of the tensile specimens at high temperature were also used; the tensile temperature was set by the supporting high temperature furnace, the hydraulic pressure was set at 300 MPa, the tensile strain rate was 2×10^{-4} s^{-1}, the same tensile rate as that of the normal temperature tensile experiments was used, and the values of the tensile strength were read directly from the tensile machine, while the values of the yield strength were obtained by the load-displacement curves recorded by the tensile machine through the reference standard-displacement curve.

https://doi.org/10.3390/cryst16010013

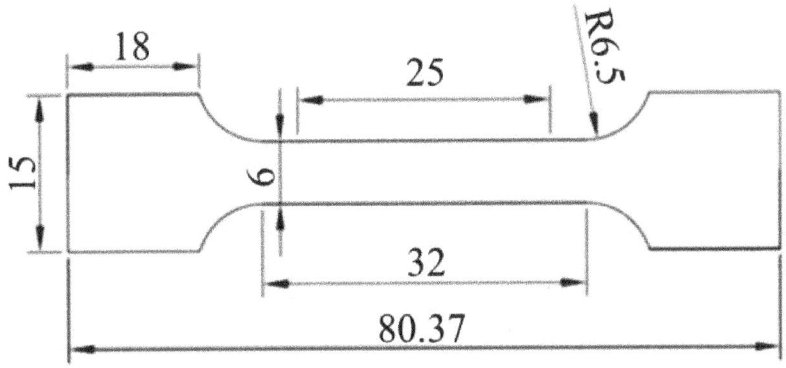

Figure 3. Size specification of tensile specimen.

3. Results and Discussion

3.1. As-Cast Microstructures

The as-cast microstructure of the experimental alloy produced by different processes is shown in Figure 4. It can be seen that there are reticulated intermetallic compounds in the structure of aluminum alloys cast in conventional molds, as well as a small number of porosity and microcrack defects (Figure 4a). The intermetallic compounds in the microstructure of the aluminum alloy produced by the TRC process show obvious bands, which generally run through the whole thickness direction, and the density and width gradually decrease from the surface to the center, and basically no pores and micro-cracks are observed. According to the analysis, the melt is easy to be sucked in during MC, resulting in volume shrinkage and defects, such as pores and microcracks. The TRC is a process that integrates continuous casting and rolling deformation. During the solidification process, the cast-rolled plate is always tightly rolled and subjected to the rolling force to compensate for the shrinkage and porosity generated during the casting process, thereby obtaining a dense plate without pores. Therefore, the plate prepared by the TRC process has a denser dendrite structure.

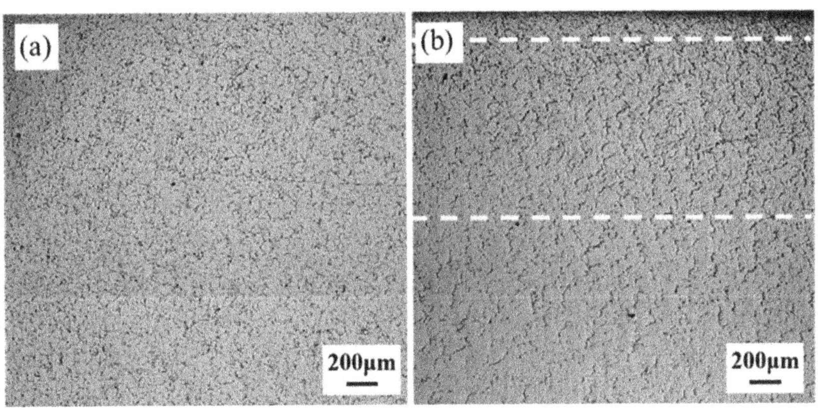

Figure 4. Microstructure of 2139 aluminum alloy prepared under different process conditions: (**a**) MC; (**b**) TRC.

The intermetallic compounds in the sheets produced by different processes are shown in Figure 5. It can be seen that the microstructure of the sheets produced by the MC process consists of an aluminum matrix and a large number of intermetallic compounds distributed

https://doi.org/10.3390/cryst16010013

along the boundary of the dendrite walls (Figure 5a). The grains of the sheets produced by the TRC process have obvious rolling microstructure characteristics, accompanied by a certain inclination angle (Figure 5b).

Figure 5c,d show the detailed information of intermetallic compounds in the plates produced by the two casting processes, respectively. There are two different intermetallic compounds at the dendrite interface. The dark region indicated by the green ellipse in Figure 5c is a pit left by the detachment of an Al_2CuMg (S-phase) particle during sample preparation, not the phase itself. In contrast, the bright intermetallic compound circled in the yellow ellipse is the Al_2Cu phase. This distinction is definitively established by the EDS point analysis presented in (Figure 5e,f). The key evidence is the presence of magnesium in the EDS spectra corresponding to the S-phase regions and its absence in the spectra from the Al_2Cu particles. The measured atomic percentages of Cu and Mg in the S-phase correspond to the Al_2CuMg stoichiometry, whereas the high Cu content and lack of Mg in the bright particles are characteristic of the Al_2Cu phase. These phases are continuously distributed on the dendrite boundaries and grain boundaries, which severely impairs the strength and toughness of the alloy. Therefore, these intermetallic compounds at dendrite boundaries and grain boundaries need to be dissolved into the matrix by homogenizing heat treatment to achieve a uniform distribution of the elements.

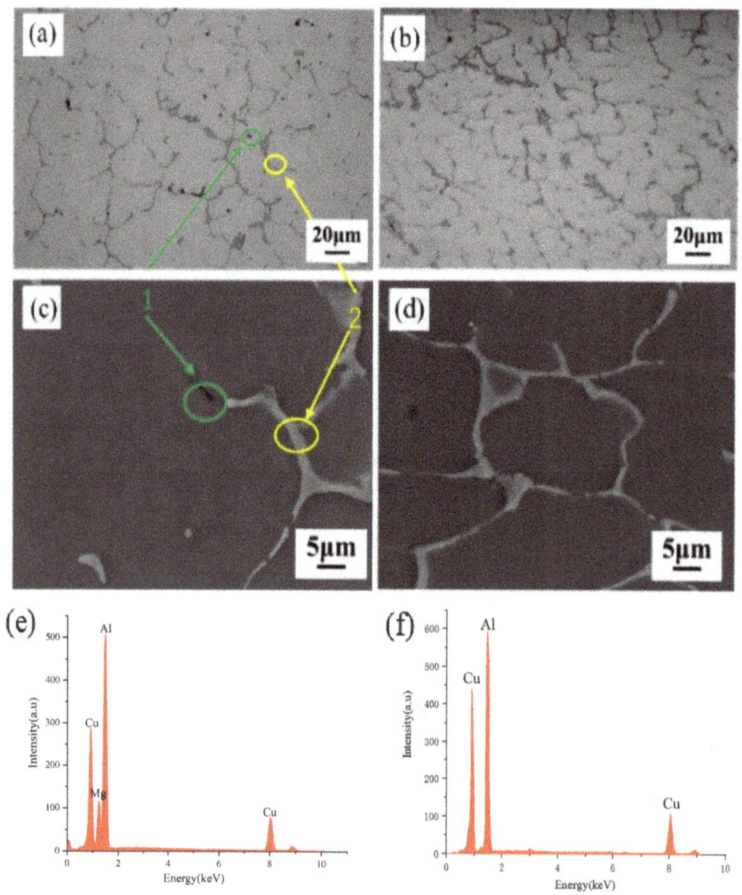

Figure 5. Microscopic morphology of 2139 aluminum alloy produced by different processes: (**a**,**c**) MC; (**b**,**d**) TRC-surface layer; (**e**) the EDS image of MC; (**f**) the EDS image of TRC.

https://doi.org/10.3390/cryst16010013

3.2. Microstructures After Homogenization

The structure and morphology of the sheet after holding at 510 °C for 24h are shown in Figure 6. It can be seen that most of the intermetallic compounds undergo dissolution after the homogenization heat treatment, and the reticulated dendritic structure disappears (Figure 6a,b). Compared to the MC samples, the changes in the 1/4 layer of the TRC samples are more significant due to the lower content of eutectic phases (Figure 6c,e).

Since the content of intermetallic compounds under the TRC process varies from the surface layer to the center layer, the surface, 1/4, and middle layers were partially enlarged (Figure 6d–f). It can be seen that the intracrystalline precipitated phases of the samples in MC after homogenization heat treatment are basically dissolved back, and very few precipitated phases that are not dissolved back remain on the grain boundaries. However, there is a small amount of fragmented intermetallic compounds, a large number of coarse banded intermetallic compounds and intermediate bulk intermetallic compounds, and a large number of insoluble phases (S-Al_2CuMg and Fe-rich phases) at the grain boundaries of the TRC process aluminum alloy after homogenization heat treatment, which prolongs the dissolution time of the crystal phase. The presence of a small amount of fragmented intermetallic compounds is attributed to localized strain concentration during the TRC, which promotes fragmentation but insufficiently dissolves these phases due to their initial coarse size and low solubility. In general, the dissolution of intermetallic compounds in the TRC samples after homogenization was not as good as that of MC, which is due to the initial large-sized intermetallic particles in the as-cast condition of the TRC sample.

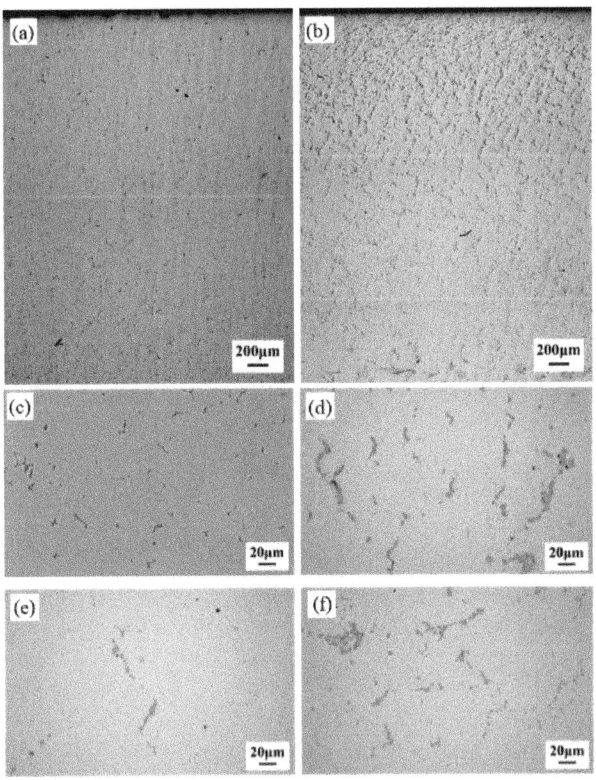

Figure 6. The microstructures of 2139 alloy after homogenization: (**a**,**c**) MC; (**b**) TRC; (**d**) TRC-surface layer; (**e**) TRC-1/4 layer; (**f**) TRC-central layer.

https://doi.org/10.3390/cryst16010013

The content and distribution of residual crystalline phases of the sheets produced by different processes after homogenization heat treatment are shown in Figure 7. It can be seen that under the same homogenization state, the different preparation processes resulted in certain differences in the type, content, and distribution of residual crystalline phases.

Typically, homogenizing heat treatment can effectively dissolve the eutectic structure, improve the micro-element distribution of the alloy, reduce the content of residual crystalline phases, and the residual crystalline phases become finer and more discretely distributed. This is because homogenization heat treatment can promote the diffusion of solute atoms from the eutectic phase into the aluminum matrix, thereby reducing compositional heterogeneity and facilitating the dissolution of intermetallic compounds, which is supported by classical diffusion theory and previous studies [21]. However, due to the TRC process characteristics that bring about coarse banded intergranular intermetallic compounds and central massive intermetallic compounds, which prolong the dissolution time of the crystalline phase, and even through a longer period of homogenizing heat treatment, a small amount of fragmented intermetallic compounds were still left behind which were difficult to be completely dissolved, so there were a large number of insoluble S-Al_2CuMg and Fe-rich phases existed in the cast and rolled samples (Figure 7b,c).

The residual crystalline phases in the TRC samples are mainly distributed in the original banded intergranular intermetallic compounds and the intermediate layer of massive intermetallic compounds, and the distribution is relatively concentrated. Table 2 and Figure 7d,e demonstrate the EDS spectroscopy analysis and scanning results of the 2139 alloy after homogenization, with the complex residual second phase dominated by the elements Al, Cu, Si, Fe, and Mn. The solubility of Mn and Fe in the aluminum matrix is much lower than that of Cu, so it is easier to precipitate from the melt to form second-phase particles rich in Mn and Fe phases. As solidification proceeds, the melt temperature gradually decreases, and Cu atoms are gradually deposited on these cores, eventually forming the AlCuFe and AlCuFeMn phases. These Mn- and Fe-rich phases usually have a high melting point and can only be dissolved at very high temperatures, making it difficult to eliminate them completely by homogenization.

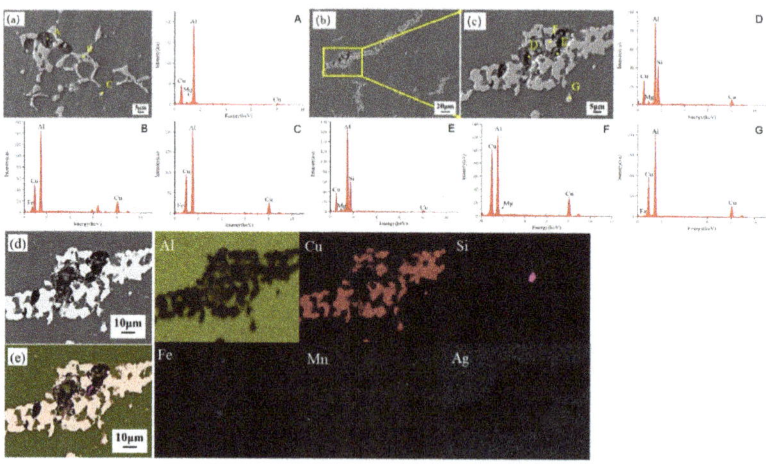

Figure 7. SEM image and EDS spectroscopy analysis and scanning results of tissue elements of 2139 alloy after homogenization: (**a**) MC; (**b**,**c**) TRC; (**d**,**e**) morphology and element distribution diagram of the insoluble second phase.

https://doi.org/10.3390/cryst16010013

Table 2. EDS spectroscopy analysis of residual phase in homogenized 2139 alloy.

Spectrogram	Atomic Percentage/%						
	Mg	Al	Si	Mn	Fe	Cu	Ag
MC-A	1.01	91.77	4.10	0.20	0.07	2.72	0.13
MC-B	0.14	70.86	0.00	1.82	7.06	20.08	0.03
MC-C	0.27	77.35	0.16	0.02	0.01	22.18	0.00
TRC-D	0.83	51.55	35.20	0.46	0.91	10.87	0.18
TRC-E	0.84	57.28	36.92	0.51	0.47	3.88	0.10
TRC-F	0.38	68.08	0.22	0.46	0.54	30.32	0.00
TRC-G	0.55	76.21	0.22	0.15	0.13	22.68	0.07

3.3. Microstructure After Rolling and Solution Treatment

The microstructure of MC and TRC samples after hot rolling, cold rolling, and solution treatment is shown in Figure 8. After cold rolling, and the second phase is aligned along the rolling direction after cold rolling, forming a second phase containing a large number of stripes. Some of the larger-sized phases were significantly broken, while some of the insoluble and incompletely dissolved phases showed relatively good ductility and were distributed in the rolling direction. With the partial dissolution of residual crystalline phases by the homogenization heat treatment process, the distribution of residual crystalline phases after cold rolling is more dispersed, and the aggregation region of coarse crystalline phases is significantly reduced.

After a short period of solid solution treatment, most of the precipitated second phases in both MC and TRC samples have been dissolved into the α(Al) matrix; only a small amount of coarse undissolved phase exists. The shape and size are only slightly reduced, but the distribution state of the cold-rolled state is basically maintained. The results of point scanning at different positions are shown in Table 3. It can be seen from the Cu content that after solution treatment, the Al_2Cu phase in the crystal is re-dissolved into the Al matrix, and the distribution of the residual crystalline phase does not change significantly. The slight reduction in the shape and size of the second phase particles after solution treatment is due to their high thermal stability and the short duration of the treatment, which limits Ostwald ripening and coalescence [22].

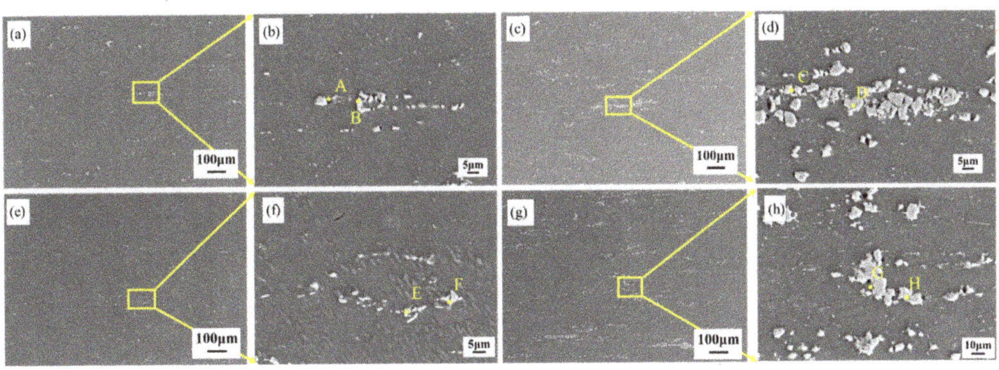

Figure 8. Evolution of crystalline phase of 2139 alloy during cold rolling and solution treatment: (**a**,**b**) cold-rolled state of MC; (**c**,**d**) cold-rolled state of TRC; (**e**,**f**) solid solution state of MC; (**g**,**h**) solid solution state of TRC.

https://doi.org/10.3390/cryst16010013

Table 3. EDS spectroscopy analysis of crystalline phase of 2139 alloy during cold rolling and solution treatment.

Spectrogram	Atomic Percentage/%						
	Mg	Al	Si	Mn	Fe	Cu	Ag
MC-A	0.15	71.96	0.09	1.80	6.81	19.14	0.05
MC-B	0.21	73.14	0.06	2.07	6.16	18.31	0.05
MC-C	0.59	89.33	0.74	0.64	0.54	7.97	0.18
TRC-D	0.53	73.61	0.03	0.32	0.19	25.14	0.19
TRC-E	0.38	89.44	0.26	1.07	1.67	7.01	0.18
TRC-F	0.33	82.85	0.44	1.61	4.50	10.25	0.02
TRC-G	0.60	91.26	0.78	1.21	1.77	4.25	0.13
TRC-H	0.34	78.81	0.03	0.11	0.10	20.49	0.11

Figure 9a,b shows that the grain size of the MC sample is 9.52 μm, and the grain size of the TRC sample is 9.33 μm, which is significantly reduced compared to the as-cast state. EBSD samples were mechanically ground and polished in accordance with standard metallographic preparation procedures ASTM E3 [23] and ISO 9042 [24], followed by appropriate surface finishing for electron backscatter diffraction analysis, ensuring reliable grain size and recrystallization characterization. Compared with the core of the cast-rolled sheet, the surface grain of the cast-rolled sheet is finer, showing the characteristics of gradient distribution. In the cold rolling process, similar to the casting and rolling process, the surface grain is subjected to greater casting and rolling force, greater plastic deformation occurs, and the accumulation of more distortion energy. According to the PSN effect, a large number of second-phase particles (such as residual crystalline phase) on the surface promotes the recrystallization nucleation mechanism of PSN and reduces the grain size. Therefore, in addition to the grain size distribution characteristics of the genetic cold-rolled TRC sample, the gradient distribution of grain size is also affected by the degree of strain and the gradient of element distribution. Due to the initial accumulation of strain energy during the TRC process, the recrystallization phenomenon with large deformation is more concentrated.

In addition, fewer low-angle and medium-angle misorientation grain boundaries can be found in both MC and TRC samples, indicating that a large number of substructures are reorganized during the solid solution process. The large angle misorientation is sparsely distributed inside the grains and is higher near the grain boundaries (Figure 9c,d). The larger the KAM value in the diagram indicates the higher the strain degree. The unidentified white part is mainly the distribution area of the residual crystalline phase. Compared with other deformation areas, the strain is concentrated in all areas where the distribution of the residual crystalline phase is concentrated. The strain concentration in regions with residual crystalline phases is attributed to the mismatch in deformation behavior between the brittle intermetallic particles and the ductile aluminum matrix, leading to stress localization and inhibited recrystallization in these areas [25].

https://doi.org/10.3390/cryst16010013

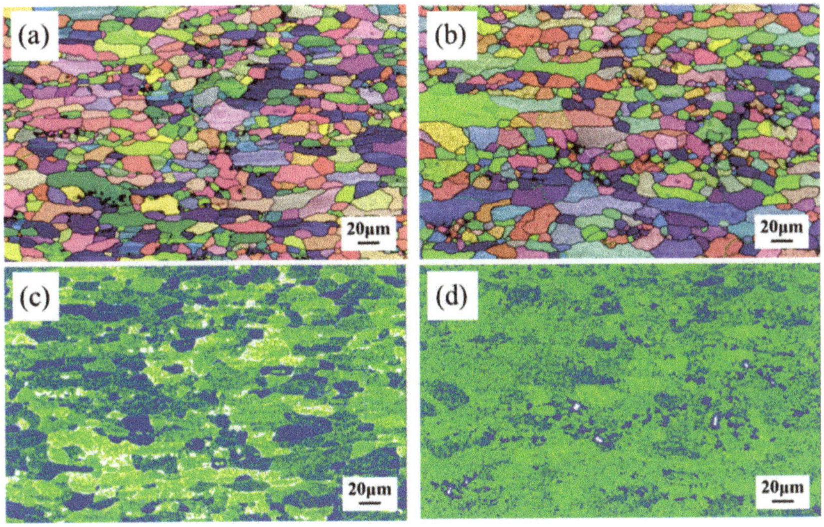

Figure 9. Observation of the solid solution microstructure and EBSD of 2139 alloy. (**a**) Grain structure dis-tribution of MC; (**b**) grain structure distribution of TRC; (**c**) KAM diagram of MC; (**d**) KAM dia-gram of TRC.

3.4. Mechanical Properties of the Sheets After Aging Heat Treatment

Normal temperature tensile experiments were conducted on the sheets after aging heat treatment, and the results are shown in Figure 10. It can be seen that at the beginning of the tensile experiment, the stress increases sharply until the peak stress is reached. After that, the stress decreases slowly until it breaks at the end of the tensile experiment. The yield strength, tensile strength, and elongation of the MC sample are 309.2 MPa, 390.0 MPa, and 11.4% after T6 (single-stage aging) treatment, respectively. The yield strength and tensile strength of the TRC sample were 354.2 MPa and 430.4 MPa, which were 10.3% higher than those of the MC sample, and the elongation of the TRC sample (12.6%) was 10.5% greater than that of the MC sample (11.4%). The tensile strength, yield strength, and elongation of the TRC specimens were increased to 385.7 MPa, 464.9 MPa, and 11.0%, respectively, after the T8 (single-stage aging) treatment. Compared with T6 (single-stage aging), the tensile strength of the alloy increased by 76.5 MPa and the yield strength by 74.9 MPa. Compared with the T6 heat treatment process, the T8 heat treatment process, which introduces 5% pre-stretch deformation before single-stage aging at 160 °C/12 h, substantially improves the strength of 2139 aluminum alloy sheet without losing too much plasticity, and enables the alloy sheet to reach a more optimal state.

The tensile fracture morphology of 2139 aluminum alloy prepared under different process conditions in different heat treatment states is given in Figure 11. The tensile fracture morphology of 2139 aluminum alloy produced by the MC process after T6 (single-stage aging) treatment has fewer dimples and a diffuse distribution, which is a mixed mode of ductile fracture and brittle fracture (Figure 11a,b). The tensile fracture morphology of the 2139 aluminum alloy produced by the TRC process after T6 (single-stage aging) treatment, compared with the MC process, the number of dimples increases significantly, but the size of the dimples is different, the size is uneven, the amount is large, and the fractures are mainly ductile fractures (Figure 11c,d). The significant increase in dimple number in TRC-T6 samples compared to MC is attributed to the finer and more uniformly distributed Ω phases and the higher dislocation density, which promote homogeneous plastic deformation and micro-void coalescence. Judging from the detailed fracture morphology diagram, the TRC

https://doi.org/10.3390/cryst16010013

samples underwent a large plastic deformation before fracture, and the alloy was highly plasticized. The tensile fracture morphology of 2139 aluminum alloy produced by the TRC process after T8 aging treatment, where larger areas of tear edges and delamination are observed, and the alloy has the worst plasticity (Figure 11e,f).

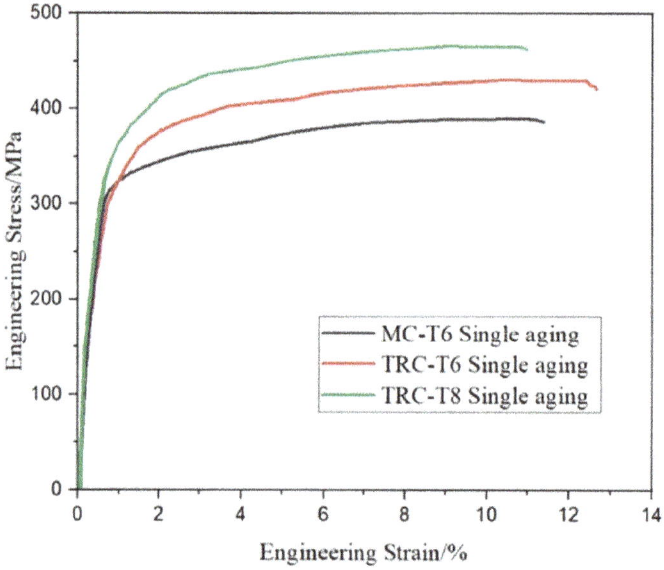

Figure 10. Tensile curves of 2139 alloy under different aging heat treatment states.

In order to investigate the high-temperature mechanical properties of TRC sheets, tensile experiments at different temperatures were carried out on specimens in T6 and T8 states, and the results are shown in Figure 12. It can be seen that the strength of the alloy decreases as the temperature increases. At 200 °C, the strength in the T6 state reaches 74% of the normal temperature strength, also 74% in the T8 state, and this reduction is further accelerated when the temperature exceeds 200 °C. At 300 °C, the strength reaches 42% and 40% of that at normal temperature, corresponding to the T6 and T8 states, respectively. In the T8 state, the high-temperature tensile and yield strengths of the alloy are higher than in the T6 state, but the elongation is lower. Therefore, 2139 aluminum alloy shows good performance at high temperatures of about 200 °C.

The fracture morphology of tensile specimens of 2139 alloy in the T6 state at 25 °C, 100 °C, 200 °C, and 300 °C is shown in Figure 13. It can be found that the normal temperature fracture of 2139-T6 alloy shows a mixed fracture with a relatively flat cross-section (Figure 13a). Despite the presence of a small number of dimples, brittle fracture predominates, which is consistent with the results of the mechanical property test. As the tensile temperature increases to 100 °C, the tensile fracture morphology of the 2139 aluminum alloy T6 state shows a significant increase in the number of dimples and an increase in the area of the dimples compared to normal temperature tensile (Figure 13b). As the temperature reaches 200 °C, the dimples become deeper at this time, and the area of ductile fracture increases significantly, which signifies that ductile fracture is dominant (Figure 13c). As the tensile temperature was further increased, the number of dimples decreased significantly, but the size of the fracture dimples increased significantly to become deeper. At 300 °C, the high-temperature fracture morphology of the T6 state of 2139 aluminum alloy shows a typical dimple-type through-crystal fracture, which is consistent with the high-temperature mechanical properties of the alloy (Figure 13d).

https://doi.org/10.3390/cryst16010013

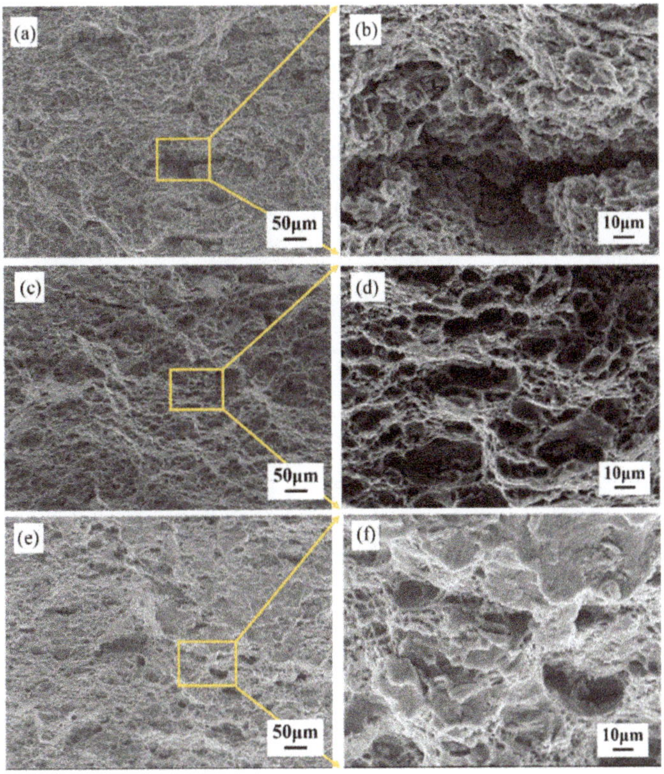

Figure 11. Tensile fracture morphology of 2139 alloy under different heat treatment states: (**a**,**b**) T6 (single-stage aging) of MC sample; (**c**,**d**) T6 (single-stage aging) of TRC sample; (**e**,**f**) T8 (single-stage aging) of TRC sample.

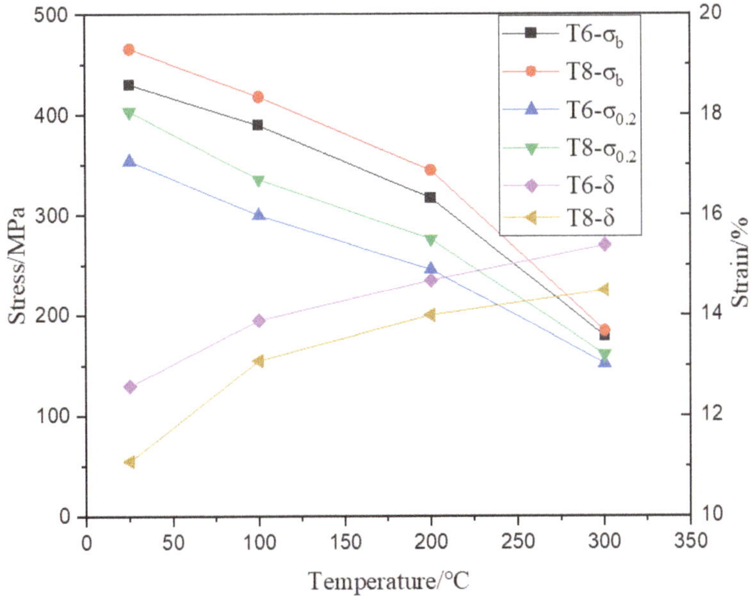

Figure 12. High-temperature mechanical properties of 2139 alloy under T6 and T8 states.

https://doi.org/10.3390/cryst16010013

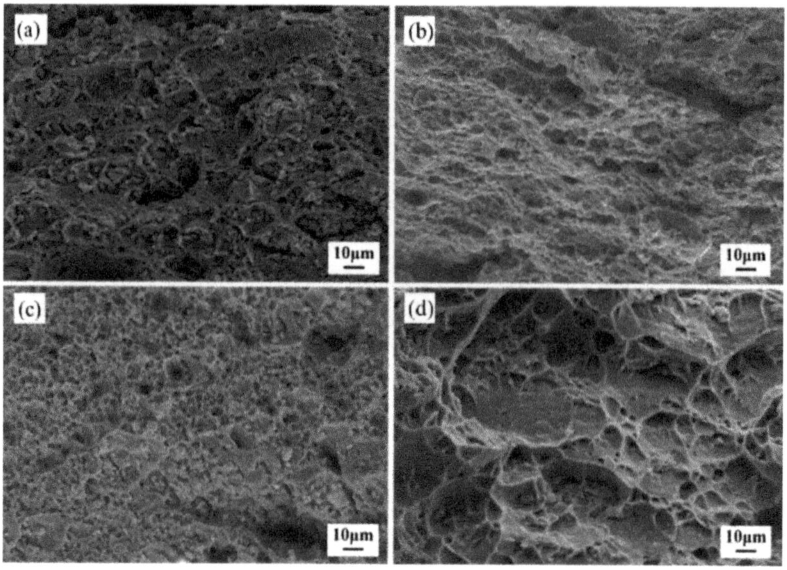

Figure 13. High-temperature tensile fracture morphology of T6 state alloy: (**a**) normal temperature; (**b**) 100 °C; (**c**) 200 °C; (**d**) 300 °C.

The high-temperature tensile fracture morphology of 2139 aluminum alloy T8 state at different temperatures is shown in Figure 14. Compared to the T6 state, the area of the fracture region increases at normal temperature, showing a mixed type of fracture. This may be attributed to the large number of dislocations introduced in the pre-deformation treatment, leading to the deformation strengthening effect (Figure 14a). The tensile fracture morphology of the T8 state of 2139 aluminum alloy at 100 °C, compared with the situation at normal temperature, the size of the dimple is smaller, the distribution of the dimple is more uniform, and the ductile fracture area is larger (Figure 14b). With the increase in tensile temperature, the dimples in the tensile fracture morphology of 2139-T8 alloy become deeper and larger, and the area of ductile fracture expands (Figure 14c). When the tensile temperature reaches 300 °C, the morphology of the high-temperature tensile fracture of 2139 aluminum alloy presents a dimple-type penetrating fracture, which is consistent with the change trend of high-temperature tensile mechanical properties (Figure 14d).

The evolution of the high-temperature fracture morphology is directly governed by thermally activated diffusion and dynamic recovery mechanisms. As the temperature rises beyond 200 °C, the enhanced atomic diffusion promotes coarsening of the Ω phase, reducing its pinning effectiveness, while simultaneously accelerating dislocation climb and rearrangement through dynamic recovery. These combined processes lead to overall material softening and an increase in uniform plastic deformation capability, which is clearly reflected in the fracture surfaces as enlarged dimple size, greater dimple depth, and expanded dimple area, as shown in Figures 13 and 14. The accelerated decline in strength observed above 200 °C corresponds to the growing dominance of dynamic recovery in the high-temperature softening behavior.

The T8 condition, with its higher dislocation density resulting from pre-deformation, retains greater strength at elevated temperatures due to its increased stored dislocation energy. However, this same high dislocation density also accelerates diffusion-assisted dislocation annihilation, leading to a more noticeable reduction in high-temperature ductility compared to the T6 condition. This mechanistic difference explains why localized

https://doi.org/10.3390/cryst16010013

tearing features appear earlier in the fracture morphology of T8 samples, as evidenced in Figure 11e,f.

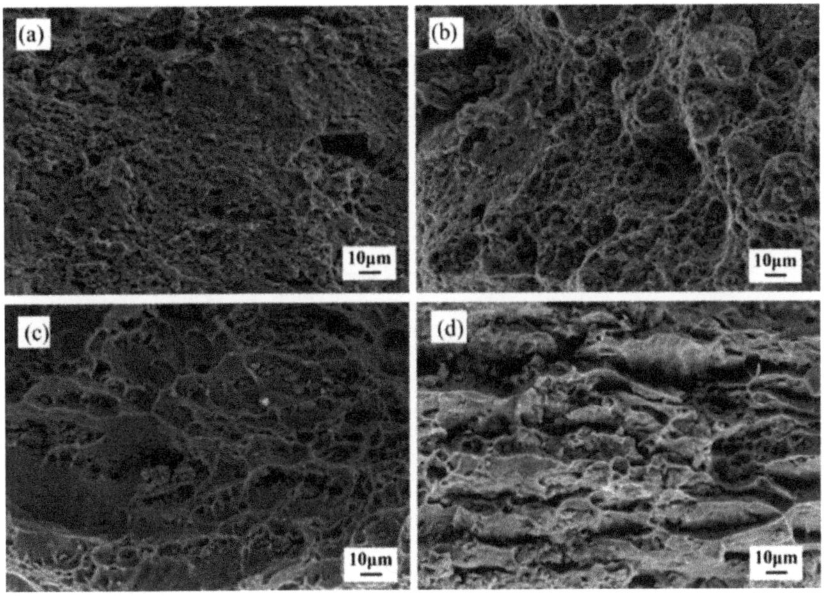

Figure 14. High-temperature tensile fracture morphology of T8 state alloy: (**a**) normal temperature; (**b**) 100 °C; (**c**) 200 °C; (**d**) 300 °C.

3.5. Analysis of Precipitated Phases After Aging Heat Treatment

Using transmission electron microscopy (TEM) technology, the distribution of precipitated phases in two alloys prepared by different processes was characterized in both T6 and T8 conditions. Figure 15a,b show the TEM images of 2139 aluminum alloy produced by MC and TRC under two different process conditions after aging treatment, and the <110>Al direction is the direction of electron incidence. After aging at 160 ° C for 12 h and T6 treatment, a small amount of needle-like Ω (Al_2Cu) phases with a large diameter and a length of about 20–80 nm can be observed in the MC sample, and the size distribution is not uniform, accompanied by a certain degree of coarsening (Figure 15a). Compared with the MC sample, the TRC sample was aged at 160 °C for 12 h and then subjected to T6 treatment. The number of Ω phases in the alloy increased significantly, the size increased slightly, the size distribution was relatively uniform, and the coarsening phenomenon was very rare (Figure 15b). The main reason is that the solid solubility in the TRC samples is high, and the phase transition driving force of the precipitated second phase in the aging process is large, which can reduce the critical nucleus size of the precipitated phase and increase the nucleation rate, thus increasing the number of precipitated phases. Therefore, after the same heat treatment, there are more and finer Ω phases precipitated in the TRC sheets, which improves the aging strengthening effect, and the strength of the TRC sheets is also improved. The microstructural distinction is governed by synergistic kinetic and thermodynamic factors. The high dislocation density in TRC samples provides potent heterogeneous nucleation sites for the Ω phase, lowering the energy barrier for its precipitation and refining its distribution. Meanwhile, the rapid solidification during TRC yields a highly supersaturated solid solution, which enhances the thermodynamic driving force for Ω-phase nucleation and simultaneously suppresses the formation of coarse equilibrium phases like Al_2Cu. This multi-mechanism enhancement parallels the approach

https://doi.org/10.3390/cryst16010013

reported by Scary et al., where 5wt% SiCp reinforcement was found to optimally improve strength, ductility, and corrosion resistance in aluminum composites through a synergistic combination of Orowan strengthening, grain refinement, and intermetallic distribution optimization [26].

Moreover, a certain amount of dislocations and dislocation tangles can be found inside the grains in the TRC samples, which is due to the deformation storage effect of the TRC process itself.

After pre-stretching the TRC sample with a deformation of 5% and then T8 treatment at 160 °C for 12 h, a large number of dislocations and dislocation tangles appeared in the alloy. The Ω phases in the alloy became fine, and the density increased significantly, and the number reached a maximum. A large area of uniform, fine, and dispersed Ω phases appeared, and then the long diameter of Ω phases became shorter, and the length further narrowed, which was about 15–50 nm (Figure 15c). The pre-stretching deformation in the T8 treatment further multiplied dislocations and created dislocation tangles. These defects served as additional nucleation sites and accelerated solute atom diffusion via pipe diffusion mechanisms. This synergistic effect of enhanced nucleation and accelerated diffusion led to the observed ultra-fine and densely distributed Ω phases in the T8 state, contributing to the peak strength achieved.

In the Al–Cu–Mg alloy system, previous studies have shown that Mg promotes Ω-phase precipitation, while Ag further modifies its precipitation kinetics by assisting GP-zone formation and solute clustering [21,27]. The Ω-phase behavior observed in the TRC-processed 2139 alloy in this work—namely its higher number density and finer, more uniform distribution in the T6 and T8 states—was consistent with these reported Mg/Ag-assisted precipitation mechanisms.

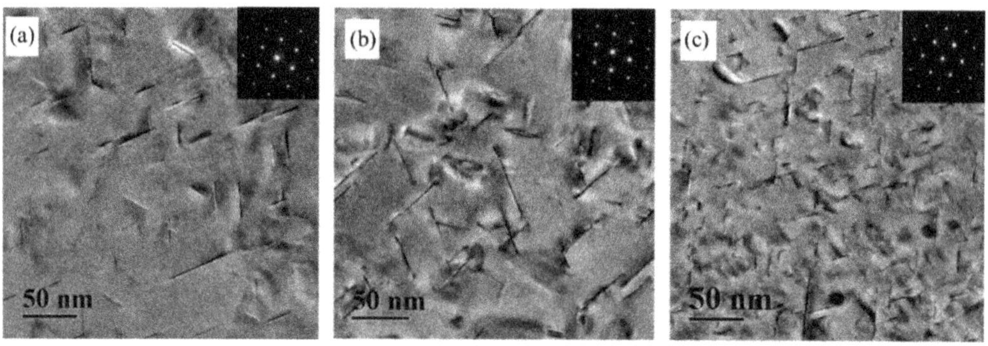

Figure 15. Microstructure of the Ω phase in MC and TRC samples after aging heat treatment: (**a**) T6 (single-stage aging) of MC sample; (**b**) T6 (single-stage aging) of TRC sample; (**c**) T8 (single-stage aging) of TRC sample.

4. Conclusions

This study investigates the use of the twin-roll casting (TRC) process for producing 2139 aluminum alloy sheets and compares the resulting microstructure and mechanical properties to those produced by the mold casting (MC) process. The main findings are as follows:

(1) The TRC process produces a microstructure where the intermetallic compounds, particularly Al_2Cu and Al_2CuMg phases, are more concentrated in the center and decrease in density and width from the surface inward. This leads to improved phase distribution and suggests that TRC enhances material homogeneity compared to MC.

https://doi.org/10.3390/cryst16010013

(2) After rolling and solid solution treatment, the residual crystalline phases (mainly undissolved intermetallic compounds) in the MC samples were evenly distributed, while the TRC samples showed localized recrystallization, particularly within second-phase and shear zones. This localized recrystallized-grain formation contributed to improved strength and ductility, indicating that TRC allowed for more controlled microstructural evolution.
(3) Mechanical testing revealed that the TRC samples in the T6 state show a 10.3% higher yield strength and 10.5% higher tensile strength compared to the MC samples, with an elongation of 12.6%, which was 10.5% higher than that of the MC samples. These improvements in strength and elongation demonstrated the effectiveness of TRC in enhancing the alloy's mechanical properties, especially at normal temperatures.
(4) The TRC-T6 sheets exhibited a significant increase in the number of Ω phases, with a more uniform distribution and minimal coarsening. After T8 heat treatment, the Ω phase became finer and more densely distributed. This enhanced phase distribution contributed to a further increase in strength and ductility, showing that TRC could optimize phase evolution for improved mechanical performance.

5. Challenges and Perspectives

In this study, we successfully applied the twin-roll casting (TRC) process to produce 2139 aluminum alloy sheets, marking a significant innovation in the production of high-performance aluminum alloys. The application of TRC allowed for remarkable improvements in both microstructure and mechanical properties, setting the stage for future advancements. However, there were notable challenges that needed to be addressed. The dissolution of iron-rich phases remained difficult, which affected the alloy's overall performance. Additionally, scaling up the TRC process for industrial production posed challenges, particularly regarding heat dissipation and material flow control, which were critical for maintaining consistent quality at larger scales. Future research will focus on optimizing the TRC process through advanced simulation tools, such as finite element analysis (FEA) and computational fluid dynamics (CFD), to refine parameters like cooling rates and roll speeds, ensuring better control over the alloy's properties. Overcoming these limitations and integrating simulation-driven optimization with large-scale adaptations will be essential to making TRC a widely adopted method for producing high-performance aluminum alloys.

Author Contributions: Conceptualization, Z.L. and H.T.; methodology, Z.L. and Y.W.; validation, Z.L., Y.W., and Q.C.; investigation, Q.C. and L.M.; resources, Z.L.; data curation, H.T.; writing—original draft preparation, Z.L. and Q.C.; writing—review and editing, Z.L., Z.Y., and X.Q.; visualization, Y.W. and Y.X.; supervision, Y.L., X.Q., and X.L.; project administration, H.T.; funding acquisition, H.T. and X.Q. All authors have read and agreed to the published version of the manuscript.

Funding: This research was funded by Guangxi Science and Technology Major Program (AA23073019), the Special Fund for Science and Technology Development of Guangxi (AD25069078), the Open Foundation of State Key Laboratory of Featured Metal Materials and Life-cycle Safety for Composite Structures (MMCS2023OF09, MMCS2023OF07), the Major Special Project of Science and Technology in Nanning (20231037), the Yanshan University Basic Innovation Research and Cultivation Project (2022LGQN015), and the Open Research Fund from the State Key Laboratory of Rolling and Automation, Northeastern University (2023RALKFKT004). Thanks to the Multi-disciplinary Integrated Innovation Experimental Teaching Center for Resources, Environment and Materials at Guangxi University.

Data Availability Statement: The original contributions presented in this study are included in the article. Further inquiries can be directed to the corresponding authors.

Conflicts of Interest: Author Yong Li was employed by the company Guangxi Advanced Aluminum Processing Innovation Center Co., Ltd. The remaining authors declare that the research was conducted in the absence of any commercial or financial relationships that could be construed as a potential conflict of interest.

References

1. Das, S.K.; Bye, R.L.; Gilman, P.S. Large scale manufacturing of rapidly solidified aluminum alloys. *Mater. Sci. Eng. A* **1991**, *134*, 1103–1106. [CrossRef]
2. Kim, Y.J.I.H. Advanced aluminum alloys for high temperature structural applications. *Ind. Heat.* **1988**, *55*, 31–34, 0019-8374.
3. Ling, K.; Chen, J.W.; Ran, H.W.; Deng, P.; Tang, C.L.; Mo, W.F.; Ouyang, Z.Q.; Wu, X.K.; Luo, B.H.; Bai, Z.H. Effect of V additions on the microstructure and mechanical properties of Al-Cu-Mg-Ag alloy. *Mater. Today Commun.* **2022**, *33*, 104197. [CrossRef]
4. Mei, Z.; Liu, Z.; Bai, S.; Wang, J.; Cao, J. Effects of yttrium additions on microstructures and mechanical properties of cast Al-Cu-Mg-Ag alloys. *J. Alloys Compd.* **2021**, *870*, 159435. [CrossRef]
5. Li, J.; An, Z.; Hage, F.S.; Wang, H.; Xie, P.; Jin, S.; Ramasse, Q.M.; Sha, G. Solute clustering and precipitation in an Al-Cu-Mg-Ag-Si model alloy. *Mater. Sci. Eng. A* **2019**, *760*, 366–376. [CrossRef]
6. Warner, T. Recently-developed aluminum solutions for aerospace applications. In *Aluminum Alloys 2006, Pts 1 and 2: Research Through Innovation and Technology*; Poole, W.J., Wells, M.A., Lloyd, D.J., Eds.; Materials Science Forum: Baech, Switzerland, 2006; Volume 519–521, pp. 1271–1278. [CrossRef]
7. Sabry, I.; El-Deeb, M.S. Enhanced structural integrity and tribological performance of Al6061–Al6082 alloys reinforced with TiB_2 and Al_2O_3 via friction stir welding. *Int. J. Adv. Manuf. Technol.* **2025**, *138*, 2893–2910. [CrossRef]
8. Rioja, R.J.; Liu, J. The Evolution of Al-Li Base Products for Aerospace and Space Applications. *Metall. Mater. Trans. A* **2012**, *43*, 3325–3337. [CrossRef]
9. Adjei-Kyeremeh, F.; Pratesa, Y.; Shen, X.; Song, W.; Raffeis, I.; Vroomen, U.; Zander, D.; Bührig-Polaczek, A. Preheating Influence on the Precipitation Microstructure, Mechanical and Corrosive Properties of Additively Built Al–Cu–Li Alloy Contrasted with Conventional (T83) Alloy. *Materials* **2023**, *16*, 4916. [CrossRef]
10. Liu, Z.-T.; Wang, B.-Y.; Wang, C.; Zha, M.; Liu, G.-J.; Yang, Z.-Z.; Wang, J.-G.; Li, J.-H.; Wang, H.-Y. Microstructure and mechanical properties of Al-Mg-Si alloy fabricated by a short process based on sub-rapid solidification. *J. Mater. Sci. Technol.* **2020**, *41*, 178–186. [CrossRef]
11. Zhang, S.-Y.; Wang, X.; Liu, X.; Mo, Y.-T.; Wang, C.; Cheng, T.; Ivasishin, O.; Wang, H.-Y. High strength-ductility synergy induced by sub-rapid solidification in twin-roll cast Al-Mg-Si alloys. *J. Mater. Res. Technol.* **2022**, *16*, 922–933. [CrossRef]
12. Neuser, M. Schaper, Mechanical and Microstructure Characterization of the Hypoeutectic Cast Aluminum Alloy AlSi10Mg Manufactured by the Twin-Roll Casting Process. *J. Manuf. Mater. Process.* **2023**, *7*, 132. [CrossRef]
13. Ullmann, M.; Stirl, M.; Prahl, U. Twin-roll casting defects in light metals. *J. Mater. Sci.* **2024**, *59*, 19003–19022. [CrossRef]
14. Sabry, I. Hybrid GRA-TOPSIS optimization of friction stir welding for SiCp-reinforced AZ31C magnesium alloy. *Int. J. Adv. Manuf. Technol.* **2025**, *139*, 4457–4478. [CrossRef]
15. Li, S.; He, C.; Fu, J.; Xu, J.; Xu, G.; Wang, Z. Evolution of microstructure and properties of novel aluminum-lithium alloy with different roll casting process parameters during twin-roll casting. *Mater. Charact.* **2020**, *161*, 110145. [CrossRef]
16. Cho, H.-G.; Kim, Y.D.; Kim, M.-S. Prediction of Grain Structure and Texture in Twin-Roll Cast Aluminum Alloys Using Cellular Automaton–Finite Element Method. *Materials* **2025**, *18*, 1075. [CrossRef]
17. Sun, B. *Theory and Technology of Strip Casting and Rolling*; Metallurgical Industry Press: Beijing, China, 2002; pp. 25–26, ISBN 978-7502430962.
18. Sabry, I. Enhanced strength, ductility, and corrosion resistance of AA6061/AA6082 alloys using Al-SiC matrix reinforcement in dissimilar friction stir welding. *Int. J. Adv. Manuf. Technol.* **2025**, *138*, 2431–2457. [CrossRef]
19. Wen, M.; Hsu, Y.; Chen, M.; Yang, W.; Lee, S. Effects of Heterogenization Treatment on the Hot-Working Temperature and Mechanical Properties of Al-Cu-Mg-Mn-(Zr) Alloys. *Materials* **2023**, *16*, 4256. [CrossRef]
20. *ASTM E8/E8M-24*; Standard Test Methods for Tension Testing of Metallic Materials. ASTM International: West Conshohocken, PA, USA, 2024. [CrossRef]
21. Cousland, S.M.; Tate, G.R. Structural changes associated with solid-state reactions in Al–Ag–Mg alloys. *J. Appl. Crystallogr.* **1986**, *19*, 174–180. [CrossRef]
22. Wang, X.-Y.; Yang, S.-P.; Wang, H.; Zhao, G.-Z. Ostwald Ripening Mechanism and Its Progress in Binary Alloys. *Mod. Phys.* **2022**, *12*, 31–37. [CrossRef]
23. *ASTM E3-11*; Standard Practice for Preparation of Metallographic Specimens. ASTM International: West Conshohocken, PA, USA, 2017. [CrossRef]
24. *ISO 9042:2021*; Industrial Thermometers—Metal Case Thermometers with Bimetal Sensing Element and Indicator. International Organization for Standardization: Geneva, Switzerland, 2021.

https://doi.org/10.3390/cryst16010013

25. Lei, T.; Hessong, E.C.; Shin, J.; Gianola, D.S.; Rupert, T.J. Intermetallic particle heterogeneity controls shear localization in high-strength nanostructured Al alloys. *Acta Mater.* **2022**, *118347*, 1359–6454. [CrossRef]
26. Sabry, I. Exploring the effect of friction stir welding parameters on the strength of AA2024 and A356-T6 aluminum alloys. *J. Alloys Metall. Syst.* **2024**, *8*, 100124. [CrossRef]
27. Huang, B.P.; Zheng, Z.Q. Independent and combined roles of trace Mg and Ag additions in properties precipitation process and precipitation kinetics of Al-Cu-Li-(Mg)-(Ag)-Zr-Ti alloys. *Acta Mater.* **1998**, *46*, 4381–4393. [CrossRef]

https://doi.org/10.3390/cryst16010013

crystals

MDPI

Article

Effect of Al-5Ti-2B on the Microstructure and Mechanical Properties of Recycled Al-7Si-0.3Mg-1Fe Alloy

Weihe Shi [1,2,3], Lin Chen [1,2,3], Bing He [1,2,3], Biwang Lu [1,2,3,*] and Jianbing Yang [4,*]

1 Guangxi Key Laboratory of Green Manufacturing for Ecological Aluminum Industry, Baise University, Baise 533000, China; shiweihe@bsuc.edu.cn (W.S.)
2 College of Ecological Aluminum Industry, Baise University, Baise 533000, China
3 Engineering Research Center of Advanced Aluminium Matrix Materials of Guangxi Province, Baise University, Baise 533000, China
4 Urban-rural Construction College, Guangxi Vocational University of Agriculture, Nanning 530007, China
* Correspondence: biwanglu@bsuc.edu.cn (B.L.); yangjianbing2002@163.com (J.Y.)

Abstract

This study systematically investigates the influence of grain refinement on the microstructural evolution and mechanical properties of recycled Al-7Si-0.3Mg-1Fe alloy through the addition of varying concentrations (0–1.25 wt.%) of Al-5Ti-2B master alloy. The synergistic effects of Al-5Ti-2B on the α-Al phase, eutectic Si, and Fe-rich intermetallics were characterized using metallographic analysis, XRD, SEM-BSE imaging, and EDS. In the unrefined alloy, the microstructure consisted of an α-Al solid solution with coarse plate-like eutectic Si, while Fe primarily formed needle-like β-Al_5FeSi phases that either surrounded or penetrated the eutectic Si. Increasing the Al-5Ti-2B addition refined both the α-Al dendrites and eutectic Si, while the β-Al_5FeSi phase transitioned from coarse to fine needles. The optimal refinement was achieved at a 1% Al-5Ti-2B addition, yielding a tensile strength of 149.4 MPa and elongation of 4.3%. However, excessive addition (1.25%) led to eutectic Si aggregation and β-Al_5FeSi coarsening, resulting in mechanical property deterioration and brittle fracture behavior. These findings provide insights into optimizing grain refinement for enhancing the performance of recycled Al-Si-Mg-Fe alloys.

Keywords: Al-5Ti-2B; recycled Al-7Si-0.3Mg-1Fe alloy; refinement treatment; microstructure; mechanical properties

1. Introduction

As global carbon neutrality strategies advance, the recycled aluminum industry has become a key pathway for achieving energy conservation and emissions reduction in the non-ferrous metals sector. Statistical data show that the energy consumption of recycled aluminum production is only 5% of that of primary aluminum, while approximately 40% of global aluminum alloy demand is met through recycled sources. Among various recycled aluminum alloys, Al-Si-Mg alloys are widely used in automotive components, electronic housings, and other fields due to their excellent casting performance and mechanical properties. However, during the recycling process, recycled aluminum inevitably becomes contaminated with iron-based impurities, leading to iron enrichment. Research data indicate that the iron content in scrap aluminum can reach 3–5 times that of primary aluminum, with the iron content in recycled Al-7Si-0.3Mg alloys typically exceeding 0.5 wt.% [1]. Al-Si-Mg alloys with higher iron content exhibit larger eutectic silicon areas and more severe eutectic silicon aggregation, making eutectic silicon prone to peeling, reducing the casting performance. Iron-rich intermetallic compounds may also serve as nucleation sites for porosity, increasing porosity rates and causing casting defects [2]. Iron primarily exists in aluminum alloys as needle-like β-Al_5FeSi phases. These hard and brittle phases disrupt the continuity of the matrix, becoming the primary crack sources in tensile fractures, promoting crack initiation and propagation [3], as well as significantly reducing the alloy's strength and elongation. When the Fe content exceeds 0.8%, the alloy's elongation may decrease by over 50% [4]. More critically, traditional purification processes struggle to economically and effectively remove iron, rendering the mechanical properties of high-iron recycled aluminum alloys (Fe > 0.8%) unable to meet industrial standards. This has become the core bottleneck constraining the high-value utilization of recycled aluminum.

The modification research on iron-rich aluminum alloys primarily focuses on the following three aspects: (1) Control of the iron-phase morphology, with the addition of elements such as Mn and Cr, of which Mn preferentially combines with Fe to form the α-Al(Fe,Mn)Si phase, replacing the brittle β-Al_5FeSi phase (needle-like or plate-like), while Cr tends to form independent intermetallic phases, such as $Al_{15}(Fe,Cr)_3Si_2$ [5,6]. Studies have shown that adding 0.5–1.0 wt.% Mn can reduce the β-phase proportion from 25% to below 5%, while controlling the α-phase size within the range of 5–10 μm [7]. Cr promotes the non-uniform nucleation of primary Si and restricts its growth, thereby refining the grain size [8]. (2) Rapid solidification technology inhibits iron-phase growth. During rapid solidification (cooling rate of 10^5–10^7 K/s), Fe elements in the aluminum matrix are "frozen" within the α-Al lattice due to an insufficient diffusion time, significantly reducing Fe segregation at the grain boundaries [9]. (3) Melt overheating treatment promotes iron-phase spheroidization [10], but it has high energy consumption costs. Although these methods have made some progress, they still have obvious limitations, as follows: Mn addition requires strict control of the Fe/Mn ratio, as excess Mn can lead to the formation of enriched phases such as $Al_{15}(Fe,Mn)_3Si_2$ or $Al_6(Fe,Mn)$ [11], whose crystal structures are predominantly cubic or hexagonal [12], resulting in poor compatibility with the aluminum matrix, and they can easily become stress concentration points [13]. Rapid solidification technology is constrained by equipment investment, and melt overheating treatment is only applicable to specific composition systems. Recent studies have found that grain refinement can indirectly improve the microstructural uniformity of iron-rich aluminum alloys. Grain refinement increases the intergranular area per unit volume, providing more heterogeneous nucleation sites for the β-Al_5FeSi phase. This spatial restriction effect significantly reduces the anisotropic growth trend, resulting in a more uniform short rod-like or granular morphology of the β phase [14]. The three-dimensional network-like grain-boundary structure (grain-boundary spacing <10 μm) formed by the fine-grained microstructure alters the diffusion kinetics of Fe elements. Molecular dynamics simulations show that the grain-boundary diffusion coefficient is 2–3 orders of magnitude higher than bulk diffusion, effectively suppressing the localized enrichment of Fe elements in grain-boundary regions. Experimental measurements indicate that the Fe segregation degree can be reduced from 15% in the original state to below 5%, thereby suppressing the preferential growth of the β phase along specific crystallographic directions [15]. Modalavalasa et al. [16] noted that while the addition of Co does not alter the morphology of the β phase, it refines its distribution. The fine-grained microstructure provides more grain boundaries as nucleation sites for phase formation, thereby disrupting its oriented growth. Simultaneously, it causes multiple deflections in crack propagation paths, consuming more fracture energy and enhancing the matrix's resistance to stress concentration [17]. The above provides new insights into regulating the performance of iron-rich recycled aluminum alloys through grain refiners. However, the mechanism of the grain refinement behavior of traditional Al-Ti-B grain refiners in iron-rich aluminum–silicon–magnesium alloy systems remains poorly understood, particularly regarding potential interface reactions, element segregation, or chemical interactions between Fe and TiB_2 particles, which may significantly influence the heterogeneous nucleation efficiency. This critical scientific issue has not been systematically explored in existing research and urgently requires in-depth analysis to guide the microstructural regulation of high-performance aluminum alloys.

This study systematically controlled the addition of Al-5Ti-2B refiner (0–1.25%) to investigate its effect on the as-cast microstructure and mechanical properties of iron-rich (1%) recycled Al-Si-Mg alloy, with a focus on quantitatively analyzing the effect of TiB_2 particles on the efficiency of α-Al heterogenous nucleation, establishing a structure–property relationship between the addition amount and grain size, revealing the regulatory effect of Al-5Ti-2B on the morphology and spatial segregation behavior of the β-Al_5FeSi phase and determining the optimal addition amount to suppress the harmful effects of the Fe phase through a combination of microstructural characterization and performance testing. By conducting a comprehensive analysis of the "refining agent dosage–microstructure–performance" chain, this study aims to advance the development of processes for the high-value utilization of recycled aluminum.

2. Materials and Methods

The experimental materials for this study were obtained from a company in Baise City and consisted of recycled Al-Si-Mg aluminum alloy (Si: 7.51%; Mg: 0.31%; Fe: 1%; Al balance), hereinafter referred to as recycled Al-7Si-0.3Mg-1Fe alloy. By adding different proportions of Al-5Ti-2B intermediate alloy, the base material was remelted to prepare the corresponding experimental samples. The specific process is as follows: (1) Melting process—Melting was performed in a crucible melting furnace at 750 °C→Al-5Ti-2B intermediate alloy was added→the melt was maintained at temperature and allowed to settle for 15 min. (2) Refining treatment—Refining was carried out by rotating stirring with high-purity nitrogen introduced from the bottom (10 min)→after slag removal, the melt was allowed to settle for another 10 min. (3) Casting and forming—When the melt temperature dropped to 720 °C, it was poured into a preheated steel mold (300 °C) to prepare the tensile test specimens. For the Al-Si-Mg-Fe alloys, due to their high eutectic reaction temperature (approximately 577 °C), the Al-5Ti-2B grain refiner was prone to poisoning by Si in a silicon-rich environment, promoting the precipitation of the $TiSi_2$ phase, thereby weakening the refining effect. Additionally, the coarsening of the iron-rich phases (such as β-Al_5FeSi) further deteriorated the alloy's mechanical properties, particularly the ductility and toughness. Therefore, to suppress the formation of harmful phases and to optimize the microstructure, a rapid cooling strategy was required to minimize the alloy's residence time in the high-temperature zone (>500 °C). Based on this, the demolding time in this experiment was strictly controlled to within 45 s to ensure that the casting rapidly separated from the mold after basic solidification, thereby maximizing the grain refinement effect of Al-5Ti-2B and minimizing the adverse effects of the brittle phases.

For different Al-5Ti-2B addition ratios (as shown in Table 1), five tensile specimens were prepared in parallel for each group and tested. The specimen dimensions were as follows (Figure 1): gauge length, L_0 = 50 mm, and original diameter, d_0 = 10 mm. Concurrently, matching the metallographic specimens, the SEM-BSE analysis specimens were prepared as follows: metallographic specimens were mechanically ground and polished, then etched with Keller's reagent for 10–20 s; the SEM-BSE specimens were only polished (no etching required). After the tensile test, samples were taken 5 mm below the fracture surface for the SEM fracture morphology observation. The specific correspondences between the specimen numbers and Al-5Ti-2B addition ratios are detailed in Table 1.

Table 1. Specimen ID and Al-5Ti-2B (wt.%).

Specimen ID	Al-5Ti-2B Additions (wt.%)
1	0
2	0.25%
3	0.5%
4	0.75%
5	1%
6	1.25%

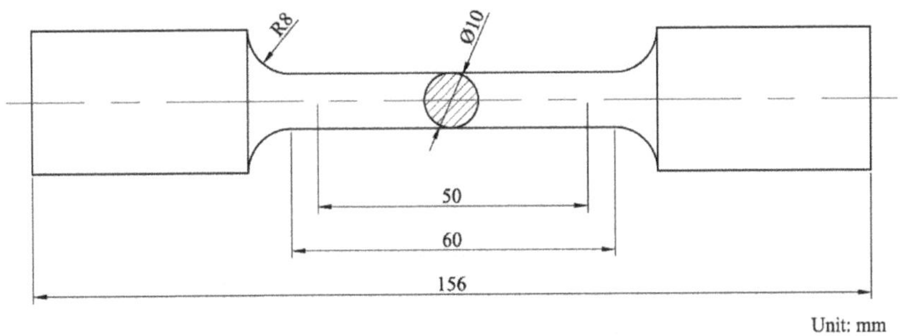

Figure 1. Tensile specimen dimensions.

Phase identification was performed using an X-ray diffractometer (XRD, Rigaku SmartLab 9, Tokyo, Japan). Backscattered electron (BSE) images, tensile fracture morphology, elemental distribution, and composition analysis were obtained using a tungsten-filament scanning electron microscope (SEM, Hitachi SU5000, Tokyo, Japan) equipped with energy-dispersive spectroscopy (EDS). Room-temperature tensile tests were conducted using a universal testing machine (CMT-4015, Shenzhen Junrui Instrument, Co., Ltd., Shenzhen, China) to determine the ultimate tensile strength and elongation. Round specimens were employed for testing, with five replicates performed for each sample type. The maximum load (F_{max}, unit: N) sustained during the tensile testing was recorded for each trial, and the average value was calculated. The original cross-sectional area (A_0) of the specimens was measured, and the ultimate tensile strength (R_m, unit: MPa) was calculated using Equation (1). Elongation was determined as the ratio of permanent elongation after fracture to the original gauge length, expressed as a percentage. The elongation (A) was calculated using Equation (2), where L_0 represents the original gauge length, and L_u denotes the final gauge length after fracture and reassembly.

$$R_m = \frac{F_{max}}{A_0} \quad (1)$$

$$A = \frac{L_u - L_0}{L_0} \times 100\% \quad (2)$$

The microstructure of the specimens was observed using an optical microscope (OM, Axio Observer 3m, Beijing Precise Instrument Co., Ltd., Beijing, China), and micrographs were taken. Metallographic images at the same magnification ratio (×100) were selected, and ImageJ 1.52 image software was used to perform binarization processing on the metallographic images to highlight the grain boundaries. Based on the processed metallographic images, the average grain boundary length, $\overline{L}$, was measured using the straight-line intersection method. The grain boundary length refers to the length of the line segment between the intersection points of the test line and the grain boundary. By calculating the average value of a large number of grain boundary lengths, the grain size can be reflected. The specific method is as follows: (1) Overlay a set of parallel straight lines on the processed metallographic images, ensuring that the total length of the lines covers multiple grains. (2) Count the total number of intersection points P between the test lines and the grain boundaries and measure the total length, L_T, of the test lines. (3) Calculate the average intercept using the formula shown in Equation (3).

$$\overline{L} = \frac{L_T}{P} \tag{3}$$

3. Results

3.1. Metallographic Analysis

Figure 2 shows the as-cast microstructure of the recycled Al-7Si-0.3Mg-1Fe alloy at different Al-5Ti-2B addition levels. The microstructure of the alloy without Al-5Ti-2B addition is shown in Figure 2a. The matrix is primarily composed of an α-Al solid solution, with eutectic Si forming coarse plate-like structures. The Mg elements are mainly dissolved in the matrix or form nanoscale Mg_2Si strengthening phases [1,2,18], but due to their small size, they are difficult to distinguish in the figure. Fe primarily exists in the alloy as needle-like β-Al_5FeSi phases, which grow from the eutectic silicon boundaries and penetrate the entire eutectic silicon structure [3]. Iron is a harmful impurity in aluminum–silicon alloys. When its content is too high, it forms coarse, brittle intermetallic compounds (such as β-Al_5FeSi). These compounds tend to segregate at grain boundaries, acting as stress concentration points that trigger microcracks and significantly reduce the strength and plasticity of the alloy [4,19]. For recycled aluminum alloys, the presence of Fe is very common; however, completely separating iron from the aluminum melt is quite challenging. Current research primarily focuses on regulating the morphology of iron phases [5–8], inhibiting their growth [9], and promoting spheroidization [10–13]. This experiment attempted to add Al-5Ti-2B intermediate alloy to refine the matrix structure while improving the morphology of the iron phases, aiming to enhance the performance of recycled aluminum alloys. Figure 2b–f show the micrographs of the alloy after adding Al-5Ti-2B, with addition levels of 0.25%, 0.5%, 0.75%, 1%, and 1.25%. As shown in the figures, with the increasing addition of Al-5Ti-2B, the primary α-Al dendrites gradually decreased in size, and the eutectic Si morphology transitioned from plate-like to granular and fibrous. Figure 3 shows the XRD pattern of the alloy with 1% Al-5Ti-2B added. The experimental results indicate that the XRD patterns of the base alloy remain unchanged after adding different amounts of Al-5Ti-2B. This may be due to the low addition amount of Al-5Ti-2B, whose diffraction signals are masked by the base alloy, or because the elements dissolve into the alloy without forming new phases. Combining the XRD patterns in Figure 3, as the addition of Al-5Ti-2B increases, the number of TiB_2 particles per unit volume of the aluminum melt increases. These particles serve as heterogeneous nucleation substrates for α-Al, with Al_3Ti forming a thin film on their surfaces, promoting α-Al heterogeneous nucleation and increasing the nucleation rate, thereby refining the matrix grains [20,21]. As the addition of Al-5Ti-2B increases, $TiAl_3$ accumulates or adsorbs onto the surface defects of the β-Fe phase, hindering its growth and causing its morphology to transition from needle-like to short, rod-like structures. When the Al-5Ti-2B addition reaches 1.25%, compared to a 1% addition, the alloy grain size increases, showing a trend toward grain coarsening. Lin et al. [22] and Ke et al. [23] found that Si in the alloy reacts with Ti to form Ti-Si compounds, consuming free Ti and reducing the grain refinement effect, known as the "silicon poisoning" phenomenon. Additionally, Chowwanonthapunya & Peeratatsuwan [24] investigated the effect of Fe on the grain refinement of Al-5Ti-1B in recycled A356 alloy, finding that as the number of cycles increased, the concentrations of Ti and B decreased, and the grain refinement effect weakened. This may be due to the presence of Fe promoting the consumption of Ti or forming complex Fe-Ti-B phases, thereby affecting the effectiveness of TiB_2.

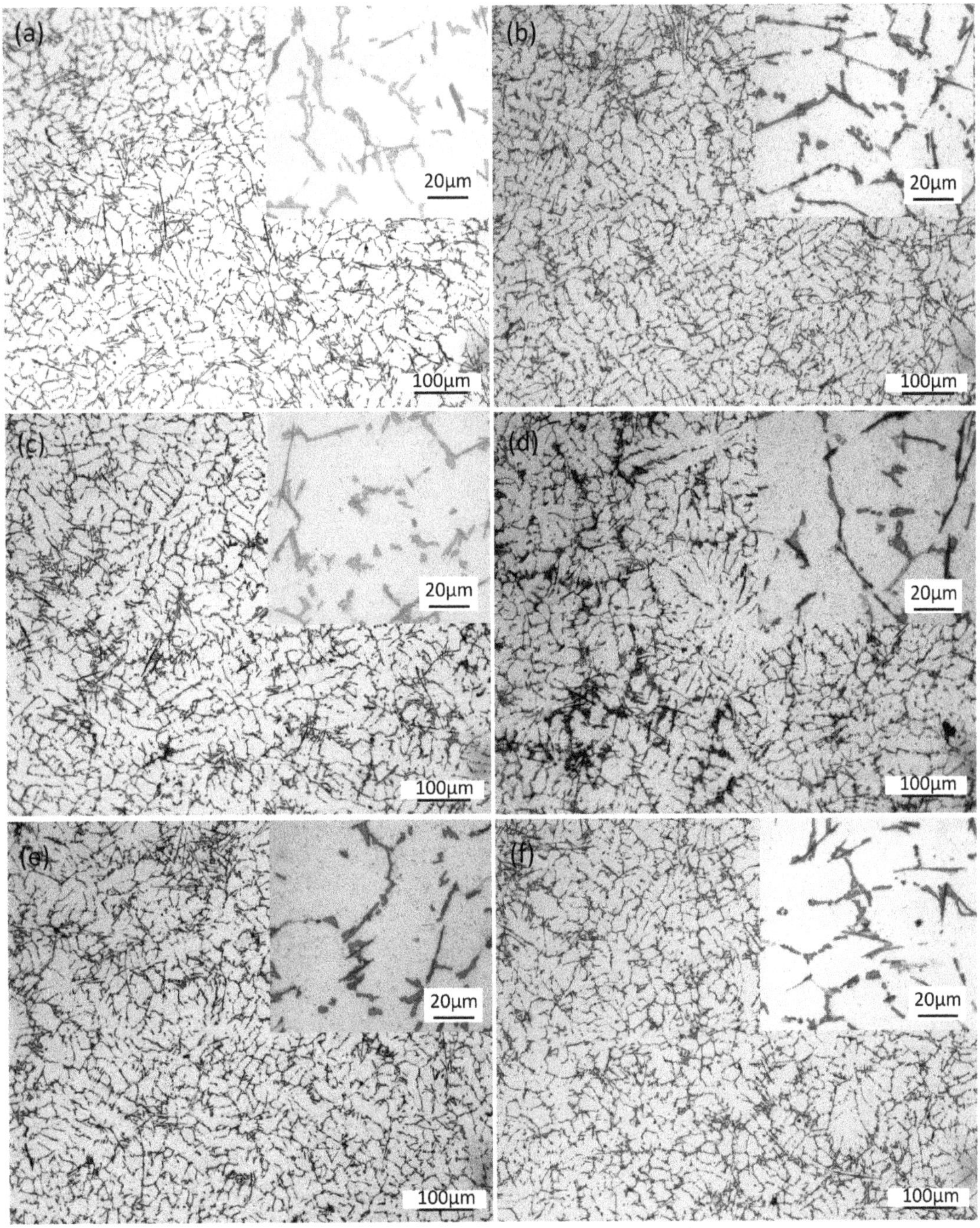

Figure 2. Metallographic organization of regenerated Al-7Si-0.3Mg-1Fe alloy with different Al-5Ti-2B additions: (**a**) 0% addition of Al-5Ti-2B; (**b**) 0.25% addition of Al-5Ti-2B; (**c**) 0.5% addition of Al-5Ti-2B; (**d**) 0.75% addition of Al-5Ti-2B; (**e**) 1% addition of Al-5Ti-2B; (**f**) 1.25% addition of Al-5Ti-2B.

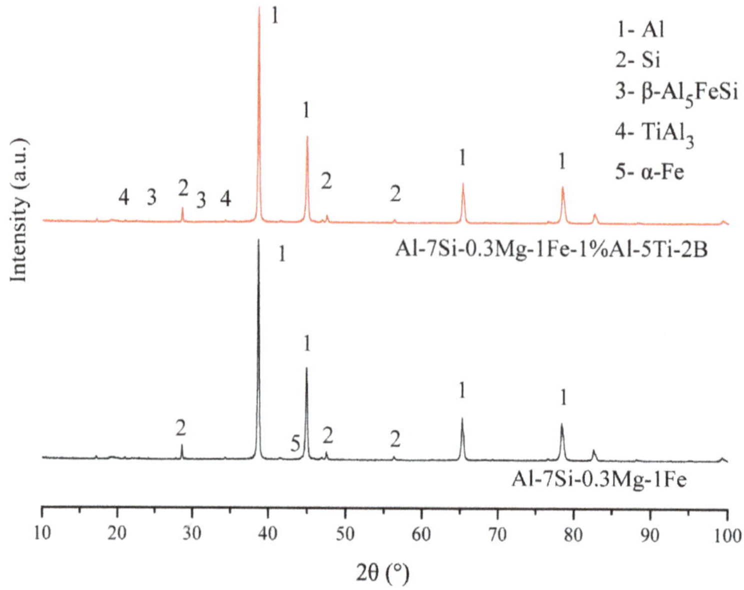

Figure 3. XRD patterns of Al-7Si-0.3Mg-1Fe alloy with 0 and 1 wt% Al-5Ti-2B additions.

To further investigate the effect of Al-5Ti-2B addition on grain refinement in the alloy, an Axio Observer 3m metallographic microscope was used to measure the grain size in the microstructure of the alloy using the straight-line intercept method. Figure 4 shows the relationship between the addition of Al-5Ti-2B and the grain intercept. As shown in the figure, with an increase in the addition of Al-5Ti-2B, the average grain intercept of the alloy exhibits a trend of first decreasing and then increasing. When no Al-5Ti-2B is added, the average grain intercept is 18.12 μm; when the addition is 0.25%, the average grain intercept is 16.96 μm; when the addition is 0.5%, the average grain intercept is 16.78 μm; when the addition is 0.75%, the average grain size is 16.5 μm; when the addition is 1%, the average grain size is 16.42 μm; and when the addition is 1.25%, the average grain size is 17.06 μm. The grain size did not decrease further but, instead, showed a trend toward increasing, confirming the previous discussion.

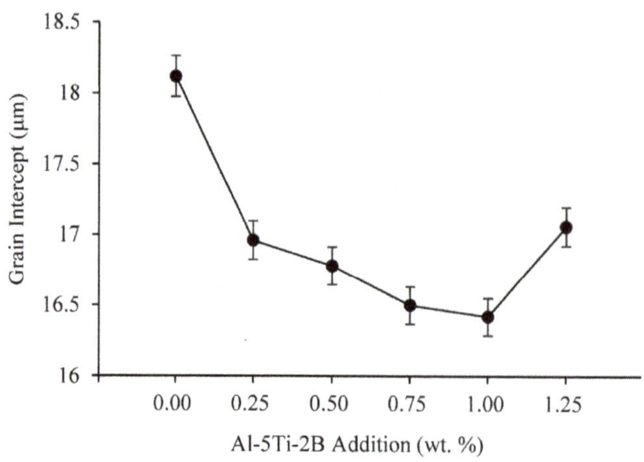

Figure 4. Influence of the Al-5Ti-2B addition amount on the grain intercept.

3.2. BSE Imaging and EDS Analysis

To investigate the regulatory effect of Al-5Ti-2B on the morphology and spatial segregation behavior of the β-Al5FeSi phase, SEM and EDS analyses were conducted on recycled Al-7Si-0.3Mg-1Fe alloys containing different amounts of Al-5Ti-2B. Figure 5a–f show the SEM morphology of the recycled Al-7Si-0.3Mg-1Fe alloy, with Al-5Ti-2B addition levels of 0%, 0.25%, 0.5%, 0.75%, 1%, and 1.25%. Figure 6 shows the EDS elemental layer map of the recycled Al-7Si-0.3Mg-1Fe alloy. As shown in the figure, Fe elements are attached to the edges of the eutectic silicon, while Mg elements are uniformly distributed throughout the matrix. Figure 6a shows the EDS elemental layer map of the alloy without Al-5Ti-2B addition, and Figure 6b shows the EDS elemental layer map of the alloy with 0.5% Al-5Ti-2B addition. Combining Figures 5–7, it can be seen that the dark gray elongated strips in Figure 5 represent eutectic Si, and TiB_2 and $TiAl_3$ are dispersed around the eutectic silicon. As the addition of Al-5Ti-2B increases, $TiAl_3$ aggregates on the surface of the eutectic silicon, inhibiting its growth, thereby refining the alloy grains. Mg forms the Mg_2Si strengthening phase with Si, which is dispersed throughout the alloy matrix, enhancing the alloy's mechanical properties. The bright white, elongated, and needle-like structures are β-Fe phases, which grow around the eutectic silicon, disrupting the alloy matrix, and severely impairing the alloy's mechanical properties. As shown in Figure 5b–f, with an increase in the addition of Al-5Ti-2B, the number of β-Fe phases in the matrix structure decreases. When the addition of Al-5Ti-2B reaches 1%, the refinement effect of the β-Fe phase is significant, with the elongated needle-like structures becoming finer and smaller. The morphology of the eutectic silicon also changes, transitioning from strip-like to granular. When the addition of Al-5Ti-2B reaches 1.25%, both the Fe phase and the Si phase exhibit obvious coarsening phenomena. The β-Fe phase appears as coarse, elongated needles intertwined within the grains and tending to aggregate, thereby weakening the mechanical properties of the alloy. In Figure 5f, the eutectic silicon exhibits severe aggregation. Chen et al. [2] and Song et al. [25] found that alloys with a higher iron content have larger eutectic silicon areas and more severe eutectic silicon aggregation. Additionally, iron-rich intermetallic compounds may serve as nucleation sites for pores, increasing the porosity and leading to casting defects.

Figures 7 and 8 show the EDS spectra of the alloy without addition and with an Al-5Ti-2B addition of 0.5%. As can be seen from the figures, the main elements in the alloy are Al and Si, with a relatively low content of Mg. Based on the metallographic microscopy, SEM morphology, and XRD analysis, the limited Mg content primarily dissolves into the Al matrix to form Mg_2Si strengthening phases, which are difficult to detect in the XRD pattern due to their weak peaks. As shown in Figure 3, the XRD pattern of the alloy without Al-5Ti-2B reveals a faint peak at $2\theta = 44.7°$, corresponding to the α-Fe phase. In contrast, after adding 1% Al-5Ti-2B, a distinct diffraction peak of the β-Al_5FeSi phase appears at $2\theta = 32.5°$. However, the XRD analysis results demonstrate that increasing the Al-5Ti-2B addition from 0.25% to 1.25% does not lead to significant changes in the diffraction peaks. Therefore, only the XRD patterns of the alloy without Al-5Ti-2B and with 1% Al-5Ti-2B addition were compared in the subsequent analysis. The EDS elemental mapping further reveals that Mg is predominantly distributed at the interface between α-Al and eutectic Si, forming Mg_2Si particles that disperse throughout the alloy matrix. These particles impede dislocation motion, thereby enhancing the alloy's strength. The key secondary phases in the Al-5Ti-2B grain refiner are TiB_2 and Al_3Ti. According to Wang et al. [21], the (0001) plane of TiB_2 provides a low-energy nucleation substrate for α-Al, while the Al_3Ti phase forms an "Al_3Ti/TiB_2" composite structure by encapsulating TiB_2, further reducing the nucleation barrier. Even in Fe-rich alloys, this dual-phase synergistic effect remains dominant.

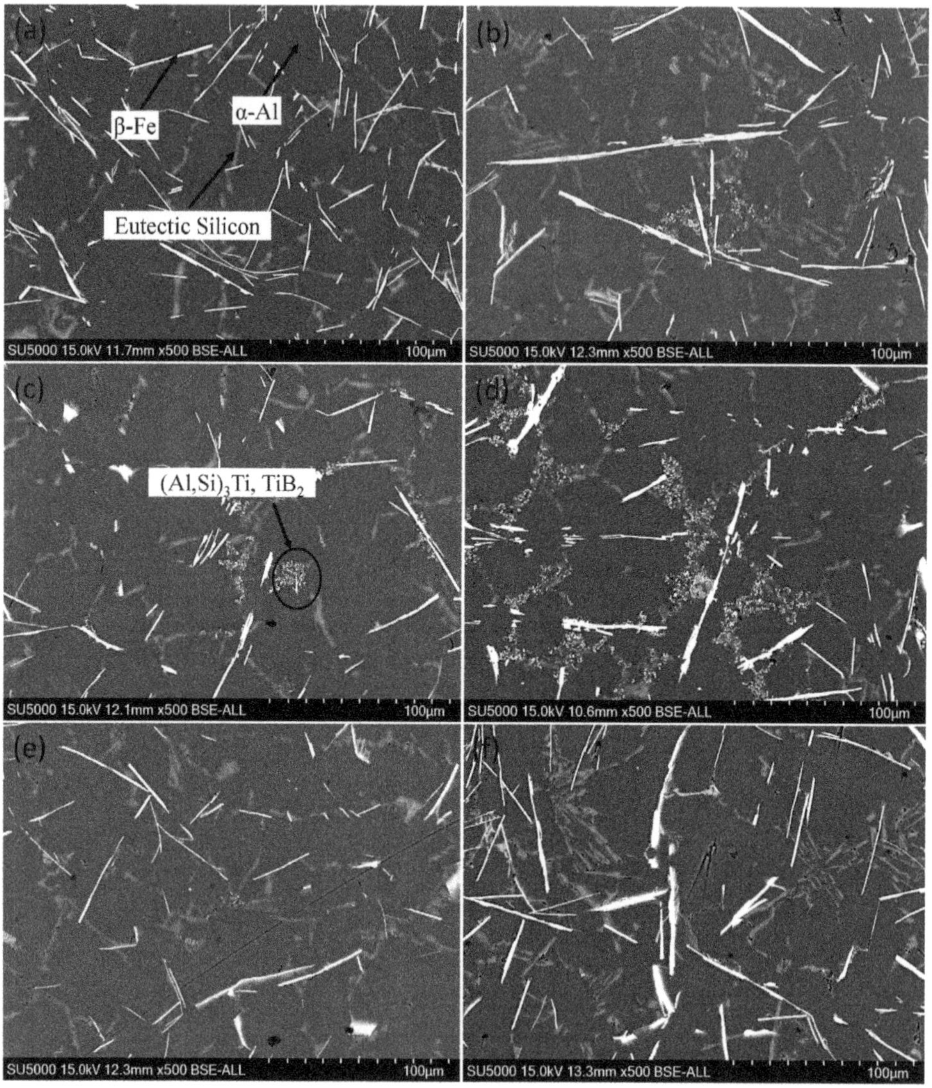

Figure 5. BSE of Al-7Si-0.3Mg-1Fe alloys with different Al-5Ti-2B additions: (**a**) 0% addition of Al-5Ti-2B; (**b**) 0.25% addition of Al-5Ti-2B; (**c**) 0.5% addition of Al-5Ti-2B; (**d**) 0.75% addition of Al-5Ti-2B; (**e**) 1% addition of Al-5Ti-2B; (**f**) 1.25% addition of Al-5Ti-2B.

To verify the findings described above and to understand the distribution of the phases in the alloy, an EDS (energy-dispersive spectroscopy) analysis was performed on the recycled Al-7Si-0.3Mg-1Fe alloy, as shown in Figure 9. Figure 9a shows the EDS spectrum image obtained when the addition of Al-5Ti-2B was 1%, and Figure 9b shows the SEM backscatter image when the addition of Al-5Ti-2B was 0.5%. Figure 9c–f are the energy spectrum diagrams of points A, B, C, and D in Figure 9a,b, respectively. From Figure 9a,c, it can be seen that the dark gray elongated structures in the alloy microstructure are eutectic Si. According to the Al-Si alloy phase diagram, primary α-Al precipitates first from the aluminum melt, and through the eutectic reaction, Al-Si eutectic silicon forms between the dendrite arms. After cooling, Mg_2Si and Si particles precipitate. Point B in

Figure 9a contains the following four elements: Al, Si, Mg, and Fe. As shown in Figure 9d, due to the high Fe content in the matrix alloy, Mg consumes Si to form the Mg_2Si phase, indirectly affecting the composition and morphology of the Al-Fe-Si phase. Doroshenko et al. [26] found that in Al-6Mg-2Ca-2Zn alloys, the addition of Mg transforms the Fe-Al-Si phase from needle-like to Chinese-character-like, with the results similar to those of this experiment. Therefore, the microstructure at point B should be the AlFeSi phase. In this case, Mg dissolves into the Al matrix, exerting a solid solution strengthening effect and indirectly influencing the precipitation kinetics of the Fe-Si intermetallic compound [27]. Additionally, Mg enriches at the grain boundaries, inhibiting the growth of coarse Fe phases and promoting the formation of nanoscale α-AlFeSi phases [28]. According to the energy spectrum diagram at point C in Figure 9e, the bright white stripes in Figure 9a may be α-Fe($Al_{12}Fe_3Si$) or β-Fe(Al_5FeSi or $Al_9Si_2Fe_2$), but the Si content in the α-Fe phase is relatively low. Combined with the XRD patterns in Figure 3, the structure is identified as the β-Al5FeSi phase. Its morphology is needle-like or plate-like, growing along the eutectic silicon boundaries and penetrating the entire eutectic structure. If Fe elements are locally enriched due to uneven stirring during melting or other factors, the Fe-rich phase will serve as an excellent nucleation site for pores during solidification, leading to increased porosity in the casting and a reduced casting performance of the alloy.

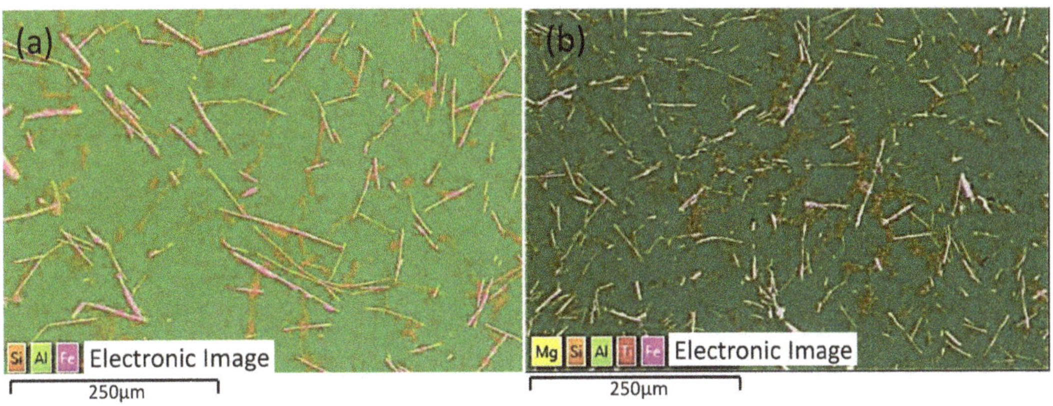

Figure 6. Elemental EDS layering of Al-7Si-0.3Mg-1Fe alloys with different Al-5Ti-2B additions: (**a**) 0% addition of Al-5Ti-2B; (**b**) 0.5% addition of Al-5Ti-2B.

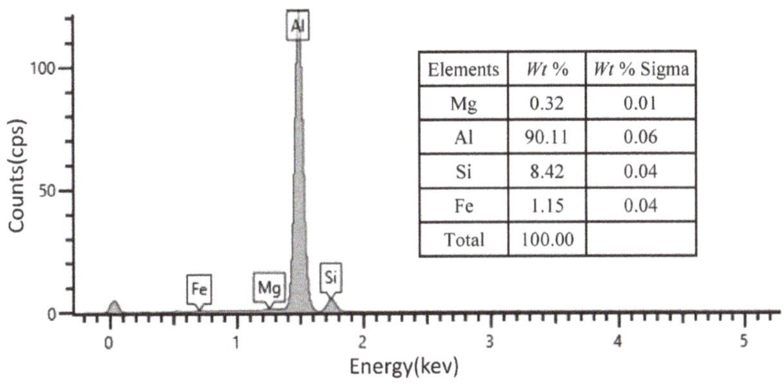

Elements	*Wt* %	*Wt* % Sigma
Mg	0.32	0.01
Al	90.11	0.06
Si	8.42	0.04
Fe	1.15	0.04
Total	100.00	

Figure 7. EDS energy spectra of Al-7Si-0.3Mg-1Fe alloy without added Al-5Ti-2B.

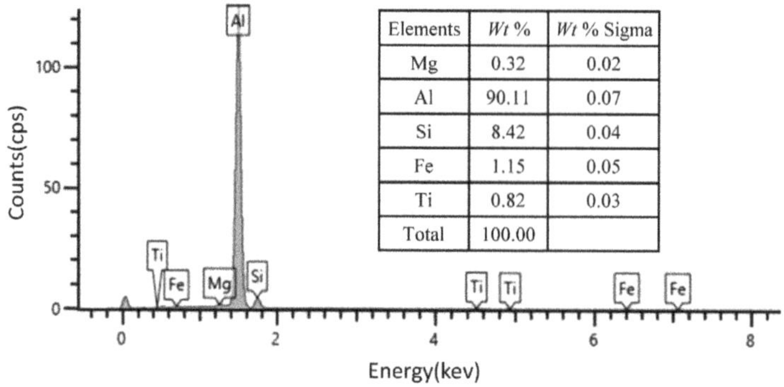

Figure 8. EDS energy spectra of Al-7Si-0.3Mg-1Fe alloy with 0.5% Al-5Ti-2B addition.

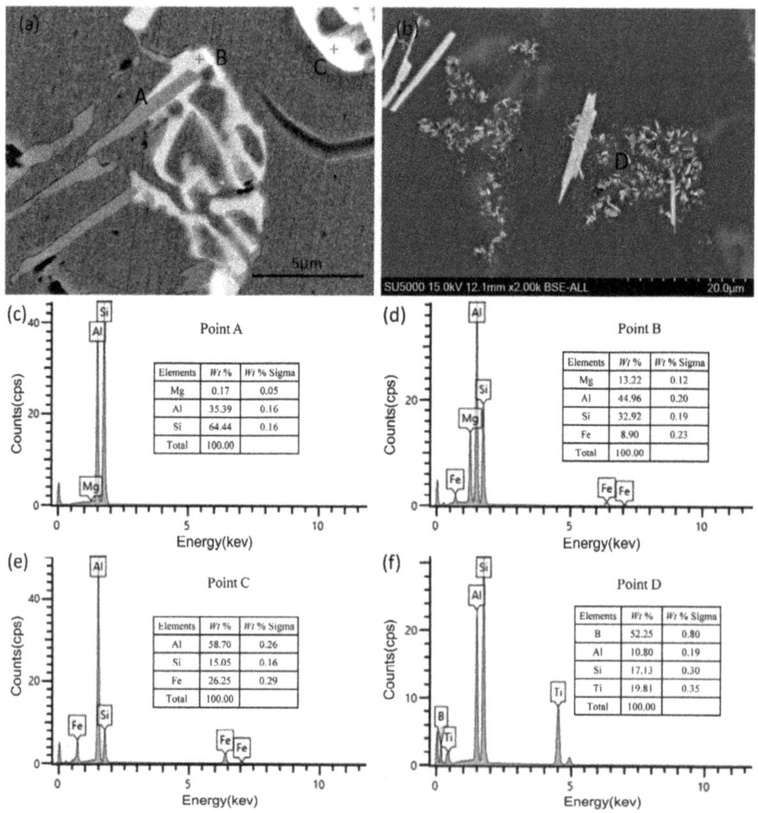

Figure 9. SEM map and EDS analysis results of the regenerated Al-7Si-0.3Mg-1Fe alloy after the addition of Al-5Ti-2B.

In summary, as shown in Figure 9a, when the addition of Al-5Ti-2B increases to 1%, the Fe-rich phase exhibits a fine needle-like or fibrous morphology, primarily in the form of a β-Fe phase. As shown in Figure 5e, the needle-like or plate-like β-Fe morphology becomes finer and shorter, and it is insufficient to penetrate the dendrites, resulting in improved mechanical properties of the alloy. From the energy spectrum at point D in Figure 9f, it can be seen that this point contains the following four elements: Al, Si, Ti, and B. This is because the addition of Al-5Ti-2B to the regenerated Al-7Si-0.3Mg-1Fe alloy forms particles such as Al_3Ti and TiB_2, which act as heterogeneous nucleation or nucleation promotion agents, refining the alloy grains. As the addition of Al-5Ti-2B increases, Al_3Ti aggregates at defects on the eutectic silicon surface, inhibiting its growth and achieving grain refinement of the eutectic silicon. Meanwhile, TiB_2 promotes the nucleation rate of Al_3Ti, thereby achieving grain refinement.

3.3. Tensile Properties Analysis

As shown in Table 2 and Figure 10, the tensile strength of the recycled Al-7Si-0.3Mg-1Fe alloy first increases and then decreases with an increase in the Al-5Ti-2B addition. When the Al-5Ti-2B addition is 1%, the tensile strength and elongation of the alloy reach their maximum values at 145.7 MPa and 3.3%, respectively. Fe forms β-Fe phases in Al-Si-Mg alloys, which are relatively coarse and reduce the fluidity of the aluminum alloy melt, affecting the feeding of the melt and leading to shrinkage porosity in castings, thereby degrading the casting performance of the alloy [2]. Additionally, the hard and brittle, elongated, and needle-like Fe phase causes cracks to propagate preferentially along the Fe phase/silicon interface, severely damaging the alloy's mechanical properties [29]. Bacaicoa et al. [3] found that the interface between the β-Al_5FeSi phase and the matrix is prone to microcracks due to differences in thermal expansion coefficients (β phase: 19.5×10^{-6}/K, while the Al matrix has a coefficient of 23.6×10^{-6}/K), microcracks are prone to form. Eutectic Si particles act as obstacles during crack propagation, causing cracks to deflect and branch [30]. At Al-5Ti-2B addition levels between 0 and 1%, combined with SEM images in Figure 5, eutectic Si primarily distributes at the α-Al grain boundaries, and bright white β-Al_5FeSi phases are distributed around the eutectic Si, with some penetrating the entire eutectic structure. As the Al-5Ti-2B addition increases, the distribution of the bright white β-Al_5FeSi phases with coarse, elongated, and needle-like morphology decreases per unit area, and the morphology of the eutectic Si transitions from rod-like to granular, with the. distribution becoming more uniform, reducing interfacial stress. TiB_2 and $TiAl_3$ are distributed around the eutectic Si, and the pinning dislocation effects are enhanced, further improving the alloy's tensile strength and mitigating the adverse effects of β-Fe on the alloy's mechanical properties. When the Al-5Ti-2B addition reaches 1.25%, the grain size increases again, grain refinement strengthening weakens, and the β-Fe phase forms coarse needle-like structures penetrating the entire eutectic Si structure, disrupting the aluminum alloy matrix. Concurrently, eutectic Si exhibits enrichment phenomena, leading to a decrease in the alloy's mechanical properties. Song et al. [25] investigated the microstructure and fracture behavior of recycled Al-Si-Mg alloys, finding that the Fe-Si phase serves as an excellent nucleation site for pores, resulting in an increase in pores within the alloy castings and an increase in inter-dendritic defects. The hard and brittle β-Fe phase causes the alloy to develop a series of microcracks, with the stress concentration at the crack tips, promoting crack initiation and reducing the alloy's tensile strength.

Table 2. Tensile strength and elongation of Al-7Si-0.3Mg-1Fe alloys.

Specimen Number	Tensile Strength (MPa)	Elongation (%)
1	117.7	2.5
2	121.1	2.0
3	139.2	1.9
4	145.7	3.3
5	149.4	4.3
6	142.6	2.5

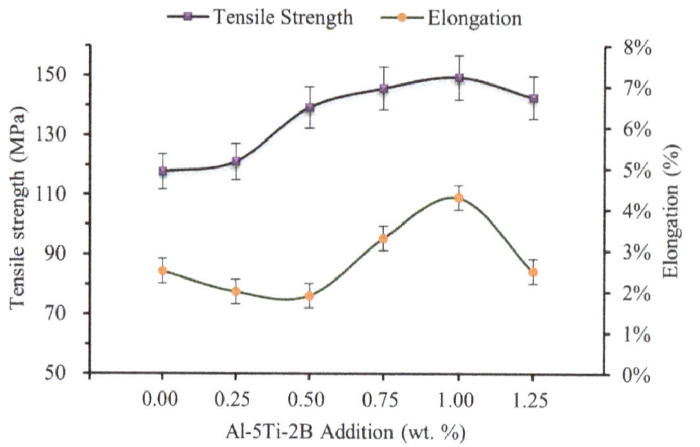

Figure 10. Trends of tensile strength and elongation of the regenerated Al-7Si-0.3Mg-1Fe alloys.

3.4. Fractography Analysis

Figure 11 shows the fracture surface morphology of the alloy after adding different amounts of Al-5Ti-2B. According to the tensile test results, as the addition of Al-5Ti-2B increases, the mechanical properties of the alloy first improve and then deteriorate. The fracture mechanism is analyzed through fracture surface morphology analysis. As reported in the literature [31–33], in unmodified iron-rich aluminum–silicon–magnesium alloys, the primary crack propagates along the β-Al_5FeSi phase/α-Al interface, forming an intergranular fracture path; eutectic silicon particles act as stress concentration points, inducing secondary cracks, and the fracture surface exhibits extensive cleavage planes. Based on the test results, the macro fracture surfaces of all samples show no obvious yield or necking phenomena, and the tensile fracture surfaces exhibit typical brittle fracture characteristics. The tensile fracture surface of the recycled Al-7Si-0.3Mg-1Fe alloy without the addition of Al-5Ti-2B is shown in Figure 11a. From the figure, it can be seen that the fracture surface morphology is primarily river-like, exhibiting a cleavage fracture surface morphology. From the fracture path analysis, the fracture surface is an intergranular fracture, belonging to brittle fracture. This is because β-Fe grows along the eutectic Si boundary, penetrating the entire eutectic structure and severing the alloy matrix. Under external force, fractures first occur, forming crack sources, and secondary cracks are prone to occur during crack propagation, accelerating material fracture. As the addition of Al-5Ti-2B increases, quasi-cleavage planes appear in Figure 11b–e, and the river-like patterns become shorter. Notably, when the Al-5Ti-2B addition reaches 1%, both the β-Al_5FeSi intermetallic phases and eutectic silicon undergo significant refinement, effectively mitigating stress concentration effects. This microstructural modification alters the fracture mechanism, causing crack propagation to circumvent the refined eutectic Si particles and leading to characteristic intergranular fracture behavior [29,30], as shown in Figure 11e. The fracture surface morphology exhibits minor tearing ridges (i.e., ductile pits (circled in the figure)), but the fracture mode remains predominantly brittle. When the Al-5Ti-2B addition reaches 1.25%, as shown in Figure 11f, the fracture surface exhibits extensive micro-porosity (indicated by arrows in the figure). Figure 11g is an enlarged view of the micro-porosity in Figure 11f, and the EDS spectrum at point A in Figure 11g is shown in Figure 11h. The Fe content at this point reaches 32.16%, and the O element is likely introduced due to improper sample handling, indicating that the microstructure at point A corresponds to the β-Fe phase. The formation of shrinkage porosity is attributed to Fe enrichment, which has a segregation effect on Al. Based on the research results in Reference [2], it is known that under external force, shrinkage porosity leads to crack initiation and propagation, resulting in a decline in the mechanical properties of the alloy.

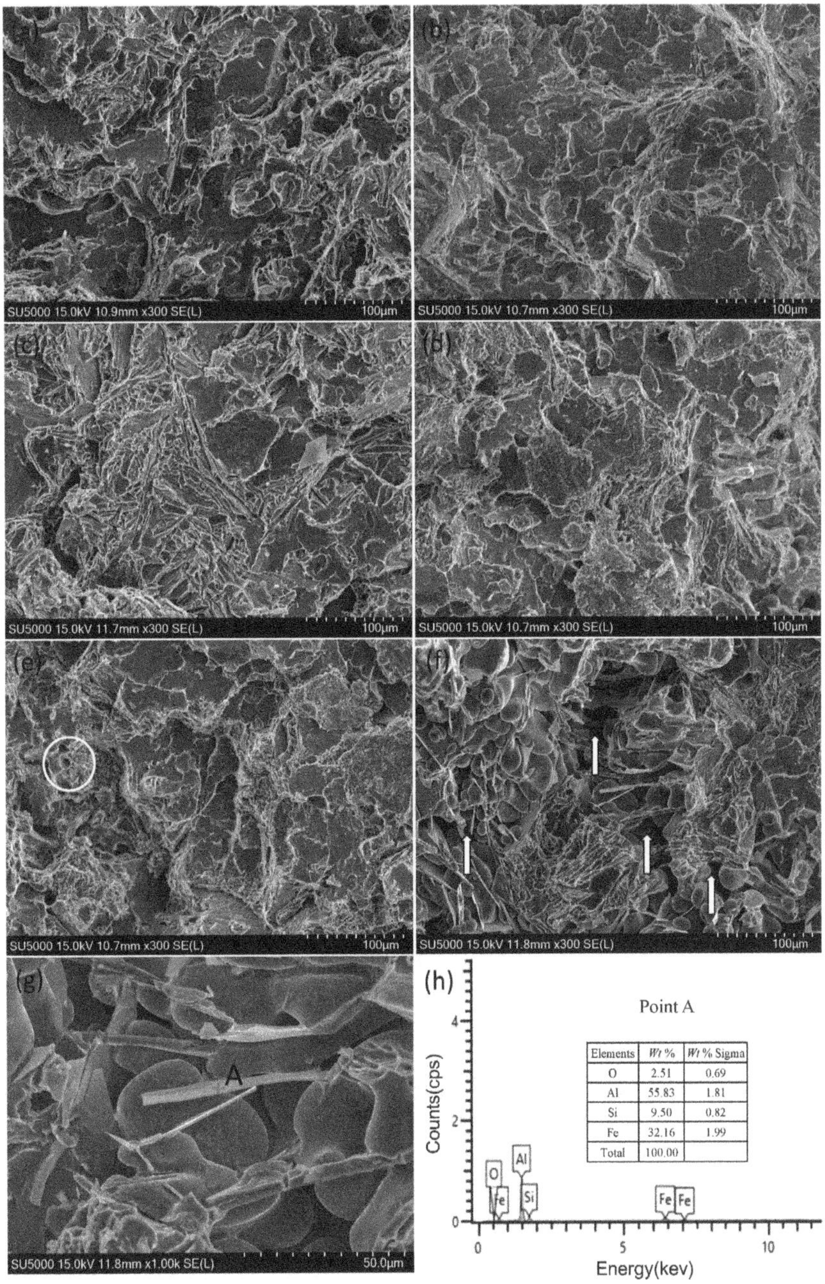

Figure 11. Tensile fracture SEM images of alloy specimens with different Al-5Ti-2B additions and the results of the energy spectrum analysis at point A: (**a**) 0% addition of Al-5Ti-2B; (**b**) 0.25% addition of Al-5Ti-2B; (**c**) 0.5% addition of Al-5Ti-2B; (**d**) 0.75% addition of Al-5Ti-2B; (**e**) 1% addition of Al-5Ti-2B; (**f**) 1.25% addition of Al-5Ti-2B; (**g**) 1.25% addition of Al-5Ti-2B (×1000); (**h**) results of the energy spectrum analysis at point A.

4. Conclusions

(1) In the recycled Al-7Si-0.3Mg-1Fe alloy, Fe primarily exists in the form of the β-Al_5SiFe phase, appearing as coarse needle-like structures distributed around the eutectic Si, with some penetrating the entire Al-Si eutectic structure and disrupting the matrix structure. As the addition of Al-5Ti-2B increases, the α-Al primary phase and eutectic Si in the matrix alloy structure become refined, and the structure tends to homogenize. When the addition of Al-5Ti-2B reaches 1%, the alloy properties reach their optimal state, with the average grain size decreasing to 16.42 μm, tensile strength increasing to 149.4 MPa, and the elongation increasing to 4.3%. When the addition of Al-5Ti-2B continues to increase, the eutectic Si aggregates, the β-Al_5SiFe phase coarsens, and micro-porosity appears in the alloy structure, leading to a decline in the alloy's performance.

(2) Iron (Fe) elements adhere to the eutectic silicon (Si), forming the β-Al_5FeSi phase. Magnesium (Mg) enriches at grain boundaries, inhibiting the growth of coarse Fe phases. Simultaneously, Mg reacts with silicon to form Mg_2Si, achieving solid solution strengthening and enhancing the alloy's performance. Al-5Ti-2B serves as a grain refiner, with its effectiveness primarily dependent on TiB_2 and Al_3Ti, which act as nuclei for the heterogeneous nucleation of α-Al. When the addition of Al-5Ti-2B is 1%, the refinement effect on the primary α-Al phase and eutectic Si in the matrix structure is significant, but its effect on altering the morphology of the β-Al_5FeSi phase is not prominent. However, after the matrix structure is refined, it inhibits the growth and distribution of the β-Al_5FeSi phase, thereby mitigating the adverse effects of the β-Al_5FeSi phase on the matrix alloy. When the addition of Al-5Ti-2B increases to 1.25%, eutectic Si aggregates, and the β-Al_5FeSi phase regrows into coarse needle-like structures, severely impairing the alloy's mechanical properties. Additionally, iron-rich intermetallic compounds may serve as nucleation sites for pores, increasing the porosity and leading to casting defects. Therefore, when improving the properties of iron-rich regenerated Al-Si-Mg alloys through the refinement effect of Al-5Ti-2B, the optimal addition amount is 1%.

(3) Based on the relevant literature and experimental results, it is known that due to the difference in thermal expansion coefficients at the interface between the β-Al_5FeSi phase and the matrix, microcracks are prone to form. Under external force, these microcracks propagate rapidly along the interface between the β-Al_5FeSi phase and the α-Al matrix. Meanwhile, unmodified eutectic Si particles act as obstacles during crack propagation, causing crack deflection and branching, which severely impair the mechanical properties of the alloy. The presence of Fe elements affects the distribution of eutectic Si, thereby reducing the alloy's casting performance. In tensile tests, the fracture surface shows no obvious yield or necking phenomena, with brittle fracture being the primary mode of failure.

5. Outlook

(1) Interfacial Behavior of TiB_2/Al and Fe Segregation Mechanisms

The current understanding confirms that Fe impurities in recycled aluminum (typically forming β-Al_5FeSi intermetallics) detrimentally affect the heterogeneous nucleation efficiency of TiB_2 particles in Al-5Ti-2B refiners. To address this fundamental issue, future investigations should prioritize the following:

(i) In situ TEM characterization of Fe segregation kinetics at TiB_2/Al interfaces, particularly focusing on the atomic-scale interaction between Fe and the (0001)TiB_2 nucleation substrate;

(ii) Quantitative assessment of nucleation potency deterioration through interfacial energy measurements (e.g., AFM-based nanomechanical testing), establishing a predictive model correlating Fe concentration with undercooling requirements.

Such studies would provide critical insights into impurity-mediated nucleation inhibition mechanisms.

(2) Rare-Earth-Enhanced Impurity Tolerance in Recycled Al-Si-Mg Alloys

A promising research direction involves developing RE-modified grain refinement systems, as follows:

(i) The synergistic effects of Ce/La multi-component additions could mitigate Si poisoning by competitively binding with Ti (as verified by ab initio MD simulations of Ti-Si-RE cluster stability);
(ii) Advanced correlative microscopy (combining APT for nanoscale Mg mapping and synchrotron μ-XRD for phase identification) is essential to quantify Mg partitioning behavior in Fe-containing systems;
(iii) Phase-field modeling incorporating experimental inputs from high-throughput XRD-EBSD could elucidate the stabilization mechanism of α-Al(Fe,Mn)Si phases during solidification.

This integrated approach may redefine impurity management strategies in recycled alloys.

(3) Thermomechanical Processing Optimization

Key knowledge gaps remain regarding the following:

(i) The critical strain threshold for Fe-intermetallic fragmentation during multi-pass rolling, which could be determined through high-temperature EBSD coupled with crystal plasticity finite element modeling (CPFEM);
(ii) The role of stored energy distribution (characterized by 3D-EBSD and TEM dislocation analyses) in governing precipitation kinetics during post-deformation aging.

Establishing these process–structure–property relationships would enable microstructure-informed manufacturing of high-performance recycled alloys.

Author Contributions: Conceptualization, W.S., B.L. and J.Y.; methodology, W.S., L.C., B.H. and B.L.; validation, W.S., L.C. and B.H.; investigation, W.S., L.C. and B.H.; resources, B.L. and J.Y.; data curation, W.S., L.C. and B.H.; writing—original draft preparation, W.S.; writing—review and editing, B.L. and J.Y.; visualization, W.S.; supervision, B.L. and J.Y.; project administration, B.L. and J.Y.; funding acquisition, B.L. All authors have read and agreed to the published version of the manuscript.

Funding: This research was funded by the National Natural Science Foundation of China (grant number: 52206002); The Science and Technology Program of Guangxi (grant number: AD23026045); The Science and Technology Research Project of Baise (grant numbers: 20221472 and 20222007); Guangxi Vocational University of Agriculture in 2023 (grant number: XKJ2340); Guangxi Key Technologies R&D Program (grant number: AB23075104).

Data Availability Statement: The original contributions presented in this study are included in the article. Further inquiries can be directed to the corresponding authors.

Acknowledgments: The authors thank Peilin Qing and Ming Qin, Baise University, and Xingzhi Pang, Guangxi University, for their kind suggestions.

Conflicts of Interest: The authors declare no conflicts of interest.

References

1. Taylor, J.A. Iron-Containing Intermetallic Phases in Al-Si Based Casting Alloys. *Procedia Mater. Sci.* **2012**, *1*, 19–33. [CrossRef]
2. Chen, Z.H.; Xu, C.; Hu, X.L.; Liu, Z.G.; Yamagata, H.; Ma, C.L. Effect of Iron-Rich Intermetallic and Eutectic Si Accumulation on Al-Si-Mg Alloy. *Mater. Sci. Forum* **2016**, *848*, 633–641. [CrossRef]
3. Bacaicoa, I.; Luetje, M.; Wicke, M.; Geisert, A.; Zeismann, F.; Fehlbier, M.; Brueckner-Foit, A. 3D Morphology of Al5FeSi Inclusions in High Fe-Content Al-Si-Cu Alloys. *Procedia Struct. Integr.* **2016**, *2*, 2269–2276. [CrossRef]
4. Ma, Z.; Samuel, A.M.; Doty, H.W.; Valtierra, S.; Samuel, F.H. Effect of Fe Content on the Fracture Behaviour of Al–Si–Cu Cast Alloys. *Mater. Amp. Des.* **2014**, *57*, 366–373. [CrossRef]
5. Liu, B.; Ma, C.; Li, L.; Yang, C.; Yu, N.; Wang, W.; Gao, T. Morphologies and Compositions of α–$Al_{15}Fe_3Si_2$-Type Intermetallics in Al–Si–Fe–Mn–Cr Alloys. *Int. J. Met.* **2022**, *17*, 1156–1164. [CrossRef]
6. Hou, L.G.; Cai, Y.H.; Cui, H.; Zhang, J.S. Microstructure Modification and Related Mechanism of Spray-formed Fe-bearing Hypereutectic Al-Si Alloys. *Mater. Und Werkst.* **2010**, *41*, 613–619. [CrossRef]
7. Kakitani, R.; Rodrigues, A.V.; Silva, C.; Garcia, A.; Cheung, N. The Roles of Solidification Cooling Rate and (Mn,Cr) Alloying Elements in the Modification of β-AlFeSi and Hardness Evolvements in near-Eutectic Al-Si Alloys. *J. Alloys Metall. Syst.* **2023**, *1*, 100005. [CrossRef]
8. Li, C.; Liu, F.; Yu, F.; Qiao, H.; Zheng, D.; Le, Q. The Growth Mechanism and Morphology Evolution of Primary Si during Slow Cooling Solidification of High Purity Al–15Si Alloy with Cr Additions. *J. Mater. Res. Technol.* **2023**, *23*, 1204–1213. [CrossRef]
9. Rajabi, M.; Vahidi, M.; Simchi, A.; Davami, P. Microstructural Evolution of Al-20Si-5Fe Alloy during Rapid Solidification and Hot Consolidation. *Rare Met.* **2009**, *28*, 639–645. [CrossRef]
10. Haque, M.M.; Ismail, A.F. Effect of Superheating Temperatures on Microstructure and Properties of Strontium Modified Aluminium–Silicon Eutectic Alloy. *J. Mater. Process. Technol.* **2005**, *162–163*, 312–316. [CrossRef]
11. Luo, L.; Tang, Y.; Liang, X.; Su, Y.; Zhang, Y.; Xie, H. Optimizing the Morphology and Solidification Behavior of Fe-Rich Phases in Eutectic Al-Si-Based Alloys with Different Fe Contents by Adding Mn Elements. *Materials* **2024**, *17*, 4104. [CrossRef] [PubMed]
12. Li, Y.J.; Muggerud, A.M.F.; Olsen, A.; Furu, T. Precipitation of Partially Coherent α-Al(Mn,Fe)Si Dispersoids and Their Strengthening Effect in AA 3003 Alloy. *Acta Mater.* **2012**, *60*, 1004–1014. [CrossRef]
13. Zhu, X.; Ji, S. The Formation of Al_6 (Fe, Mn) Phase in Die-Cast Al-Mg Alloys. *IOP Conf. Ser. Mater. Sci. Eng.* **2019**, *529*, 012011. [CrossRef]
14. Shen, X.; Liu, S.; Wang, X.; Cui, C.; Gong, P.; Zhao, L.; Han, X.; Li, Z. Effect of Cooling Rate on the Microstructure Evolution and Mechanical Properties of Iron-Rich Al–Si Alloy. *Materials* **2022**, *15*, 411. [CrossRef] [PubMed]
15. Yang, Y.; Tang, P.; Lu, J. Unveiling the Refinement Effect of the Coarse Fe-Rich Phase in Al-Si-Fe Alloy under Controlled Diffusion Solidification Treatments. *J. Mater. Eng. Perform.* **2024**, *34*, 9759–9771. [CrossRef]
16. Modalavalasa, K.; Namburi, G.M.K.; Battula, S.R.S.S.; Hasan, M.; Koppula, N.R.; Ramavajjala, A.K.; Rajulapati, K.V.; Rao, A.K.P. Investigation of the Microstructure and Dry-Sliding Wear Behavior of Co Containing Al-Si-Fe Scrap Alloy. *Tribol. Trans.* **2024**, *67*, 765–774. [CrossRef]
17. Gencalp Irizalp, S.; Saklakoglu, N. Effect of Fe-Rich Intermetallics on the Microstructure and Mechanical Properties of Thixoformed A380 Aluminum Alloy. *Eng. Sci. Technol. Int. J.* **2014**, *17*, 58–62. [CrossRef]
18. Lin, C.-W.; Hung, F.-Y.; Lui, T.-S. Microstructure Evolution and Microstructural Characteristics of Al–Mg–Si Aluminum Alloys Fabricated by a Modified Strain-Induced Melting Activation Process. *Metals* **2017**, *8*, 3. [CrossRef]
19. Lin, B.; Li, H.; Xu, R.; Shi, Y.; Xiao, H.; Zhang, W.; Liu, K. Precipitation of Iron-Rich Intermetallics and Mechanical Properties of Al–Si–Mg–Fe Alloys with Al–5Ti–B. *Mater. Sci. Technol.* **2018**, *34*, 2145–2152. [CrossRef]
20. Wang, S.C.; Li, J.L.; Gan, C.L.; Zheng, K. Grain Refinement Mechanism and Effective Nucleation Phase of Al-5Ti-1B Master Alloy. *Mater. Sci. Forum* **2017**, *898*, 1231–1235. [CrossRef]
21. Wang, Z.; Wang, S.; Yang, Q.; Liu, X.; Dong, C.; Liu, L. Refinement Properties and Refinement Mechanism of a New Master Alloy Al-5Ti-1B-1RE Refiner. *Coatings* **2024**, *14*, 1485. [CrossRef]
22. Lin, L.; Zhou, L.; Xie, Y.; Bai, W.; Li, F.; Xie, Y.; Lu, M.; Wang, J. Study on the Effect of Pre-Refinement and Heat Treatment on the Microstructure and Properties of Hypoeutectic Al-Si-Mg Alloy. *Materials* **2022**, *15*, 6056. [CrossRef] [PubMed]
23. Qiu, K.; Wang, R.; Peng, C.; Wang, N.; Cai, Z.; Zhang, C. Effects of Mn and Sn on Microstructure of Al–7Si–Mg Alloy Modified by Sr and Al–5Ti–B. *Trans. Nonferrous Met. Soc. China* **2015**, *25*, 3546–3552. [CrossRef]
24. Chowwanonthapunya, T.; Peeratatsuwan, C. The Influence of Fe on Grain Refinement of Recycled A 356 Alloy Initially Refined by Al-5Ti-1B Master Alloy. *Trends Sci.* **2022**, *20*, 6393. [CrossRef]
25. Song, L.Y.; Liu, H.X.; Nie, S.N.; Yu, B.Y.; Wang, S.C.; Zheng, L.; Li, R.X. Microstructure and Fracture Behavior of Al-Si-Mg Alloy Prepared with Recycled Alloy. *Mater. Und Werkst.* **2019**, *50*, 462–470. [CrossRef]
26. Doroshenko, V.; Shurkin, P.; Sviridova, T.; Fortuna, A.; Shkaley, I. Phase Composition and Microstructure of Cast Al-6%Mg-2%Ca-2%Zn Alloy with Fe and Si Additions. *Metals* **2023**, *13*, 1584. [CrossRef]

27. Khan, M.H.; Das, A.; Khan, T.I.; Li, Z.; Kotadia, H.R. Microstructural Modification in an Al-Mg-Si (6082) Alloy with High Fe Content Using Inoculation and/or High-Intensity Ultrasonication. *J. Alloys Compd.* **2025**, *1012*, 178500. [CrossRef]
28. Kim, J.; Kim, D. Effects of Fragmented Fe Intermetallic Compounds on Ductility in Al–Si–Mg Alloys. *J. Nanosci. Nanotechnol.* **2018**, *18*, 2137–2139. [CrossRef]
29. Bogdanoff, T.; Lattanzi, L.; Merlin, M.; Ghassemali, E.; Seifeddine, S. THE Influence of Copper Addition on Crack Initiation and Propagation in an Al–Si–Mg Alloy during Cyclic Testing. *Materialia* **2020**, *12*, 100787. [CrossRef]
30. Elsebaie, O.; Samuel, A.M.; Samuel, F.H.; Doty, H.W. Impact Toughness of Al–Si–Cu–Mg–Fe Cast Alloys: Effects of Minor Additives and Aging Conditions. *Mater. Amp. Des.* **2014**, *60*, 496–509. [CrossRef]
31. Xia, L.; Zheng, J.; Liu, Y.; Song, J.; Jiang, B. The Influence of Heat Treatment on the Heterogenous Structure and Deformation Behavior of a High Pressure Die-Cast Al-10Si-0.5Mn-0.3Mg Alloy. *Sci. Sin. Technol.* **2025**, *55*, 617–629. [CrossRef]
32. Basak, S.; Biswas, P.; Patra, S.; Roy, H.; Mondal, M.K. Effect of TiB_2 and Al_3Ti on the Microstructure, Mechanical Properties and Fracture Behaviour of near Eutectic Al-12.6Si Alloy. *Int. J. Miner. Metall. Mater.* **2021**, *28*, 1174–1185. [CrossRef]
33. Liu, Y.; Wu, C.; Tu, H.; Lu, X.; Wang, J.; Su, X. Microstructure and Mechanical Properties of Al-10Si Alloy Modified with Al-5Ti. *China Foundry* **2018**, *15*, 405–410. [CrossRef]

crystals

MDPI

Article

Research on the Efficient Desilication Process of Low-Grade Bauxite in Guangxi

Guoxian Hu [1], Anmin Li [1,2,*], An Xia [1], Dongjie Zhang [1], Liwen Pan [1,2], Xiaolian Zhao [1,2] and Xingzhi Pang [1,2]

1 School of Resources, Environment and Materials, Guangxi University, Nanning 530004, China; 15732973106@163.com (G.H.); 15375284514@163.com (A.X.); 15578488783@163.com (D.Z.); plw988@163.com (L.P.); xiaolianbsh@126.com (X.Z.); pxzh2006@126.com (X.P.)

2 China Guangxi Key Laboratory of Processing for Non-Ferrous Metals and Featured Materials, Guangxi University, Nanning 530004, China

* Correspondence: lianmin@gxu.edu.cn

Abstract

With the continuous exploitation of bauxite mineral resources, Guangxi bauxite faces many difficulties in alumina production due to its characteristics of high silicon content, high iron content, and a low Al-Si ratio. In view of this, this study is closely related to the key link of bauxite pre-desiliconization and strives to break free from the status quo to improve the aluminum/silicon ratio and help optimize the subsequent alumina-refining process. In the work presented in this paper, the unique mineralogy of Guangxi bauxite was comprehensively considered, covering its complex mineral composition and fine distribution characteristics. The barium hydroxide pre-desilication technology was first used for in-depth experimental exploration, and the silicon removal efficiency under different working conditions was systematically compared. The system compared the silicon removal effect and the associated aluminum loss under different working conditions. The results of this study will lay a solid foundation for the rational and efficient development of bauxite in Guangxi, which is expected to reduce the cost of alumina production, improve the economic benefits for the Guangxi aluminum industry, simultaneously strengthen the efficiency of resource recycling, accelerate the sustainable development of the industry, and provide a useful reference example for subsequent similar studies.

Keywords: Guangxi bauxite; desilication; Bayer process; alumina

1. Introduction

Bauxite is the main raw material for primary aluminum production. The essence of the whole chain of aluminum industrial production is to achieve the whole process conversion from mining bauxite, alumina, electrolytic aluminum, and aluminum primary products to deep aluminum processing according to the needs of the product terminal. Bauxite resources play a decisive role in the construction and development of aluminum industry. Guangxi is rich in bauxite resources, has good resource endowment, and has good geological conditions for bauxite mineralization. It is one of the four provinces and regions with the largest bauxite reserves in China and occupies an important position in the country. According to the mineral resources survey results, by the end of 2021, the total amount of bauxite resources in Guangxi, accounted for more than 25% of the country's bauxite resources. More than 90% of these resources are accumulative bauxite, and less than 10% are sedimentary bauxite [1]. However, with the decrease in domestic bauxite

https://doi.org/10.3390/cryst15080675

reserves and grade, the import volume of bauxite is increasing year by year, and it is urgent to strengthen the utilization of low-grade bauxite [2].

Silicon has a significant impact on the utilization of bauxite resources. In the production process of primary aluminum, the silicon element exists in the form of silicate and reacts with an alkali solution to produce sodium silicate, which consumes a large amount of alkali, and increases production costs [3]. During the leaching process of bauxite, sodium silicate forms a co-precipitate with sodium aluminate, resulting in the loss of aluminum, reducing the leaching rate of alumina, and causing stuttering on the surface of the equipment, thus affecting its normal operation and increasing maintenance costs and energy consumption [4]. In addition, almost all silicon enters the red mud during the production process, resulting in an increase in the amount of red mud. At present, there are not many good treatments for red mud, only some methods of recycling and reusing parts of red mud such as oxides [5].

In the process of bauxite purification and primary aluminum production, desilication technology is one of the core links that determine the economic efficiency of the process. The current mainstream methods for desilication include physical [6], chemical [7], and biological [8] methods. Physical methods mainly utilize the differences in mineral density and surface properties to separate silicate impurities, such as flotation, gravity separation, and magnetic separation techniques [9]. These methods have certain advantages in processing high-silica bauxite, but usually require complex equipment and multi-step operations. Massola et al. [10] used reverse flotation (a physical method) with ether amine as the collector and corn starch as the inhibitor of desilication of bauxite under pH 10 conditions, and finally obtained a concentrate with an alumina content of 42.3% and an alumina/silica ratio of 12.6. After magnetic separation, the alumina content was further increased to 54.0%. Abroon et al. [11] tested the effects of cationic collectors such as DDA, CTAB, and DTAL on flotation efficiency within the pH range of 6–7, and found that DTAL performed the best in terms of the Al_2O_3 recovery rate and A/S ratio. Chemical methods use chemical reactions to remove silicon impurities, such as the alkaline leaching process in the Bayer process [12]. Zhengshuai J et al. [13] used lime for the desilication of the silicon-containing alkaline solution after alkaline leaching, achieving the recycling of the alkaline solution, and the aluminum/silicon ratio of the concentrate after recycling was still greater than 7. Sun Y et al. [14] successfully extracted Al_2O_3 and SiO_2 from high-silica bauxite using a two-step process of sodium carbonate sintering combined with water leaching and acid leaching. The total recovery rate of Al_2O_3 was about 97% and that of SiO_2 was about 86%, achieving efficient resource utilization. The biological method utilizes the reaction between microorganisms and minerals containing silicon, which exists in soluble substances to reduce the silicon content and has good environmental potential. Shah S S et al. [15] used organic acids produced by the filamentous fungus Penicillium simplicissimum to acid hydrolyze silicates in bauxite, leaching aluminum and achieving an environmentally friendly desilication pathway without the generation of red mud. Qi Wang et al. [16] used a biological method to remove silicon from high-silica bauxite. Through the action of two efficient silicon-solubilizing bacteria (Arthrobacter pascens H19 and Burkholderia anthena G21), the silicon content in the bauxite was significantly reduced, and its aluminum/silicon ratio was significantly increased to 9.15–9.64, meeting the requirements for producing alumina by the Bayer process.

The work of this article is to optimize the alkali leaching pre-desilication treatment in the Bayer process for producing primary aluminum. The Bayer process has the advantages of low energy consumption, simple process, and high product quality, and is currently the main method for producing aluminum from bauxite in the world [17]. Despite significant success in industrial production, the desilication process remains a complex and critical step. The presence of silicon can seriously affect the quality and production efficiency of alumina, so it is of great significance to study how to effectively remove silicon. Sodium hydroxide is commonly used as a desilication agent in the Bayer process [18]. Bauxite reacts with the sodium hydroxide solution under high-temperature and pressure conditions, and silicon dioxide reacts with sodium hydroxide to form sodium silicate, which then forms hydrated aluminum silicate precipitate to remove silicon impurities. However, this process has high requirements pertaining to the aluminum/silicon ratio of bauxite, and there are problems such as high alkali consumption and a large amount of red mud.

For this reason, barium hydroxide ($Ba(OH)_2$) was selected as the desiliconization agent in this study due to its strong alkalinity and high reactivity with silicate.

2. Materials and Methods

2.1. Experimental Principle

Barium hydroxide, as the desilication agent in this study, exhibited the following core reaction during the experimental process:

$$Na_2SiO_3 + Ba(OH)_2 = BaSiO_3 + 2\ NaOH \quad (1)$$

During the desilication process of barium hydroxide, the stable precipitation of barium silicate ($BaSiO_3$) generated by the reaction can achieve the efficient separation of silicon, while regenerating NaOH further reduces alkali consumption. This reaction can also reduce the loss of aluminum to a lower level during the desilication process. After barium hydroxide treatment, the ratio of aluminum to silicon in the slurry is higher than the 6–7 control line required by the Bayer process, which reduces the burden of desilication in the subsequent Bayer process, is conducive to the subsequent Bayer process, and greatly reduces the cost of alumina production process. This method provides the possibility for the achievement of efficient desiliconization technology, which is more in line with the needs of Guangxi alumina production and promotes the rapid development of local economy.

2.2. Experimental Raw Materials and Reagents

The raw ore of bauxite used in the experiment was taken from a place in Guangxi. Its components are listed in Table 1.

Table 1. The chemical composition of raw bauxite (wt. %).

Al_2O_3	SiO_2	CaO	Fe_2O_3	TiO_2	ω (Al/Si)
50.00	8.60	0.25	22.87	3.62	5.82

During the experiment, bauxite and limestone were blended with a circulating mother liquor at a mass ratio of 6.85:1. This mixture was charged into an autoclave and maintained at 270 °C for 60 min. In the ore-blending stage, solids were added to achieve a target solid concentration of 370 g/L, utilizing 2 L of the circulating mother liquor. The composition of the mother liquor was as follows: total alkali (NT) = 261.7 g/L, caustic alkali (NK) = 250 g/L, and carbon alkali (Nc) = 4.5 g/L. Lime was incorporated according to a target CaO/SiO_2 (C/S) molar ratio of 1.65. In the bauxite slurry prepared by the above operation, the solubility of the harmful substance TiO_2 is relatively small, and the solubility of Al_2O_3 is sufficient [19]. The specific composition is shown in Table 2.

Table 2. The chemical composition of raw slurry composition (wt. %).

Al_2O_3	SiO_2	Na_2O	K_2O	BaO	ω (Al/Si)
31.28	10.29	39.96	14.38	1.15	3.04

The barium hydroxide powder used in the experiment is a pure reagent, the mass fraction of hydrochloric acid is 30%, and the carbon dioxide gas is a pure gas.

The equipment used in the test includes a constant-temperature water bath heating pot (to provide a stable-temperature environment for the experimental reaction), a vacuum drying oven (for drying the precipitate obtained during the experiment), and an electronic balance (for accurately weighing the mass of the experimental raw materials).

2.3. *Experimental Methods*

2.3.1. Pre-Desiliconization of Bauxite Slurry

The pre-desiliconization test of the slurry was carried out in an orderly manner in the constant-temperature water bath heating pot. Alumina slurry (100 mL) was transferred to a beaker. A measured quantity of barium hydroxide powder was added, and the mixture was homogenized using a magnetic stirrer to ensure complete contact. The beaker was sealed and immersed in a thermostatic water bath preheated to the target temperature. Stirring commenced immediately, and recording the reaction time was began. After the predetermined reaction time, the beaker was rapidly removed. The slurry underwent liquid–solid separation via filtration. The separated solids were vacuum-dried. Chemical composition analysis was performed on the dried solids, and their phase composition was characterized by XRD. The pre-desilication rate of the bauxite ore was calculated based on the chemical composition of the slurry following precipitation and acid dissolution, using the formula below:

$$\text{Pre-desiliconization rate} = M_{\text{Si removed}} / M_{\text{Si in original solution}} \times 100\% \quad (2)$$

2.3.2. Dissolution of Aluminum Hydroxide

In the bauxite slurry after heat preservation treatment and solid–liquid separation in the previous stage, carbon dioxide gas is continuously injected for 30 min. When the predetermined time is reached, the solution is filtered using a vacuum filter to obtain a white precipitate and a filtrate. The resulting white precipitate is dried in a vacuum drying oven, and the final product is aluminum hydroxide. The experimental process is shown in Figure 1.

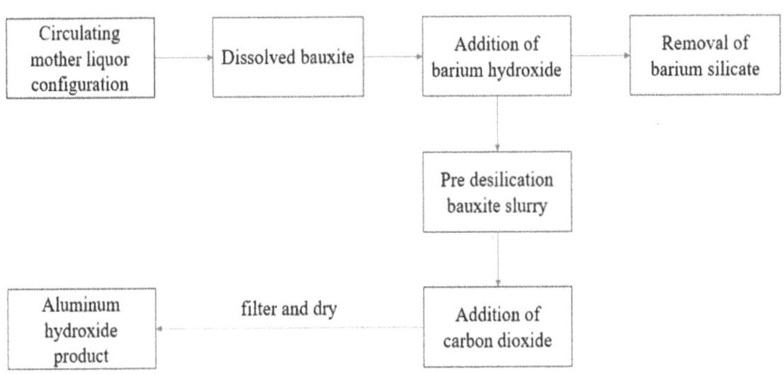

Figure 1. Experimental process.

3. Results

3.1. Pre-Desiliconization of Bauxite Slurry by Barium Hydroxide Powder

During the experiment, different conditions such as the amount of barium hydroxide added, temperature gradient, and insulation time gradient were set to systematically investigate the effect of barium hydroxide on the pre-desilication rate of the original slurry. The corresponding pre-desilication rate data under different conditions are shown in Table 3. As shown in Table 3, under the condition of a constant insulation temperature of 25 °C, increasing the amount of barium hydroxide added from 2 g to 3 g resulted in a certain improvement in the desilication rate of the samples at different insulation times. At the same time, under the conditions of 2 g and 3 g of barium hydroxide addition, extending the insulation time resulted in an increase in the desilication rate of the samples, and the desilication effect was significantly improved. However, when the amount of barium hydroxide added remains constant and the insulation temperature is increased, the desilication rate decreases to a certain extent, and the desilication effect deteriorates, showing a negative correlation trend, as shown in Table 3. Taking into account the desilication effect, under the experimental conditions of group 6, namely, with a $Ba(OH)_2$ addition of 3 g, a holding temperature of 25 °C, and a holding time of 50 min, the desilication effect is optimal.

Table 3. Pre-desilication rates corresponding to different conditions.

Number	1	2	3	4	5	6	7	8	9	10	11	12
Amount of barium hydroxide added/g	2	2	2	3	3	3	3	3	3	3	3	3
Temperature/°C	25	25	25	25	25	25	25	50	60	70	80	90
Time/min	30	40	50	30	40	50	30	30	30	30	30	30
Desilication rate/%	96.77	96.87	97.13	97.30	97.02	97.53	96.94	96.70	96.62	96.91	96.44	96.41

From the principle of chemical reaction, it is known that barium hydroxide is continuously consumed during the desilication process. Only by ensuring sufficient barium hydroxide can the silicon element be fully reacted and completely removed. In addition, extending the insulation time appropriately makes the reaction more complete, helps the crystallization of barium silicate to be more complete, and makes the reaction more thorough, thereby improving the precipitation efficiency of barium silicate, enhancing the desilication effect, and improving the pre-desilication rate.

3.2. Influence of Insulation Time on Leaching Performance of Aluminum Hydroxide

To investigate the effect of holding time on the leaching performance of aluminum hydroxide, experiments were conducted under conditions where the amount of barium hydroxide added remained constant while the holding time was varied. When 3 g of barium hydroxide was added, three time intervals were set, i.e., 30 min, 40 min, and 50 min. Based on the data, the curve depicting the silicon content percentage over time was plotted, as shown in Figure 2. It was clearly observed that, for the same amount of barium hydroxide added, the silicon content percentage gradually decreased as the holding time increased from 30 min to 50 min. This indicated that the holding time significantly impacted the leaching performance of barium hydroxide on aluminum hydroxide. During longer holding times, longer contact and reaction times between barium hydroxide and silicates were allowed. This facilitated the continuous progression of the desilication reaction, promoted a more complete crystallization of barium silicate, improved the precipitation efficiency, and thereby enabled a more effective removal of silicon. Consequently, the purity was enhanced, and the desilication rate was increased. Therefore, in practical production, appropriately extending the holding time to facilitate the desilication reaction is recommended to improve the desilication rate.

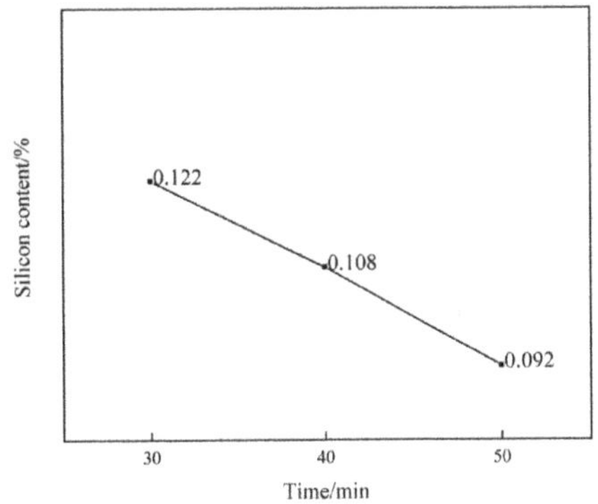

Figure 2. The proportion of silicon content changes with time under the addition of 3 g of $Ba(OH)_2$.

3.3. XRD and SEM Analyses of Dissolved Aluminum Hydroxide

The dissolved aluminum hydroxide precipitate was subjected to XRD phase analysis, as shown in Figure 3. From the figure, it can be seen that the leached precipitate is indeed identified to be aluminum hydroxide, with only a small amount of the impurity phase. This is attributed to the inclusion of sodium during the precipitation of aluminum hydroxide, resulting in impurities [20].

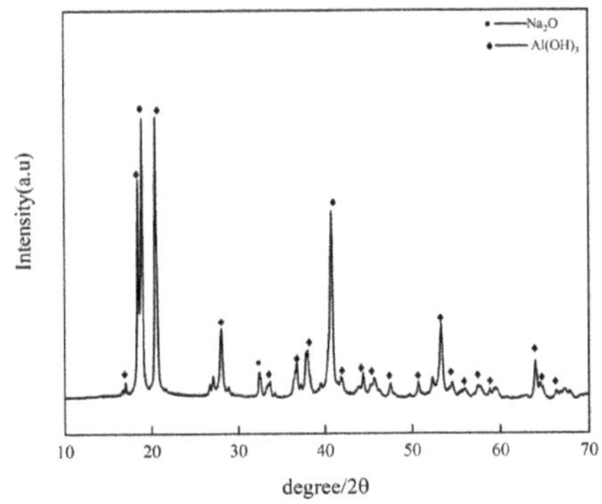

Figure 3. Dissolved aluminum hydroxide.

The scanning electron microscope (SEM) characterization was performed on the acquired aluminum hydroxide powder. Figure 4 presents the morphologies observed for the resulting aluminum hydroxide powder. The analysis reveals that the aluminum hydroxide particles exhibit a small particle size and homogeneous distribution.

The selected aluminum hydroxide powder underwent localized SEM and EDS investigations, with the results illustrated in Figures 5 and 6.

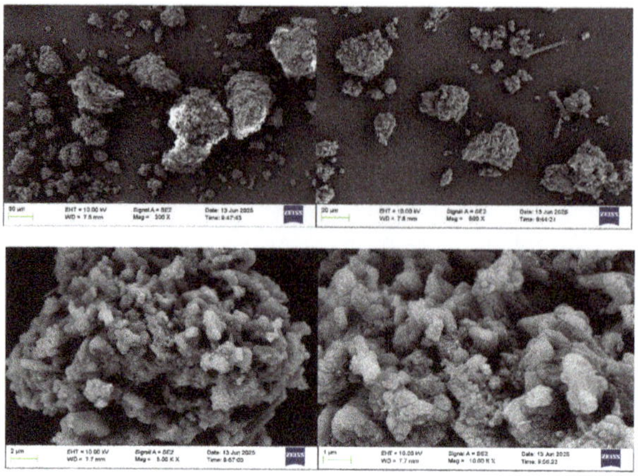

Figure 4. SEM of aluminum hydroxide powder at different magnifications.

Figure 5. SEM of localized aluminum hydroxide powder.

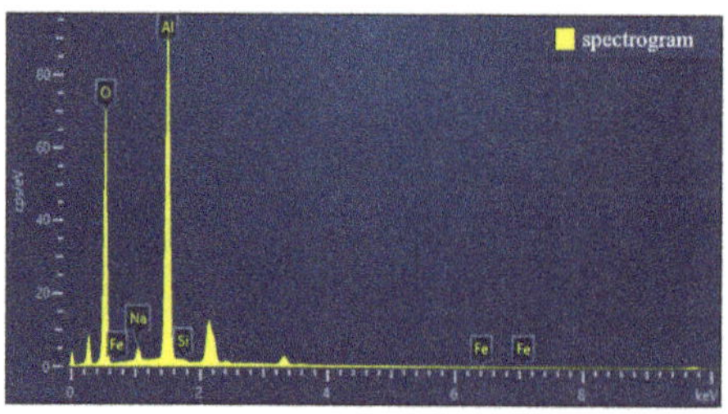

Figure 6. EDS analysis of aluminum hydroxide powder.

According to the SEM and EDS data analyses, the sample contains a very low concentration of silicon, which exists as fine particles dispersed in the solid gaps of aluminum hydroxide and is classified as a trace impurity. The slightly elevated sodium content can be attributed to its introduction in the previous process. Meanwhile, the iron content is extremely low and has almost no influence.

Compared with the traditional lime-based desilication, barium hydroxide desilication has significant advantages in multiple aspects.

In terms of reaction rate, the advantages of the barium hydroxide method are particularly prominent. The lime method for desilication mainly utilizes the reaction of calcium hydroxide ($Ca(OH)_2$) with the silicon element to generate hydrated calcium silicate precipitate for desilication [21]. However, the solubility of calcium hydroxide in water is relatively low, and the concentration of Ca^{2+} ions it can provide in aqueous solutions is limited. Chemical reactions are limited by the concentration of Ca^{2+} ions, resulting in a slower reaction rate. On the contrary, barium hydroxide ($Ba(OH)_2$) has a high solubility in water and can quickly produce a large amount of Ba^{2+} ions when dissolved in water, increasing the concentration of Ba^{2+} ions. These Ba^{2+} ions can quickly react with the silicon element in the solution to form a barium silicate precipitate, greatly improving the rate of desilication reaction.

Barium hydroxide desilication also has significant advantages in aluminum loss. During the process of desilication by the lime method, the generated hydrated calcium silicate precipitate has a large specific surface area, which makes it easier to adsorb aluminum ions, resulting in the chemical loss of aluminum. The precipitation properties of barium silicate generated by the desilication of barium hydroxide are relatively stable, and its adsorption effect on aluminum ions is weak, which can effectively reduce the chemical loss of aluminum.

From the point of view of energy consumption, barium hydroxide desiliconization also has certain advantages. Because the barium hydroxide method has a fast reaction rate, it can achieve a better desiliconization effect in a short time, so the time and energy consumption required for the reaction can be reduced. In contrast, due to the slow reaction rate of lime desiliconization, in order to achieve the ideal desiliconization effect, it often needs a longer reaction time and a higher reaction temperature, which undoubtedly increases the energy consumption [22].

In addition, compared with acid-based desilication, barium hydroxide desilication also has environmental advantages. Acid-based desilication usually uses strong acids such as sulfuric acid and hydrochloric acid, which are costly. During the desilication process, a large amount of acidic wastewater is generated, which is highly corrosive and makes it difficult to separate aluminum later, causing serious pollution to the environment [23]. In the process of desilication by the barium hydroxide method, no acidic wastewater is generated, and the precipitation of barium silicate produced by the desilication reaction can also be reused as building materials, fully meeting the requirements of green environmental protection.

4. Conclusions

(1) This study systematically explored the pre-desilication process of barium hydroxide based on the characteristics of high silicon content and a low aluminum/silicon ratio in a low-grade bauxite mine in Guangxi, aiming to improve the aluminum/silicon ratio and optimize the subsequent alumina production process. The research results indicate that, to a certain extent, the desilication effect is positively correlated with the amount of barium hydroxide added and the holding time, and negatively correlated with the reaction temperature. The research results provide the possibility of achieving efficient desilication technology for low-grade bauxite and improving the utilization rate of mineral resources, thereby better meeting the demand for alumina production in Guangxi and promoting economic development.

(2) In this paper, adding 3 g of barium hydroxide powder to 100 mL bauxite slurry and keeping it at a temperature of 25 °C for 50 min can achieve the best desilication effect. The experimental results showed that, under this condition, the desilication rate reached 97.53%, and the aluminum/silicon ratio was significantly improved, exceeding the control line required for the production of alumina by the Bayer process. At the same time, the properties of the barium silicate precipitate generated by the

barium hydroxide method for desilication are relatively stable, and it has a weak adsorption effect on aluminum ions, which can effectively reduce the loss of aluminum during desilication, reduce aluminum loss, and effectively solve the problems of high aluminum loss (5–10%) in the traditional lime method and serious environmental pollution in the acid method.

(3) The research results have demonstrated certain value and advantages in terms of economy and resource utilization. In terms of economy, after desilication treatment with barium hydroxide, the aluminum/silicon ratio of the slurry significantly increases, exceeding the control line required by the Bayer process for producing alumina. This advantage greatly reduces the burden of desilication in the subsequent process of producing alumina using the Bayer process, and correspondingly reduces the equipment investment, chemical reagent consumption, and energy consumption required in the subsequent desilication process, thereby effectively reducing the cost of alumina production and improving the economic efficiency of the enterprise.

(4) Future research can further focus on the development of barium salt cycle process, the collaborative removal mechanism of fine silicon particles, and the optimal design of industrial continuous reaction devices, in order to deepen the application potential of technology and help the efficient development and industrial upgrading of bauxite resources in China.

Author Contributions: Conceptualization, G.H. and A.L.; methodology, A.X.; formal analysis, G.H.; investigation, D.Z.; resources, A.L.; data curation, D.Z.; writing—original draft preparation, G.H. and A.X.; writing—review and editing, A.L. and X.P.; supervision, L.P. and X.Z.; funding acquisition, A.L. All authors have read and agreed to the published version of the manuscript.

Funding: This research was funded by the 2022 Guangxi Key Research and Development Program, grant number GuiKeAB22080015, the Special Fund for Science and Technology Development of Guangxi, grant number AD25069078, 2022 Specific Research Project of Guangxi for Research Bases and Talents, grant number GuiKeAD21238010 and 2021 Special funds for central guidance of local scientific and technological development funds, grant number GuiKeZY21195030.

Data Availability Statement: Data are contained within the article.

Conflicts of Interest: The authors declare no conflicts of interest.

References

1. Zhang, H.; Hu, P.; Jiang, J.; Cheng, X.; Wang, J.; Liu, J.; Xiang, P. Distribution, genetic types and current situation of exploration and development of bauxite resources. *Geol. China* **2021**, *48*, 68–81.
2. Chen, Y.; Long, F.; Cao, X.; Li, Y.; Zhang, W.; Zhang, T.; Lv, G. Exploration of Large-Scale Application of Efficient and Clean Utilization of Low-Grade Bauxite. *Separations* **2023**, *10*, 336. [CrossRef]
3. Rao, D.S.; Das, B. Characterization and beneficiation studies of a low grade bauxite ore. *J. Inst. Eng. Ser. D* **2014**, *95*, 81–93. [CrossRef]
4. Modi, M.; Dewangan, P. Beneficiation of Bauxite Ore Characterized by Low-Grade and High Silica Content Using Crushing and Scrubbing Technique. *Int. J. Eng. Trends Technol.* **2024**, *72*, 207–213. [CrossRef]
5. Papadopoulos, N.; Karayianni, H.-S.; Hristoforou, E.; Metaxa, E. Valuable Products Obtained from Red Mud. GR2010000003. 15 July 2010.
6. Gibson, B.; Wonyen, D.G.; Chelgani, S.C. A review of pretreatment of diasporic bauxite ores by flotation separation. *Miner. Eng.* **2017**, *114*, 64–73. [CrossRef]
7. Zhao, X.; Liu, Y.; Wang, L.; Hua, Y.; Cheng, T.; Zhang, T.; Zhao, Q. Co-Extraction of Aluminum and Silicon and Kinetics Analysis in Carbochlorination Process of Low-Grade Bauxite. *Materials* **2024**, *17*, 3613. [CrossRef] [PubMed]
8. Shah, S.S.; Palmieri, M.C.; Sponchiado, S.R.P.; Bevilaqua, D. Environmentally sustainable and cost-effective bioleaching of aluminum from low-grade bauxite ore using marine-derived Aspergillus niger. *Hydrometallurgy* **2020**, *195*, 105368. [CrossRef]
9. Ren, S.P.; Xiong, T.; Zhou, Y.C. Study and application of desilication process in mineral processing. *Min. Metall.* **2013**, *10*, 195–197.

10. Massola, C.P.; Lima, J.R.B.; Andrade, C.F. Separation of silica from bauxite via froth flotation. *Miner. Eng.* **2009**, *22*, 315–318. [CrossRef]
11. Abroon, A.; Rahimdad, A. Desilication and flotation techniques for separating them from the bauxite. *Adv. Environ. Biol.* **2014**, *8*, 1280–1288.
12. Xu, Y.; Chen, C.; Li, J. A Novel Sustainable and Facile Method in Alumina Industry: Pre-Treatment of High Silica Bauxite by Cyclic Alkaline Leaching. *J. Sustain. Metall.* **2025**, *11*, 645–656. [CrossRef]
13. Jiang, Z.; Xia, F.; Zhang, S.; Zhang, Q.; Zhang, M. Study on efficient desulfurization and desilication process of low-grade bauxite. *Nonferrous Met. Sci. Eng.* **2022**, *13*, 26–34.
14. Sun, Y.; Pan, A.; Ma, Y.; Chang, J. Extraction of alumina and silica from high-silica bauxite by sintering with sodium carbonate followed by two-step leaching with water and sulfuric acid. *RSC Adv.* **2023**, *13*, 23254–23266. [CrossRef] [PubMed]
15. Shah, S.S.; Palmieri, M.C.; Sponchiado, S.R.P.; Bevilaqua, D. A sustainable approach on biomining of low-grade bauxite by *P. simplicissimum* using molasses medium. *Braz. J. Microbiol.* **2022**, *53*, 831–843. [CrossRef] [PubMed]
16. Wang, Q.; Sheng, X.-F.; He, L.-Y.; Shan, Y. Improving bio-desilication of a high silica bauxite by two highly effective silica-solubilizing bacteria. *Miner. Eng.* **2018**, *128*, 179–186. [CrossRef]
17. Paz, S.P.A.; Angélica, R.S.; Kahn, H. Optimization of the reactive silica quantification method applied to Paragominas-type gibbsitic bauxites. *Int. J. Miner. Process.* **2017**, *162*, 48–57. [CrossRef]
18. Wu, Y.; Pan, X.; Han, Y.; Yu, H. Novel Process and Mechanism for Chemical Desilication from Kaolinite-Rich Diasporic Bauxite. *J. Northeast. Univ. (Nat. Sci.)* **2019**, *40*, 1424–1429.
19. Yin, Z.; Gu, S. Influence of lime adding method on scaling process in Bayer preheating process of diasporic bauxite slurry. *Chin. J. Nonferrous Met.* **2001**, *11*, 910–914.
20. Dobra, G.; Garcia-Granda, S.; Iliev, S.; Cotet, L.; Negrea, P.; Duteanu, N.; Boiangiu, A. Aluminum hydroxide impurities occlusions and contamination sources. *Rev. Chim.* **2020**, *71*, 65–76. [CrossRef]
21. Zhang, Z.Y. Experimental study on digestion of gibbsite and boehmite mixture. *Light Met.* **2019**, *1*, 10–13.
22. Smith, P. The processing of high silica bauxites—Review of existing and potential processes. *Hydrometallurgy* **2009**, *98*, 162–176. [CrossRef]
23. Wang, Y.; Liu, W.; Liu, W.; Chen, M.; Hu, L. Two-step acid leaching extraction of Fe&Al from karst bauxite and optimization experimental study. *Chem. Eng. Process.-Process Intensif.* **2025**, *208*, 110131.

MDPI AG
Grosspeteranlage 5
4052 Basel
Switzerland
Tel.: +41 61 683 77 34

Crystals Editorial Office
E-mail: crystals@mdpi.com
www.mdpi.com/journal/crystals

www.ingramcontent.com/pod-product-compliance
Lightning Source LLC
Chambersburg PA
CBHW040806250926
51850CB00007BA/1093
9783725878970